Euphemism and Dysphemism

Euphemism

&

LANGUAGE USED AS

New York · Oxford · OXFOR

Dysphemism

SHIELD AND WEAPON

Keith Allan
Kate Burridge

UNIVERSITY PRESS · 1991

Oxford University Press

Oxford   New York   Toronto
Delhi   Bombay   Calcutta   Madras   Karachi
Petaling Jaya   Singapore   Hong Kong   Tokyo
Nairobi   Dar es Salaam   Cape Town
Melbourne   Auckland

and associated companies in
Berlin   Ibadan

Copyright © 1991 by Keith Allan and Kate Burridge

Published by Oxford University Press, Inc.
200 Madison Avenue, New York, NY 10016

Oxford is a registered trademark of Oxford University Press

Library of Congress Cataloging-in-Publication Data
Allan, Keith, 1943–
Euphemism and dysphemism : language used as shield and weapon /
by Keith Allan, Kate Burridge.
p. cm.   Includes bibliographical references and index.
ISBN 0-19-506622-7
1. English language—Euphemism.   2. Dutch language—To 1500—Euphemism.
3. English language—Grammar, Comparative—Dutch.
4. Dutch language—Grammar, Comparative—English.
5. Euphemism.   I. Burridge, Kate.   II. Title.
PE1585.A37   1991   427'.09—dc20
90-49602

1 3 5 7 9 8 6 4 2

Printed in the United States of America
on acid-free paper

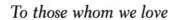

To those whom we love

Preface

The form of language a person uses can cause offense to other human beings and to gods—even to dangerous animals. The result of misusing language can be dire: according to Leviticus (24:16) God told Moses 'He that blasphemeth the name of the LORD shall be put to death.' To shield a speaker from the consequences of giving offence, all languages have euphemisms. Euphemistic expressions trade on illusion: the bluntly profane (and therefore in some eyes blasphemous) 'Jesus Christ!' is euphemistically transmuted into 'Jeepers Creepers!' but the same person is referred to by both, and if the former is profane, even blasphemous, so should the latter be, too. Similarly, the euphemistic 'Your dog went to the bathroom in my driveway!' describes an event that is equally well captured by 'Your dog shat in my driveway!' Euphemisms have existed throughout recorded history; they are used among preliterate peoples, and have probably been around ever since recognizably human language developed. Likewise has offensive language, what we call 'dysphemism.'

How are euphemism and dysphemism to be defined? Is euphemism necessarily tied in with taboo? What topics are taboo? Why are women's bodies so strongly tabooed? Do euphemism and dysphemism reflect the intrinsic conflict between intellect and body within human beings? What sorts of euphemisms are there? Where do conventional euphemisms and dysphemisms come from? How are euphemism and dysphemism related to one another? How are they related to neutral terms, if there are such things? Why is it that the euphemism of yesteryear (e.g., 'toilet') is replaced by a new one ('bathroom')? How do euphemism and dysphemism relate to slang and jargons? What makes people swear? How many euphemisms are there in English for the genital organs? All these questions and many more are broached in this book.

Like many books, this one came to be written as the result of an accident. A student drew Keith Allan's attention to the following passage in *Introduction to Language* by Fromkin et al.:

> What is surprising is that two words or expressions can have the identical linguistic meaning and one can be acceptable for use and the other strictly forbidden or the cause of embarrassment or horror. In English, words which we have borrowed

from Latin or French seem to carry with them a "scientific" connotation and thus appear to be technical terms and "clean", while good old native Anglo-Saxon words are taboo. This seems to reflect the view that the vocabulary used by the upper classes was clearly superior to that used by the lower classes, a view that was, of course, held and propagated by the upper classes. Peter Farb points out that this distinction must go back at least as far as the Norman conquest in 1066, when "a duchess *perspired* and *expectorated* and *menstruated*—while a kitchen maid *sweated* and *spat* and *bled*."   (Fromkin et al. 1984:271f)

The freshman student took this to imply that speakers of old and medieval Germanic dialects did not use euphemisms. Whether or not this is an appropriate inference from the preceding text, it is historically false. Because Kate Burridge had worked with, among other things, the language of Middle Dutch medical texts, we got together to compare euphemism in present-day English and in Middle Dutch. The more deeply we delved into euphemism, the greater the project grew. We found it necessary to distinguish euphemisms from dysphemisms, then to sort them into types, next to tease out the effects of style and jargon, and so forth. Before long, what began as a short paper expanded into the present book.

In the course of our investigations we discovered that, when it comes to largely oral traditions such as slang and obscenity, different dialect groups will frequently have different interpretations for a common form—which suggests that the common written language has a levelling effect on English dialects. We also came to suspect that there is a wealth of unexplored vocabulary in oral culture; for instance, what Australian author Nancy Keesing has called 'Shielaspeak' (i.e., colloquial language used among women on topics to do with women). One of the most striking effects of our journey into oral culture was to be sharply reminded of the delightful inventiveness of so much of its terminology; the overwhelming majority of language examples discussed in linguistic textbooks are sterile fragments of the rich language that they supposedly identify. There can be no doubt at all that Lakoff and Johnson were correct in observing that

> most people think they can get along perfectly well without metaphor. We have found, on the contrary, that metaphor is pervasive in everyday life, not just in language but in thought and action. Our ordinary conceptual system, in terms of which we both think and act, is fundamentally metaphorical in nature.
>
> (Lakoff and Johnson 1980:3)

There is plenty of confirmatory evidence within this book.

We had a lot of fun gathering material for the book; and a lot of help getting it together. Our great pleasure is to express special gratitude to John Haiman, Bill McGregor, Heidi Weber Platt, Cynthia Read, Ross Weber, and the anonymous reader for Oxford University Press: without their help the text would have been much worse than it is. They are, course, all blameless for the remaining flaws. We were also offered useful chunks and snippets of information by a large number of supportive friends, students, and colleagues to whom we hereby offer our heartfelt thanks; they are: Wendy Allen, Larry Audette, Tony Barta, Barry Blake, Bill Bright, Carol Budge, Sarah Castle, Hilary Chappell, Sandra Cootes, Susanna Cumming, Dick Demers, Charlotte Dihoff, Bruce Donaldson, Diana Eades, Sue Favret,

George Gunter, Liza Harris, Marion Harris, Lliam Harrison, Linden Hilgendorf, Greg Horseley, Wim Hüskin, Rhys Isaac, Sakarepe Kamene (for information on Zia), Judy Kermode, Tomiko Kodama (Japanese), Terry Langendoen, Adrienne Lehrer, Lise Menn, Jack Murphy, Suzanne Murphy, Peng Long (Chinese), Sugu Pillai (Malaysian Tamil), John Platt (whose untimely death we mourn), David Potts, Ellen Prince (Yiddish), Doug Saddy, Paul Saka, Mel Scullen, Pornpimol Senawong (Thai), Jae Song, Nik Suriana Nik Yusoff (Bahasa Malaysia), Putu Wijana (Bahasa Indonesia), Monty Wilkinson, Lesley Wright, Zhang Shi (Chinese). Keith Allan is very grateful to the Department of Linguistics at the University of Arizona for generous access to facilities during the time he spent there. Perhaps we should add our thanks for the use of the facilities provided by our regular employers, Monash University and La Trobe University, respectively. Finally, we pay tribute to Don Laycock who was especially generous in lending us his collection of *Maledictas* and one or two other texts; we were anticipating his advice on our final draft, but unfortunately he died long before it was finished. We think that Don might have enjoyed this book, even though he would not have agreed with everything in it. Thanks Don, wherever you are.

We offer *Euphemism and Dysphemism* to our readers with the wish that they will concomitantly enjoy the book and benefit, as we have done, from the study of its subject matter. An interesting perspective on the human psyche is to be gained from the study of euphemisms used as a shield against the disapprobation of our fellows or at worst death, and from the examination of dysphemisms used as a weapon against those we dislike or as a release valve against the vicissitudes of malign fate. Many euphemisms and dysphemisms demonstrate the poetic inventiveness of ordinary people: they reveal a folk culture that has been paid too little attention by lexicographers, linguists, and literaticians—and, indeed, by the very folk who use them: you, me, and our friends and relatives. We hope the reader will come to share our respect for 'la masse parlante,' as Saussure (the father of modern linguistics) called them. Yet perhaps we should rather be referring here to the masses for whom Shakespeare's comedy was written, those among 'la masse parlante' who would appreciate the double-meanings in, for instance, what Margaret says in the following girl-talk:

HERO	God give me joy to wear it, for my heart is exceeding heavy.
MARGARET	'Twill be heavier soon by the weight of a man.
HERO	Fie upon thee, art not ashamed?
MARGARET	Of what, lady? Of speaking honourably? Is not marriage honourable in a beggar? Is not your lord honourable without marriage? I think you would have me say, saving your reverence, 'a husband'. And bad thinking do not wrest true speaking, I'll offend nobody. Is there any harm in 'the heavier for a husband'? None, I think, and it be the right husband, and the right wife; otherwise 'tis light, and not heavy. Ask my Lady Beatrice else; here she comes.
	Enter Beatrice
HERO	Good morrow, coz.
BEATRICE	Good morrow, sweet Hero.

HERO	Why, how now? Do you speak in the sick tune?
BEATRICE	I am out of all other tune, methinks.
MARGARET	Clap's into 'Light o' Love'; that goes without a burden. Do you sing it, and I'll dance it.
BEATRICE	Ye light o' love, with your heels! Then, if your husband have stables enough, you'll see he shall lack no barns.
MARGARET	O illegitimate construction! I scorn that with my heels.
BEATRICE	'Tis almost five o'clock, cousin; 'tis time you were ready. By my troth, I am exceeding ill—heigh-ho!
MARGARET	For a hawk, a horse, or a husband?
BEATRICE	For the letter that begins them all, H.
MARGARET	Well, and you be not turn'd Turk, there's no more sailing by the star.
BEATRICE	What means the fool, trow?
MARGARET	Nothing I, but God send everyone their heart's desire!
HERO	These gloves the Count sent me, they are an excellent perfume.
BEATRICE	I am stuffed, cousin, I cannot smell.
MARGARET	A maid, and stuffed! There's goodly catching of cold.
BEATRICE	O, God help me, God help me, how long have you professed apprehension?
MARGARET	Ever since you left it. Doth not my wit become me rarely?
BEATRICE	It is not seen enough, you should wear it in your cap. By my troth, I am sick.
MARGARET	Get you some of this distilled *carduus benedictus*, and lay it to your heart, it is the only thing for a qualm.
HERO	There you prick'st her with a thistle.
BEATRICE	*Benedictus!* why *benedictus?* You have some moral in this *benedictus.*
MARGARET	Moral! no, by my troth, I have no moral meaning, I meant, plain holy-thistle.

(SHAKESPEARE *Much Ado About Nothing* III.iv.23–74)

This is one text we shall not be explaining; should you feel you need help interpreting it, we can recommend you get hold of *Shakespeare's Bawdy* by Eric Partridge, and use it as a primer. But be assured, dear reader, that you will not need help with the rest of this book (though we do provide a Glossary). And, as you turn the pages, we entreat you to enjoy the products of the human mind as it confronts the problem of how to talk in different contexts about body parts, bodily functions, sex, lust, disapproval, anger, hate, disease, death, fear, and God.

Melbourne, Australia   K. A.
September 1990   K. B.

P.S. We shall be glad to hear from readers who wish to offer observations on any of the matters we touch on in this book, or who wish to enlighten our ignorance with information we have omitted and s/he believes we ought to have included. Write to us in Australia at the Linguistics Department at either Monash University, Clayton, Victoria 3168 or La Trobe University, Bundoora, Victoria 3083.

Contents

Contents

Euphemism and Dysphemism

Introductory Remarks on Language
Used as Shield and Weapon

Knowing ignorance is strength.
Ignoring knowledge is sickness.
(Lao Tsu *Tao Te Ching* 71)

The principal purpose of this book is to give an expository and explanatory account of the kinds of expressions people use for two quite different purposes: euphemism and dysphemism. Euphemism is characterized by avoidance language and evasive expression; that is, Speaker uses words as a protective shield against the anger or disapproval of natural or supernatural beings. If this seems too negative, we can alternatively describe euphemism as "expression that seeks to avoid being offensive." Dysphemism is, roughly speaking, the contrary of euphemism. Investigating dysphemism, we examine the verbal resources for being offensive, being abusive, or just letting off steam. We approach the subject in the spirit demonstrated in classical works like Durkheim's *Incest: The Nature and Origin of the Taboo* (1963 [1897]), Frazer's *The Golden Bough Part II: Taboo and The Perils of the Soul* (1911), and Freud's *Totem and Taboo* (1953 [1913]). But because our background is linguistics, not sociology, anthropology, or psychology, we—unlike these particular authors— show that euphemism and dysphemism are not merely a response to taboo: they also function where Speaker avoids using, or, on the contrary, deliberately decides to use, a distasteful expression and/or an infelicitous style of addressing or naming. And we also set forth a classification of euphemisms and dysphemisms according to their formal and functional characteristics. We include examples like the following: the ancient Greeks called The Furies "the well-minded ones"—*Eumenides*; a Middle Dutch physician talks of *figs in the secret passage* to denote "piles"; in the nineteenth century, the Victorian moral code prevented those in polite society from uttering words like *legs, trousers* or *underclothing*; some fishermen still avoid words alluding to women and certain kinds of land animals during a fishing expedition; people say *jiminy cricket* instead of *Jesus Christ*; and newspapers still occasionally

3

print *f*— for you-know-what; a politician speaks of a *categorical inaccuracy* or *misspeaking* himself instead of admitting he lied; people get *selected out* rather than *dismissed* or *fired*; the cricketer (but, we regret to admit, neither of us) can make sense of *a man at deep fine leg and another at wide mid-on:* this is jargon, and jargon can be dysphemistic to outgroupers; for instance, when doctors refer to a *circumorbital hematoma* instead of *black eye* they are using jargon, but if a doctor advises a terminally ill patient *Perhaps you had better get your affairs in order* it is to avoid mentioning death, a topic strongly tabooed almost everywhere; in the United States, the brand name of a pencil was changed from *Mikado* to *Mirado* during World War II; and during the 1914–1918 war, *machine gun* occasionally became *sewing machine*; the push for nonsexist usage has rendered words like *manhole cover* and *chairman* taboo for many people; and so forth. The list is extremely heterogeneous and seemingly endless; but in the chapters that follow we attempt to impose order on apparent chaos.

When we seek to explain the sources of the metaphors and figures that have been created for euphemistic or dysphemistic purposes, we step into lexicology. But we emphasize that our book is NOT (and was never intended to be) a dictionary of euphemism and dysphemism; there are already many such volumes in existence (e.g., Grose 1811, Farmer and Henley 1890–1904; Partridge 1961, 1970, 1984; Fryer 1963; Wentworth and Flexner 1975; Read 1977; Rawson 1981, 1989; Spears 1982; Neaman and Silver 1983; Paros 1984; and McDonald 1988, to cite just a few). Our aim is not to outdo or even replicate these treasure troves of euphemism and dysphemism, so you, the reader, will probably be able to think up many more examples of euphemistic and dysphemistic usage than we shall write about. Words are powerful things, and the reticence and fear that some taboo terms arouse in us can be strong. The word *cancer,* for example, has come to be equated with malignancy, corruption, and death; so it is generally shunned. People generally prefer to talk about *growths* and *tumors,* both of which CAN be benign, whereas a view prevails that cancer cannot. Different connotations motivate the choice between these different vocabulary items. If this seems to imply that euphemism and dysphemism are matters of purely lexical choice, that is not in fact the case. We shall show, in the course of this book, that euphemism and dysphemism are principally determined by the choice of expression within a given context: both world spoken **of,** and the world spoken **in.** We cannot properly judge something as euphemistic or dysphemistic without this information—which is why dictionaries of euphemisms are never entirely successful. The proof of this should become evident in various of the chapters that follow.

As we have said, we offer an analysis of the motivations behind these different kinds of language use, a classification of them, an occasional history, and a functional account of the various expressions of euphemism and dysphemism. And we develop the notion that euphemism and dysphemism are defined by reference to concerns about face, concerns that are of immense significance on any occasion of language interchange (see the Glossary, and also Allan 1986a; Brown and Levinson 1987). Talking is one of the most pervasive of human social activities, and like all the others, can only successfully take place if participants mutually recognize that certain conventions govern their own actions and use of language, and also their interpretations of the actions and utterances of others. For instance, there are conventions governing how close to one another interlocutors should stand; how to begin and end phone conversations; how to show interest in what Speaker is saying, how to make requests politely or sharply, or how to offer thanks. There is a presumption that both Speaker and Hearer will act reasonably and cooperatively, and we interpret Speaker's utterances accordingly. Furthermore, Speaker does not randomly chose the forms and style to use in making an utterance; s/he normally has some reason for selecting the particular ones used—a reason sought by Hearer (not necessarily consciously) when interpreting Speaker's utterance. Grice (1975) recognized four different kinds of cooperation governing the most typical forms of social behavior. He formulated them as maxims prescribing the norms of behavior that participants are expected to observe during a linguistic exchange, although they may occasionally choose not to do so. We have adapted the Gricean originals to cover a greater range of data than he envisaged. There is the maxim of quantity: SPEAKER SHOULD MAKE THE STRONGEST CLAIM POSSIBLE CONSISTENT WITH HIS/HER PERCEPTION OF THE FACTS, WHILE GIVING NO MORE AND NO LESS INFORMATION THAN IS REQUIRED TO MAKE HIS/HER MESSAGE CLEAR TO HEARER. It would normally be dysphemistic to say things like *My neighbor, who is a woman, is pregnant*, because it contains an unnecessary relative clause; we know that if the neighbor is pregnant, it MUST be a woman. Second, there is the maxim of quality: SPEAKER SHOULD BE GENUINE AND SINCERE. That is, Speaker should state as facts only what s/he believe to be facts; make offers and promises only if s/he intends carrying them out; pronounce judgments only if s/he is in a position to judge, and so on; thus, we, the authors, would be violating the maxim of quality if we were to say *We believe this book is suitable bedtime reading for six year olds*. Violations of the quality maxim are normally dysphemistic. Third, is the maxim "be relevant," which Grice called relation: IN GENERAL, AN UTTERANCE SHOULD NOT BE IRRELEVANT TO THE CONTEXT IN WHICH IT IS UTTERED, BECAUSE THAT MAKES IT DIFFICULT FOR HEARER TO COMPREHEND; and we presume that Speaker has some reason for making this utterance in this context, in the particular form in which it occurs, rather than maintaining silence or uttering something different. (Sperber and Wilson 1986 argue that 'relevance' is the fundamental principle of language interaction, but we disagree.) Generally speaking, to be irrelevant, is to be dysphemistic. The last maxim of cooperation is manner: WHERE POSSIBLE, SPEAKER'S MEANING SHOULD BE PRESENTED IN A CLEAR, CONCISE MANNER THAT AVOIDS AMBIGUITY, AND AVOIDS MISLEADING OR CONFUSING HEARER THROUGH STYLISTIC INEPTITUDE. Thus, for example, one should ordinarily avoid saying things like *There is a male adult human being in upright stance using his legs as a means of locomotion to propel himself up a series of flat-topped structures some fifteen centimeters high* rather than . . . [we leave it to the reader]. Wasting Hearer's time and mental effort is dysphemistic.

Every time we open our mouths, we have to consider whether what we say is likely to maintain, enhance, or damage our own face, as well as considering the effect of our utterance on others. We have to work to create the effect we intend to create; Goffman refers to this as "face-work". Social interaction is generally oriented towards maintaining (= saving) face, and one of the ground rules in an encounter is a tacit agreement between the different parties that everyone should operate with exactly this in mind. Just as we look after our own face (self-respect), we are expected to be considerate of and look after the face-wants of others, turning a tactful blind eye, perhaps, or telling a white lie. How often have some of us bought a product simply to save the face of the salesperson—with a sneaking suspicion that the sales-pitch was engineered for precisely this outcome! Depending on our own social attributes and on the situation itself, we will have to adjust the line we take during an exchange and adopt different face-saving strategies accordingly; that is, we will have to choose our language expressions, tone and quality of speaking, looks and gestures to produce just the desired face effects. Those who are skilled in face-work are described as having social savoir faire; they are said to be perceptive, and diplomatic. But of course not everyone is going to be the same. Different groups play by different rules, and conventions vary considerably between individuals even within the same community. Cultures can have very different norms, too, and consequently quite different repertoires of face-work strategies. Nonnative speakers are often unaware of these differences and because of this may occasionally inadver-tently cause offense. This becomes critical where representatives of whole nations are involved and it is not just one person's face that is at stake. The reason that euphemism is the stock in trade of politicians and diplomats is just that they must pay special attention to face effects.

Mutual consideration (being mindful of one another's face) can also create a conflict of wants. Protecting someone else's face can lead to loss of one's own, just as defending one's own face can sometimes threaten the face of others. It is a difficult game to play, requiring just the right amount of attention and perceptiveness. We do not want to be accused of being thick-skinned because we pay too little attention to our own face; but neither do we want to be seen as thin-skinned because we pay too much attention to it. But being overgracious and oversolicitous of others' faces can be seen as dysphemistic, and a person who does that risks being condemned as ingratiating and unctuous!

As we remarked earlier, the default situation for nonhostile social interaction is a mutual expectation that the participants will try to avoid any potential face affront to the others. All societies have strategies for avoiding encounters that are potentially face-threatening, for example by using go-betweens or intermediaries. Avoidance relationships in the kinship systems of certain Australian Aboriginal societies, for example, have led to the development and use of a special language for communicat-ing with tabooed kin (cf. Frazer 1911; Dixon 1980). Where affront cannot be prevented, there are conventions of one kind or another for ameliorating it (see Allan 1986a; Brown and Levinson 1987; Leech 1983). For instance, Speaker might apologize for the potential dysphemism, or point out that s/he is not personally responsible, or that it is really in Hearer's long term interest. A critical comment can be ameliorated with some sort of hedging phrase to soften the potential face affront to the hearer; something underplaying Speaker's own ideas or intelligence is effec-tive, like the humiliatives *I've probably completely missed the point, but . . . ; It's probably me, but . . . ; I maybe don't understand properly, but . . . ;* and so on. In making a request Speaker can minimize an imposition by giving, or at least appear-ing to give, the option of refusal by hinting or asking that Hearer do something, rather than ordering that s/he do it; compare *Shut the door!* with *Will / Can / Could /*

Would you shut the door? or *Would you mind shutting the door?*, or even more obliquely *My goodness, it's cold in here!*. In response to an invitation, an expression of regretted inability is generally preferable to downright refusal. Compare some of the possible responses to the offer *Do you want to come out for a meal tonight?*

1. *No, I don't.*
2. *I'm sorry, I don't.*
3. *I'm sorry, (I'd love to but) I can't.*

Refusals and denials of any sort are potentially hazardous to face and either of the first two responses can be used to hurt or offend (i.e., be dysphemistic to) the offerer—although (2) is slightly better than (1) because at least Speaker apologizes. The third response, expressing inability, is more considerate of face and therefore more polite. Generally speaking, the greater the oncoming face-affront, the greater is the politeness shown, and the greater the degree of euphemism required. The legal profession has ritual face-savers down to a fine art; for example, in Britain or Australia the formula *In answer to my learned friend's erudite submission* is generally understood to mean "you are wrong"; in court, criticisms and arguments are always given *with the greatest of respect* (see Ch. 8). In sum, it is conventional to appear to be polite, whatever one's true feelings. Greetings, apologies, and congratulations may not be quite so welcome when uttered cursorily and as a matter of 'form' as they would be if uttered with deeply felt sincerity; but it would often be downright dysphemistic to keep silent.

The cooperative principle and matters of face are relevant to the subject of this book because it is principally a study of euphemism and dysphemism, and we define both of them in terms of face effects. It is therefore important not only to know what face effects are, but also how they relate to other politeness phenomena and conventions for language behavior. Face affronts are typically dysphemistic, and therefore potential face affronts are sometimes avoided, or at least ameliorated, by the use of euphemisms. In the production of social harmony, euphemism and dysphemism are key performers.

Dividing language behavior into that which seeks to avoid being offensive, and that which doesn't, leads us to identify and distinguish the locutionary and the illocutionary functions of euphemism and dysphemism; and to identify 'dysphemistic euphemisms' and 'euphemistic dysphemisms'—which respectively label "dysphemistic locutions that have a euphemistic illocution" and "euphemistic locutions which have a dysphemistic illocution" (examples of the latter would be the expletives *Shoot!*, *Sugar!* and *Shucks!*).

By and large, euphemism and dysphemism are obverse sides of the same coin. In developing our account of them, we need to show how it is that they (particularly euphemism) have come to be so consistently and so ubiquitously maintained in human society. On the face of it, euphemism is not such a big deal in English as in, say, Austronesian societies in which violations of taboo are expected to have dire consequences. It seems to us that the real difference is in societal attitudes to violations of the prevailing social norms. The comparatively mild censure with which present day North American and European societies treat blasphemy, political dissent, and even crimes against property and persons, are comparatively recent innovations: such transgressions have been, and in some cases still are, among other things, violations of taboos. One does not need to look far back in history to find dire consequences for people foolish enough to be heard or observed violating them. Looked at this way, the social constraints on the observation of verbal taboos in places like Austronesia seem less exotic; in any case, many of their old taboos are disappearing under the influence of that hegemonous alien culture spreading down from the northern hemisphere. Sure, there are cultural differences with respect to the use of euphemism and dysphemism, but they are differences in degree rather than differences in kind. Attitudes to bodily effluvia, body parts, to notions of social status and the like, to death, disease, to dangerous animals, and to the supernatural, vary tremendously between cultures; but essentially these same parameters recur in every culture (and subculture) to motivate euphemism and dysphemism.

The point of view we have just described was determined after careful investigation of euphemism in a wide variety of languages, and that is why we shall occasionally mention languages other than present day English. In particular, of course, we compare euphemisms in English and Middle Dutch—that is, Dutch spoken during the Middle Ages. Why Middle Dutch? Well, if we had confined ourselves to early examples of English we would have been restricted to secondary sources like the *O.E.D.* To have to rely solely on secondary sources is a real drawback for a study such as this because of the lack of contextual details they provide. On the other hand, we do have at our disposal original Dutch texts, on which one of the authors (KB) has already worked extensively. So, we opted for a special consideration of Middle Dutch. Something which is not generally appreciated about Dutch is its very long written tradition, providing us with a repertoire of early texts as impressive as any to be found in a Germanic language; and there is the added advantage that many of these texts are in nonliterary genres. This is an important consideration for our study on two counts. First, we want something that gives an accurate picture of ordinary spoken language. There wasn't the same gap between the spoken and written word that we are used to today because there was no long established written tradition, and scribes were not confined by norms artificially prescribed by literary pedants. The lack of stylization and literary ambition in these texts means that, as far as we can tell, they reflect the vernacular language of their day (cf. Ch. 1 and Burridge [forthcoming] Ch. 1). Second, the nonliterary texts we chose deal particularly with sickness, death, sex, and bodily functions (i.e., with most of the topics which have been found to motivate euphemism). So, in style and thematic material, these Middle Dutch medical texts were ideally suited to provide the kind of historical perspective we sought for our study. Moreover, they gave us an appreciation of the Middle Ages which could never have been gained from consulting secondary sources. Middle Dutch might look a bit strange, but readers should not find it much more obscure than Middle English—in fact the two languages are tantalizingly similar. As a result of the Norman conquest in 1066, Middle English is significantly different from Old English. The effect was to create lots of doublets in the language; the one deriving from French or Latin would have more prestige, and was likely to be the euphemism; while the corresponding Germanic form, deriving from Old English, was more likely to be dysphemistic. In reality, the situation was far more complex than this suggests, but the overlay of Romance on Germanic that makes English such a mongrel language adds an unwelcome variable that is not present in Dutch. Although French was a prestige language for Dutch speakers also, it did not exert nearly the same influence on the Dutch language as it did on English at this time, particularly in the area of vocabulary, and for the most part Middle Dutch maintains its Germanic character from Old Dutch. For these reasons, we opted for a special consideration of Middle Dutch.

Much of the material we discuss comes from oral tradition in English and other languages; but we also draw heavily from written records for euphemisms, dysphemisms, and information about sociocultural arrangements and attitudes. Of course, it is true that written records are not the ideal source for those kinds of euphemism and dysphemism that blossom in the spoken language. However, to make this the reason for ignoring euphemism and dysphemism in 'dead' languages would be extraordinarily obtuse. There is plenty of relevant material available in works of art from time immemorial, ranging from great works of literature like Shakespeare's plays, through lesser writers like Catullus or the Earl of Rochester, to the occasional gem from graffiti writers; these offer many insights into the aspects of language use that are our concern in this book. And, of course, documents such as the medieval Dutch medical treatises that we report on, provide the only evidence available of the kinds of expressions used in a medieval speech community. Unless we can persuade Dr. Who (or someone else with a time machine) to take us back through history to investigate the spoken language of our forebears, their written records are all we have. Stricter conventions are observed in writing than in speech, because the domain of writing is usually public rather than private; but the readership expected for a medieval text was very much smaller than we would expect for a book today. Like speech, writing exists in many different forms; and, again like speech, different occasions will call for different conventions and different degrees of euphemism or dysphemism. The fact remains, however, that whatever drives these verbal behaviors is the same whether they are manifest in conversation, comic verse, formal courtly poetry, drama, or instructive texts such as the ones we examine here from medieval Holland. And we are concerned in this book not only with the formal expression of euphemism and dysphemism, but also with the traditions of etiquette, the aspects of social organization and culture, and the contexts which impel these verbal behaviours.

Chapter 1 defines euphemism, dysphemism, dysphemistic euphemism, and euphemistic dysphemism; and offers classifications for them. Chapter 2 looks at euphemism in naming and addressing people. Chapter 3 examines the taboos on bodily effluvia, especially the menstruation taboo, and tabooed body-parts; it seeks to explain why there are stronger taboos on the bodies and effluvia of women than those of men. Chapter 4 surveys the huge lexicon of tabooed effluvia, sexual activity, and all the associated body-parts. Chapter 5 examines the language of abuse in every field from profanity to obscenity to racism and the like. There is also a brief consideration of dysphemistic locutions used in framing nondysphemistic illocutions to one's friends. Chapter 6 studies the language of death: every society seems to euphemize at least some aspects of death. We look, too, at the deceptive language used by killers, both criminal and military. Chapter 7 takes a step back from death to tackle verbal expressions for the sick and their diseases. In Chapter 8 we examine jargon (sometimes called register): we show that jargon has euphemistic functions for in-groupers, but is too often dysphemistic towards out-groupers. Chapter 9 is on euphemism and dysphemism in works of art; in it we analyze three bawdy works: a modern ballad, a traditional ballad, and one example of Shakespeare's low comedy. Finally, in Chapter 10, we draw together the loose ends from earlier chapters and offer our concluding thoughts on euphemism and dysphemism.

We believe the content of this book you are about to embark upon, indeed that you have already embarked upon, contributes to both empirical investigation and theory. It is intended to reach a wide audience, ranging through specialists such as linguists, anthropologists, sociologists, psychologists, therapists, and nonspecialist readers, all of whom are, like us, fascinated by the rich material that we shall uncover that forges the shield and the weapon.

CHAPTER
— 1 —

Euphemism, Dysphemism, and Cross-Varietal Synonymy

Who ever stubbed his toe in the dark and cried out, 'Oh, faeces!'?
(Adams, 1985:45)

Q: How is a wombat like a man?
A. It eats$_{(,)}$ roots, shoots, and leaves.
(Traditional Australian feminist joke)

Euphemism

A painted I-won't-say-what, standing there just as the Lord made her, with only a little bit of shift to cover places that I won't mention.
(James R. Parker, *New Yorker* June 16, 1945:20)

In *The Golden Bough* Sir James Frazer wrote about the use of euphemisms by hunters in primitive (he might have said 'savage') societies:

The speaker imagines himself to be overheard and understood by spirits, or animals, or other things whom his fancy endows with human intelligence; and hence he avoids certain words and substitutes others in their stead, either from a desire to soothe and propitiate these beings by speaking well of them, or from a dread that they may understand his speech and know what he is about, when he happens to be engaged in that which, if they know of it, would excite their anger or their fear.
(Frazer 1911:417)

What Frazer says here is applicable to many of the euphemisms we shall be discussing in this book, but to encompass the enormously wide range of expressions that have been called euphemisms we need to define them as follows:

A **euphemism** is used as an alternative to a dispreferred expression, in order to avoid possible loss of face: either one's own face or, through giving offense, that of the audience, or of some third party.

11

Folk belief has it that what we are calling 'dispreferred expressions' typically denote taboo topics, and therefore might alternatively be called 'taboo terms.' In its original conception (the word *taboo* was borrowed from Tongan, an Austronesian language) taboo was prohibited behavior. It was prohibited because it was believed to be dangerous to certain individuals, or to society as a whole. To violate the taboo would automatically cause harm (even death) to the violator or his/her fellows. However, many of the taboo terms we shall discuss are avoided because their use is regarded as distasteful within a given social context: they are dispreferred, not because of any fear that physical or metaphysical harm may befall either Speaker or Hearer, but lest Speaker lose face by offending Hearer's sensibilities. Some speakers would claim that to utter the taboo terms would offend their own sensibilities, because of the supposed unpleasantness or ugliness of the taboo terms themselves (see upcoming discussion). Hock (1986:303ff) refers to 'tabooistic distortions' in the phonological forms of body-part names in the history of Indo-European languages; such distortions were probably motivated by superstitions in some cases, and by distaste in others. In Austronesian languages, such tabooistic distortions wreak havoc with the comparative phonology of once cognate terms in related languages to cause a great deal of confusion among comparative linguists (cf. Simons 1982 for exemplification of this and also extensive discussion of the linguistic effects of taboos current in Austronesia).

In fact, many euphemisms are alternatives for expressions Speaker would simply prefer not to use in executing a particular communicative intention on a given occasion or set of occasions. For example, *Time* magazine had the following to say about the U.S. House of Representatives: 'Bribes, graft and expenses-paid vacations are never talked about on Capitol Hill. Honorariums, campaign contributions and per diem travel reimbursements are' (*Time Australia* April 17, 1989:36). *Bribes* and *graft* are hardly taboo terms, even among politicians. Or take the example of a widow who prefers to say she has *paying guests* rather than *lodgers*: it is stretching credulity to claim that *lodger* is a taboo term, even for this lady; rather, it is the case that, to her mind, *paying guest* has more positive connotations—or alternatively, fewer negative connotations—than does *lodger*; similarly, the terms *honorarium*, *campaign contributions*, and so on, have much more positive connotations than *bribes*, *graft*, and *expenses-paid vacations*. Social taboos in most English-speaking communities stretch from those on bodily effluvia, reproductive processes, and the associated body-parts, which are (in some form or other) universal, to such matters of taste as (not) talking about personal income, which are rarer. In Hong Kong, Singapore, or the Philippines it is as polite to ask a new acquaintance *What is your salary?* as it is for Anglos to ask *What do you do?*, because in those countries one's income is 'free-goods', and the question will not affront Hearer's impositive (sometimes called 'negative') face. There are euphemisms without real taboo counterparts, ones that upgrade an alternative nomenclature. For instance the *paying guest* example just mentioned; the use of *corporate apparel* for "company uniform"; *revenue augmentation* for "increased earnings"—except when it is used by the government to mean "raising taxes," which is taboo; a barrister's *refresher* is "the fee for the second and each subsequent day"; *sanitation engineer* sounds more exalted than does "garbage collector," as does *vermin control officer* for "ratcatcher"; *comfortably off* does not mention wealth whereas "fairly wealthy" does; a *preloved* object sounds less worn than a "second-hand" or "used" one does; and a *starter home* or a *cozy home suitable for renovation* is so much more enticing than "small dilapidated cottage"; *underprivileged* sounds much better than "poor and needy," as does *senior citizens* rather than "old people." German *Gastarbeiter* "guest workers" sounds more welcoming than does *Fremdarbeiter* "alien workers"; *ethnic group* in place of "race" avoids contamination from *racism*; and *old and new Australians* doesn't even mention ethnic origins, though we should infer that *Old Australians* are Anglo-Celtic— not, you will notice, Aborigines—whereas new ones are not. Some of these euphemisms are motivated by pretentiousness, and are perhaps identifiable as jargon (see Ch. 8 for detailed discussion). Others are vaguely deceptive (e.g., the *cozy home suitable for renovation*, or *revenue augmentation* to mean "raise taxes"; and, even *guest workers*—a loan translation from German—can be seen that way). Militarese is replete with deceptive euphemisms (see Ch. 6).

A related kind of euphemism is illustrated in the following true story. A man we know was once the offended party in a minor traffic accident in the back blocks of a Third World country. It was suggested by the two police constables who turned up to check his vehicle that they could help him out by saving him the trouble of going to court. He expressed his gratitude for such help, and in return offered to take them out for a drink. 'Are you big drinking men, or little drinking men?' he asked. 'Little drinking men,' they replied. Being busy with other pressing matters, he couldn't promise to actually meet the policemen for a drink, so he gave them a little money to have a drink on him. The moral is that when a helpful policeman *drinks*, there is no need to mention *bribery*.

There is a whole batch of euphemisms for avoiding the mildly distasteful. For instance, consider the import of the parenthetical remark in the following quotation from *Desert Solitaire* by Edward Abbey (1968:40): 'I cross the swinging footbridge over Salt Creek pestered all the way by a couple of yellow cowflies (cattlemen call them deerflies).' *Deerfly* is a euphemism for cattlemen, but cowflies are not taboo; by calling these bloodsucking pests *deerflies*, cattlemen seek to avoid all responsibility for their numbers by associating them with feral beasts rather than their own charges. There are dozens of comparable examples in many fields of discourse; for example, the use of *gender* for "sex" (there is a distinction between them: *sex* is biological reproductive differentiation—though see Frank and Treichler 1989:11; *gender* is differentiation in a sociocultural role, it is also a grammatical (sub)category); *positive discrimination* is "discrimination that a majority within society approves of"; *After careful consideration we have regretfully concluded that your manuscript falls outside the scope of our current publishing program* is what publishers write when they mean "We don't want to publish your lousy manuscript"; *We'll have to let you go* replaces "You're fired"; *make redundant* is instead of "sack"; *take industrial action* instead of "strike"; *life insurance* "insurance for when you are dead" (delicious irony!); *disengage from the enemy* or *tactical withdrawal* for "retreat"; *misappropriate* for "steal"; *helping the police with their enquiries* for "arrested but not yet charged"; *adult videos / books* for "pornographic and/or violent videos/books"— in Tucson, Arizona, a place selling such things describes itself as *Continental Adult Shop*. The *Mature Media Group* is an Australian consortium of porn-movie makers. A prostitute may well work in a *massage parlor* and so describe herself as a *masseuse*; and the *working-girl* who used to be known as a *streetwalker* is someone who *will show you a good time*.

In short, euphemisms are alternatives to dispreferred expressions, and are used in order to avoid possible loss of face. The dispreferred expression may be taboo, fearsome, distasteful, or for some other reason have too many negative connotations to felicitously execute Speaker's communicative intention on a given occasion.

Types of Euphemism

Many euphemisms are figurative; many have been or are being the cause of semantic change; some show remarkable inventiveness of either figure or form; and some are indubitably playful. Euphemism can be achieved through antithetical means, such as by circumlocution and abbreviation, acronym or even complete omission and also by one-for-one substitution; by general-for-specific and part-for-whole substitution (terms we prefer to the more traditional 'synecdoche and metonymy'); by hyperbole and understatement; by the use of learned terms or technical jargon instead of common terms, and by the use of colloquial instead of formal terms. Many learned terms and some technical jargon is either borrowed from another language or constructed from one: for English, they are mostly derived from Latin or Ancient Greek. Most languages seem to have some euphemisms based on borrowed words or morphs. Consider some examples of these many types of euphemism.

Euphemism and Verbal Play: Figures, Metaphors, Flippancies and Remodellings

> Her *scotches*, long and slender
> Reached to her *kingdom come*,
> Her *hobsons*, low and husky
> Made my *newingtons* go numb.
>
> I took her for some *Lillian Gish*
> Down at the chippy caff.
> We squeezed into my *jam-jar*
> And drove back to my gaff.
>
> She then began removing
> Her full-length *almond rock*,
> Revealing size nine *how-de-do's*
> Which gave me quite a shock.
>
> And with a sexy *butchers*
> She murmured 'I'm all yours.'
> She then took off her *fly-be's*
> And dropped her *early doors*.
> (From 'Cockney's Lament' by
> Ronnie Barker 1979:21f)

[Key: *scotch [peg]* "leg"; *kingdom come* "bum, fanny"; *hobson's [choice]* "voice"; *Newington [Butts]* "guts"; *Lillian Gish* "fish"; *jam-jar* "car"; *almond rock* "sock/ frock"; *how-de-do's* "shoes"; *butcher's [hook]* "look"; *fly-be[-nights]* "tights, panty-hose"; *early doors* "drawers, underpants"]

We find **figurative expressions** like *the cavalry's come* for "I've got my period," and *go to the happy hunting grounds* for "die." The first of these is **metaphorical** in implicitly representing the onset of catamenia as the arrival of the redcoated cavalry (see Ch. 4 for more discussion). An even more inventive metaphor is *the miraculous pitcher, that holds water with the mouth downwards* for "vagina" (cf. Grose 1811). It seems unlikely that this lengthy example of verbal play was widely used, and its flippancy puts us in mind of euphemisms like *kick the bucket* for "die" with their real or pretended disdain for a taboo.

It is questionable whether *kick the bucket* or *get your finger out* can usefully be called euphemism rather than simply slang, which they undoubtedly are also. In Ch. 8 we will discuss the interrelation of jargons and slang with euphemism; however, at this juncture we merely note that certain slang terms, some of which might be judged euphemistic, exemplify verbal play (i.e., rhyme, quasi-reduplication, alliteration, pleasing rhythms, silly words, etc.). Examples (a couple of them archaic) are *over-shoulder boulder-holders* "bra"; '*Wham, bam, thank you ma'am!*'; *hoddy-doddy (all arse and no body)* "a short clumsy person"; *om-tiddly-om-pom* and *umpti-poo* "toilet"; *tantadlin tart* "turd"; *tallywags* or *twiddle-diddles* "testicles"; *doodle, diddle, dink, dong* "penis"; *tuzzy-muzzy* "vagina"; *rantum-scantum* "copulate"; numerous terms for "masturbate": *beat the bishop / beaver, pull the pope, pull one's pud, crank one's shank, jerkin' the gherkin, tweak one's twinkie, juice the sluice, stump-jump* (cf. Aman 1984/5:106). Then there is rhyming slang like in 'Cockney's Lament', some of which is quoted at the head of this section, also *whistle [and flute]* "suit," *jimmy-riddle* "piddle, piss," *groan and grunt* "cunt, girlfriend," *bristols [Bristol cities]* "titties, breasts," *Brahms [and Liszt]* "pissed, drunk," etc.

Remodellings like *sugar, shoot,* or *shucks* for "shit," *tarnation* for "damnation," *darn, dang,* and *drat* for "damn," *tidbits* for "titbits," *basket* for "bastard," *cripes* or *crumbs* for "christ," usually end up as one-for-one substitutions (see later) in which either the onset or rhyme of the dispreferred term is matched with that of a semantically unrelated word. In the southwestern United States one occasionally encounters a Spanish place name like *Tres Pinos* ("three pines"); the normal anglicized pronunciation of Spanish 'pinos' would be /'piinous/, making it almost homophonous with *penis*—indeed, fully homophonous if the final, unstressed, syllable is reduced to /-nəs/; to avoid the resulting embarrassment, 'pinos' is phonologically remodelled to /'painous/. Before *coney* "rabbit" dropped out of use in the late-nineteenth century, it was pronounced /'kounɪ/ to phonologically dissimilate it from the otherwise homonymous word meaning "cunt," normally pronounced / 'kʌɪ/ (see Ch. 4). Bislama, an English-based pidgin spoken in Vanuatu, has the following remodelled euphemisms: *kastom* "custom," *kasis* "cassia tree," *kasetoel* "castor oil" ll remodellings of *kan* "cunt"; *kos* "course" instead of *kok* "cock"; *bagraes* "bag of rice" instead of *bagrit* "bugger it"; *fas* "stuck" instead of *fak* "fuck" (cf.

Crowley 1989:105). Gandour (1978) describes 'khamphŭan' "word reverse" in Thai: 'In this word game, the vowel(s) and final consonant as well as the tone may be exchanged between two words or syllables' (p. 111). Thus *khray phăay lom* "who farted?" becomes *khray phŏm laay* "whose hair [is] striped?" Here the resulting string contains normal Thai words, but nonce words can result; for instance, *kháw lıng rabong* "he penis infected [= he's got the clap]" becomes *kháw long rabıng* "he down—."

A different kind of remodelling is exemplified by 'secret languages' or 'play languages' like Pig Latin, Eggy-peggy, Up-up, and so on (cf. Kirshenblatt-Gimblett 1976; Simons 1982:187). Instead of effecting one-for-one substitution on the pattern described above, there are regular morphophonological changes such as metathesis or affixation to every word in the secret language. Here is an example of Eggy-peggy in Nancy Milford's *Love in a Cold Climate*; Fanny has just been recognized as the Bolter's daughter by one of a group of women whom she'd just previously described as 'chattering like starlings in a tree':

> "Come and have your tea, Fanny," said Lady Montdore. She led me to the tea table and the starlings went on with their chatter about my mother in eggy-peggy, a language I happened to know quite well.
> "*Eggis shegge reggeally, peggoor sweggeet!* I couldn't be more interested, naturally, when you come to think of it, considering that the very first person the Bolter ever bolted with, was my husband . . .   (N. Mitford 1979:274. Italics added.)

According to our classification of euphemism, 'special languages' like the 'mother-in-law' languages of aboriginal Australia which, until recently, were used to certain categories of cross-kin (see Frazer 1911:346f; Dixon 1972:32, 1982 ch.2) are used for euphemistic purposes.

Circumlocutions, Clippings, Acronyms, Abbreviations, Omissions, and One-for-one Substitutions

There are **circumlocutions** like *little girl's room* for "toilet," and *categorical inaccuracy* or *terminological inexactitude* for "lie," or *the person I am wont to refer to by the perpendicular pronoun* for "I/me." The first is also figurative; the other two use the puffed up jargon of bureaucratese (or a pastiche of it) instead of a common term (see Ch. 8 for more discussion). In 1985 when the Australian dollar was taking a tumble, dealers and economists talked of it having *a substantial downside risk potential*. A recent Australian Broadcasting Commission program on education kept referring to *those on the lower end of the ability scale* (cf. Peterson 1986:54). They might also have said *low ability subjects* or *educationally disadvantaged / challenged groups*. The process involved here is a kind of componential analysis; the senses of taboo terms are unpacked and each of the meaning components are listed. The resulting periphrasis functions as a euphemism: *rape* becomes *criminal sexual assault* or (as the newspapers used to call it) *a serious offense against a woman*. Using this method new euphemisms can easily be created: *urine* becomes *excrementitious human kidney fluid; feces* becomes *solid human waste*, and *pus* is *viscous matter of a wound*.

There are **clippings** like *jeeze* for "Jesus," *bra* for "brassiere" (both end-clipped), and the archaic *nation* for "damnation" (cf. Grose 1811, this is foreclipped); **acronyms** like *snafu* for "situation normal, all fucked up" or *commfu* "complete monumental military fuck up"; **abbreviations** like *S.O.B.* for "son-of-a-bitch" or *pee* for "piss." Also *f---* instead of printing "fuck." (For the difference between acronyms and abbreviations see the Glossary). **Omissions** take this kind of euphemism one step further. There are quasi-omissions which substitute some nonlexical expression for the dispreferred term. For example, Grose (1811) contains the following:

> CAULIFLOWER. [. . .] Also the private parts of a woman; the reason for which application is given in the following story: A woman, who was giving evidence in a cause wherein it was necessary to express those parts, made use of the term cauliflower; for which the judge on the bench, a peevish old fellow, reproved her, saying she might as well call it artichoke. Not so my lord, replied she; for an artichoke has a bottom, but a **** and a cauliflower have none.

This, together with the appearance of **** in other entries, leaves no doubt that Grose uses four asterisks as a synonym for what he otherwise calls 'the monosyllable' (if, dear reader, you are uncertain of his meaning, be assured it will very soon be made explicit). The spoken counterparts to dashes and asterisks are things like *mhm*, *er-mm*, and so on; for example, in Pinero's *The Gay Lord Quex* (1926:116f) a lady says on discovering some French novels on her friend's table 'This is a little—h'm— isn't it?" (It is worth recording that the reply to this is 'I read those things for their exquisitely polished style; the subjects escape me.'—a likely story!) **Full-omissions** seem less common than **quasi-omissions,** but there is *There's the pot calling the kettle black* which omits *arse* from the end (cf. Grose 1811). Also *I need to go*, from which is omitted *to the lavatory*. We found one Middle Dutch writer who used *vernoy smans / swijfs* "irritation of the man's / woman's" thereby altogether omitting the name for the genital organ itself. This particular example reminds us of the Vagisil™ medication for *feminine itching*, which is discussed in the next section.

    There then are the **one-for-one substitutions** like *bottom* for "arse, ass", *casket* for "coffin." Both these illustrate meaning extensions, and are arguably figurative. Not all one-for-one substitutions are like that: for instance, Grose's *the monosyllable* for "cunt," nor the eighteenth and nineteenth century *inexpressibles* for "underclothes, breeches"—which also exemplifies the general-for-specific class of euphemisms.

General-for-Specific and Part-for-Whole Euphemisms

Euphemisms like the legal term *person* for "penis" (cf. Pannick 1985:145 for English law, and Vagrancy Act, State of Victoria (Australia), §7, 1(c)) employ a GENERAL-FOR-SPECIFIC strategy; this example is also a one-to-one substitution. (We refer to this use of *person* as the legal use, but it is occasionally found elsewhere; e.g. in Kerr's translation of Latin epigramist Martial's *penem*, though this dates back to 1919.) There are a couple of very nice examples of general-for-specific euphemisms in a 1989 magazine advert for Vagisil™ products. Vagisil Feminine Itching Medication is described as follows:

> This instantly soothing medication relieves external feminine itching as easily as aspirin relieves a headache.

> That's good news because minor feminine itching is about as common as a headache—caused by everyday things like jogging, pantyhose, even normal perspiration.

This advertisement was coupled with another:

> **Feminine moisture** end it now and stay fresher all day
> Now stay drier, feel fresher all day with VAGISIL™, the first Feminine Powder with a totally unique formula to solve wetness problems.

There are various subclasses of general-for-specific: the euphemism just mentioned is whole-for-part; *nether regions* for "genitals" invokes the **general-area-for-a-specific-area-within-it**; *go to bed* for "fuck" invokes the **usual-location-where-a-specific-event-takes-place**; U.S. President Richard Nixon's references to *prething* and *postthing* (where 'thing' = "Watergate break-in"), the use of *thingummybob* for "penis" (or whatever), and expressions like *the you-know-what* to denote almost anything that can be properly inferred from context—all these employ **the-maximally-general-for-something-specific** strategy for euphemism. Rather similar was (is?) the use of *inexpressibles* or *unmentionables* and perhaps *smalls* for "underclothing"; also Grose's use of *the monosyllable*: these employ **the-nonspecific-for-something-specific** strategy. And so on: the number of general-for-specific subclasses is probably boundless. Some of these would traditionally have been called **metonymies.**

A PART-FOR-WHOLE euphemism is demonstrated in *spend a penny* for "go to the lavatory" (from the days when public lavatories cost a penny to access); and *I've got a cough* may occasionally ignore the accompanying stuffed up nose, postnasal drip, and running eyes. Afrikaans *ghat*, originally "hole", is used in much the same way as British or Australian *bum*, or American *fanny*. But euphemisms of this kind, which would traditionally have been called **synecdoches,** seem comparatively rare.

Hyperbole and Understatement in Euphemism

Hyperboles (overstatements) are found in euphemisms like *flight to glory* meaning "death," or *villa in a premier location by the bay* referring to a "dilapidated artisan's cottage, five streets away from the bay," or *Personal Assistant to the Secretary (Special Activities)* for "cook" (cf. Rawson 1981:11, who remarks that this 'illustrates a basic rule of bureaucracies: the longer the title, the lower the rank'—presumably to upgrade the lower ranks in at least this one inexpensive respect; see Ch. 8 for further discussion).

There are euphemistic **understatements** like *sleep* for "die." Many general-for-specific euphemisms are also understatements; for instance, the maximally general ones like *thing* for whatever ("Watergate break-in", "genitals") or *deed* for "act of murder" (or whatever); but also *anatomically correct dolls* for "dolls with sexual organs"; and things like *companion, friend, this guy I'm seeing* and even *lover* for "regular sexual partner." Litotes like *He's not very bright* meaning "he's as thick as two short planks" fall into this class. At one time the French verb *baiser* meant "kiss" (*embrasser* has replaced it) but now has come to mean "screw, fuck"; the transition shows a nice euphemistic understatement that was, in this case, a part-for-whole euphemism.

"Excuse my French"—Euphemism Through Borrowing

[*Philip Swallow to Vic Wilcox, with Robyn Penrose as bystander*] 'That's why the whole business of the cuts has been such a balls-up. Excuse my French, Robyn.'
Robyn waved the apology aside.   (Lodge *Nice Work* 1989:344)

The use of Latin homonyms provides Standard English with euphemisms for bodily effluvia, sex, and the associated acts and bodily organs. The use of *perspire* instead of *sweat*, *expectorate* instead of *spit*, *defecate* and *feces* instead of *shit*, *copulate* instead of *fuck*, *anus* instead of *arsehole*, *genitals* or *genitalia* instead of *sex organs*, *vagina* instead of *cunt*, *labia* instead of *lips* [of the vagina], and so forth, is accepted practice when using Standard English. Until recently, translations of taboo terms from exotic languages, and descriptions of taboo acts, frequently caused an author to switch from English to Latin. For instance, Hollis in *The Masai: Their Language and Folklore* translates the story of the demon Konyek: at one point in the story Konyek sits beneath a tree in which a frightened woman is hiding, causing her to tremble so much *neisirisir ngulak*, which Hollis translates as 'Incipit mingere guttatim' (Hollis 1905:137). This would be more aptly rendered for an English readership as "it made her piss herself." His translation of the very brief Maasai tale ' 'L omon le-'ngai o en-gop' is as follows:

The story of the sky and the earth.
We understand that the sky once married the earth.
   Haec verba dicere volunt. Ut maritus supra feminam in coitione iacet, sic coelum supra terram. Ubi lucet sol et cadit imber, terra calorem recipit et humorem: non aliter femina hominis semine fruitur.   (Hollis 1905:279)

The Latin reads: "They say that just as a husband lies over his wife to make love, so does the sky lie above the earth. When the sun shines and the rain falls, the earth receives heat and water: in the same way a woman is fertilized by a man." The final sentence of the Maasai text translates: "If the sun beats down on it and the rain falls, what is underneath arrives there [interpret that as you will!], and so it is with a man and his wife." Even in Lewis and Short's *A Latin Dictionary* (impression of 1975, though it dates from 1879) the meaning for *cunire* is given as 'est stercus facere' instead of the English "have a shit" or even "defecate." It is probable that Latin was euphemistically used because of the author's prudery in not wishing to use everyday English terms, but with the added rationalization that the Latin text would be uninterpretable to the uneducated—and therefore to the young and innocent.
   Latin is not the only foreign language source for English euphemisms. There is also French: *po* for "chamber pot" from French *pot* (in which the final *-t* is not pronounced), *lingerie* for "women's underclothing," *masseuse* for "whore," *matériel* for "armament and ammunition," *sortie* for "a sallying forth by a military unit"; and so on. According to our definition of euphemism, *brassiere* is not a euphemism because there is no alternative. Indeed, the abbreviation *bra* is more like a euphemism than *brassiere* itself. On the other hand, the use of French is per se euphemistic, and there really is no native English term—other than such horrors as *tit-covers*, *breastplates*, *over-shoulder boulder-holders*, (cp. the Viennese *Mirabellenetui "plums-case"*). In America, particularly in proximity to New York, Yiddish is a source for euphemistic

terms such as *tush(y)* "bottom," which is really "Yinglish" from Yiddish *tokhes/tukhes* itself a euphemism derived from Hebrew *takhath* "under". *Schlong* "penis" from Yiddish *schlang* "snake" or "penis" may not be exactly euphemistic.

Using words borrowed from other languages to function as euphemisms is characteristic of many languages. For instance, among Thais fluent in English, including doctors addressing well-educated patients, English words are used in preference to Thai euphemisms for penis, vagina, menstruation (abbreviated to *men*—but probably not from *menses*) and sperm. It is interesting that other body parts and bodily effluvia are not generally rendered in English, just those connected with reproduction. The national language of Papua New Guinea is Tok Pisin, which developed out of pidgin English (*Tok Pisin* < *talk pidgin*); and Holzknecht (1988:58) reports that 'The English and now Tok Pisin items *sit* "shit" and *pak* "fuck" are often used by Adzera speakers in front of in-laws, when the Adzera items are strongly tabooed.' Adzera is a Markham (Austronesian) language spoken in Morobe Province in Papua New Guinea, and Holzknecht writes of Markham language users in general:

> In social situations where people are likely to become angry with each other, where they might transgress the taboo rules and say something unacceptable in front of an in-law, Tok Pisin is often used. When people are arguing with each other, during football matches and when men are drunk, they will speak to each other in Tok Pisin rather than their vernacular. In this context, Tok Pisin is neutral and the words have no power to harm anyone.   (Holzknecht 1988:67)

Learned Terms or Technical Jargon and Colloquial or Common Terms as Euphemisms

In the previous section we exemplified the euphemistic use of learned terms or technical jargon instead of common terms, an example is *feces* for "shit." The antithetical strategy is to use colloquial rather than more formal terms, as with the use of *period* for "menstruate."

Which is the Euphemism?

It may seem obvious that *pass away* is a euphemism whereas *die* is the dispreferred term, and so forth. But the truth is, as usual, not so straightforward. How many of us would seriously say *These flowers have passed away* rather than *These flowers have died?* Compare the terms *menstruation* and *period*. There are occasions the former is the euphemism; others when the colloquial term is more appropriate; and yet other occasions when the learned and the colloquial terms are equally appropriate—where neither is a euphemism. It follows that, although we believe all euphemisms can be analyzed as the preference for one expression over another, we do not believe that one term will always necessarily be 'dispreferred' while the other is concomitantly a 'euphemism'—in such a way that they ought to be thus characterized in the dictionary.

Normally, the choice between alternatives depends entirely on context. For instance, the choice between *menstruation* and *period* is a matter of style, and one expression will often be more acceptable than another within a given style. For instance, in Ch. 4 we will discuss 'The menstruation taboo': to refer to this as 'The period taboo' would be inappropriate; or, in other words, dispreferred in the context of this book. The choice is not always so clear: in an article on the victims of bank holdups and the like in *Time Australia* September 12, 1988:24 was the following:

> "A lot of stress starts to happen for those who have been lying on the floor with a gun at their head," says Michelle Mulvihill, a Sydney psychologist. "Women lose their periods; people develop migraines, backaches and symptoms of real anxiety." She says . . .

In this context, Mulvihill could just as well have said *Women stop menstruating*, although it would have been marginally more formal than 'Women lose their periods.' The comparative informality of the latter is consistent with Mulvihill's style of spontaneous speech (as quoted). Furthermore, *period* seems to be the most commonly used noun among contemporary British, Australian, and American women: note, however, that this does not, in our view, make *menstruate* the 'dispreferred' term. *Menstruation* is an abstract noun whereas *period* is a concrete noun; hence one can say *my period* but not **my menstruation*—from which we would understand 'my menstrual cycle', which is not synonymous with *my period*. Not also that there is an adjective *menstrual* which has no felicitous counterpart derived from *period*. These different characteristics are consistent with the greater formality of the term *menstruation*.

Euphemism and style are not the same thing, they intersect and interact: the style used defines the set of euphemisms which are conventional within that style; the euphemisms used help to define and maintain a particular style. For example, circumlocution and metaphor characterize high style, both in polite society and in allegorical literature; learned terms are used in formal styles and professional jargons like medicalese, and remodelling in colloquial styles; see Ch. 8 for more discussion.

Despite our claim that particular expressions are not necessarily euphemistic in all contexts, it would ignore reality to pretend that ordinary people do not speak as if some expressions are intrinsically euphemistic—for instance, *loo* is euphemistic, whereas *shithouse* is not. What seems to be meant by this is that, in order to be polite to a casual acquaintance of the opposite sex in a formal situation in a middle class environment, one would normally be expected to use the euphemism rather than its dispreferred counterpart(s). When we describe some expression as a euphemism, without reference to the context of its use, this is what we have in mind.

Are Some Words Intrinsically Nasty? No! But Dirt Clings

> Annum *appellas alieno nomine; cur non suo potius? Si turpe est, ne alieno quidem; si non est, suo potius.*

> "When you speak of the *anus*, you call it by a name that is not its own; why not rather call it by its own [= *culus*]? If it is indecent, do not use even the substituted name; if not, you had better call it by its own."
>
> (Cicero *Epistulae ad Familiares*, IX,xxii)

Despite the fact that every language has some vocabulary based on sound symbolism, and even though everyone within a language community must tacitly concur in using a certain form with certain meanings, the correlation between the form and the meaning of a language expression is nonetheless arbitrary. Therefore, there is no a priori basis for distinction between the mentionable euphemism and the unmentionable taboo term. It is mysterious why the euphemisms *pass away, misappropriate, We'll have to let you go,* and *I'm going to the loo* have fewer unpleasant connotations than their corresponding taboo terms *die, steal, You're fired,* and *I'm going for a piss.* The difference presumably derives from a 'naturalist hypothesis' (cf. Allan 1986a §2.8), a persistent belief that the form of an expression somehow communicates the essential nature of whatever it denotes; in Frazer's words: 'the link between a name and the person or thing denominated by it is not a mere arbitrary and ideal association, but a real and substantial bond which unites the two' (1911:318). Taboo terms are contaminated by the taboo topics that they denote; but by definition the euphemisms are not—or not yet—contaminated. In fact, a euphemism often degenerates into a taboo term through contamination by the taboo topic. For example, Cicero observes that Latin *penis* "tail" had been a euphemism for *mentula:* 'At hodie *penis* est in obscenis' "But nowadays *penis* is among the obscenities" (*Epistulae ad Familiares,* IX, xxii). English *undertaker* once meant "odd-job man" (someone who undertakes to do things), which was used as a euphemism for the person taking care of funerals; like most ambiguous taboo terms, the meaning of *undertaker* narrowed to the taboo sense alone, and is now being replaced by the euphemism *funeral director.* What often happens with euphemisms like this, is that they start off with a modifying word, 'funeral' in *funeral undertaker,* then the modifier is dropped as the phrase ceases to be euphemistic, consider: *mentally deranged > deranged* and *lunatic asylum > asylum.* It is conceivable that *funeral director* will one day be clipped to mere *director,* which will then follow *undertaker* and become a taboo term. Other examples of single word euphemisms that have degenerated are: English *hussy* originally "housewife", French *fille* and German *Dirne* both originally "girl", and Ancient Greek *hetaira* originally "female friend"—all of which were also euphemisms for *prostitute* that themselves became taboo terms. (In French, "girl" is now *jeune fille,* in German *Fräulein* or *Mädchen;* in Ancient Greek, "girl/female friend" came to be rendered by *philē.*) The once euphemistic 'toilet' (from French *toile* "cloth") in *going to the toilet* is being, or has already been, superseded by *bathroom* or *restroom* in American and *loo* in spoken British and Australian. Perhaps we should say that *toilet* is 'fading' as a euphemism and may well disappear, as has *necessary house* (see Grose 1811). We know of no converse histories in which a taboo term has been elevated to a euphemism, with the possible exception of Greek *kakos* "bad": it derives from the Indo-European root **kak(k)-* "shit", which is the source for our *cacky* and its cognates in other Indo-European languages (cf. Arbeitman 1980:78f); of course, Indo-European **kak(k)-* might have been a euphemism that once meant "bad" and later became contaminated. Certainly, some euphemisms that have degraded into taboo terms come back from the abyss after they have lost their taboo sense: during the seventeenth and eighteenth centuries the verb *occupy* meant "copulate", during which time non-taboo senses lapsed; it only re-entered the lexicon in its current sense of "inhabit, take up" after it had ceased to be used in the taboo sense. All this supports the view that taboo terms are classified as such because of a belief, be it ever so vague, that their form reflects the essential nature of the taboo topics they denote. This is exactly why the terms themselves are often said to be unpleasant or ugly sounding, why they are miscalled 'dirty words.'

There is a wealth of evidence that where a language expression is ambiguous between a taboo sense and a nontaboo sense its meaning will often narrow to the taboo sense alone. Some examples: (1) The noun *accident* once meant "that which happens, a chance event" (cf. *accidentally, by accident*), but its association with misfortune has narrowed the meaning to "chance misfortune" in *There was an accident, He had an accident*. (2) *Coney* (rhymes with *honey*) was the word for "rabbit" until the late nineteenth century, when it dropped out of use because of the taboo homonym meaning "cunt" (see Ch. 4). (3) The British still use *cock* to mean "rooster"; however, because of the taboo homonym meaning "penis," this sense of *cock* started to die out in American in the early nineteenth century; it is nowadays very rare in Australian. There has also been an effect on words containing *cock*: for example, former Mayor Ed *Koch* of New York City gives his surname a spelling-pronunciation /koč/ which rhymes with *Scotch*; the family of Louisa May *Alcott* (author of *Little Women*) changed their name from *Alcox*; *cockroach* is often foreclipped to *roach* in American; but on the other hand, *cockpit* and above all *cocktail* show no sign of being avoided. Also, although there were other factors at work too, the use of *haystack* in place of *haycock*, and the use of *weather-vane* as an alternative to *weather-cock*, were undoubtedly influenced by taboo avoidance. (5) Bloomfield (1927:228) noted that *ass* meaning "donkey" was dropping out of use in American because of the taboo homonym (meaning "arse or cunt"). (5) Cicero in *Epistulae Ad Familiares* (IX,xxii) pointed out that *ruta* "rue" and *menta* "mint" can be used without impropriety; the same is true for the diminutive of *ruta, rutula*, but not of *menta*, because the resulting *mentula* means "penis". (6) Hock (1986:295) believes that phonetic similarity to *fuck* led to the demise of the following words; *fuk* "a sail"; *feck* (cp. *feckless*); *feck* or *fack* "one of the stomachs of a ruminant"; *fac*, an abbreviation for *factotum*; *fack(s)* "fact(s)". We think he drives his argument too far, however, because the pronunciation /fæk(s)/ is still used freely, indeed so freely that 'FAX' (facsimile) machines are almost a fad. (7) Since the 1960s the adjective *gay* has been used less and less in the sense "bright, full of fun" because it also has the meaning "homosexual". Farmer and Henley (1890–1904) have the nineteenth-century slang sense of *gay* as "given to venery"; and a *gay girl* was "a strumpet", not a lesbian. Partridge (1970) reckons the sense "male homosexual" came into being 'since ca. 1930—if not much earlier.' The extension of *gay* to female homosexuals is comparatively recent, since the 1960s.

There are two reasons why language abandons homonyms of taboo terms. One is the relative salience of taboo terms. In 1935, Allen Read wrote: 'The ordinary reaction to a display of filth and vulgarity should be a neutral one or else disgust; but the reaction to certain words connected with excrement and sex is neither of these, but a titillating thrill of scandalized perturbation.' (Read 1977:9). Osgood, Suci, and Tannenbaum (1957) discovered a general tendency for any derogatory or unfavorable denotation or connotation within a language expression to dominate the interpretation of its immediate context. MacWhinney et al. (1982:315) found that 'sentences with profane and sexually suggestive language elicited responses quite different from those [without. . . .] Sentences with off-color language [Euphemism!] possess a memorability that is quite independent of their role in conversation.' The reason for this may be that obscene vocabulary is stored or accessed differently in the brain from other vocabulary; the evidence for this comes from people manifesting 'Gilles de la Tourette's syndrome, which is characterized by unusual tics progressing to involuntary outbursts of foul language (coprolalia)' (Valenstein and Heilman 1979:431). These people may lose all other language ability, which would only be possible if the means of storage and/or access were separate from that of obscene vocabulary, although why there should be such a separation is a mystery.

The other reason for abandoning the homonyms of taboo terms is that a speaker won't risk appearing to use a taboo term when none was intended. For example, there are some (mostly older) English speakers who, if they catch themselves using the adjective *gay* in its former sense will, with mild embarrassment, explicitly draw attention to this intended meaning. Their late nineteenth-century forbears, fearful of seeming impropriety, avoided the terms *leg* and *breast* even when speaking of a cooked fowl, referring instead to its *dark* (or *red*) *meat* and *white meat*.

Where there is little likelihood of being misunderstood, the homonyms of a taboo term are likely to persist in the language. This is the case for instance with *queen* "regina", which is under no threat from the homonym meaning "gay male" simply because one denotatum is necessarily female, the other is necessarily male; the converse holds, *mutatis mutandis* for the end-clipped American epithet *mother* "motherfucker". Similarly we experience no constraint in saying *It's queer* but we generally avoid saying *He's queer* if we mean "He's peculiar" preferring *He's eccentric* or *He's a bit odd*. *Bull* meaning "bullshit" is dissimilated from *bull* "male, typically bovine, animal" because it heads an uncountable noun phrase instead of a countable one.

Dissimilation, however, is not always a protection for the innocent language expression. For instance, *regina* makes some people feel uncomfortable because of its phonetic similarity to *vagina*. It is quite usual for speakers to avoid expressions that are phonetically similar to taboo terms (cf. Keesing & Fifi2i 1969:162f and Ch. 2). Then, again, one author recalls a nonnative graduate student presenting a paper in which he several times used the phrase 'my testees' to refer to "those subjected to a test": this appellation provoked a good deal of barely suppressed mirth in part of the audience. Pornpimol Senawong tells us that bilingual Thais may get apprehensive about using the Thai word *fuk* "gourd, pumpkin" in the hearing of other Thais fluent in English. *Fuk* is used for the name of the main character in the award-winning Thai novel *Kham Phi Phaksa* (*The Judgement*) by Chart Kobjitti, and there was much speculation about how the name would be transliterated when the novel was translated into English. We can report that the translator called him 'Fak'. Thai English-teachers experience some embarrassment, and their students some amusement, with the English word *yet*, which is the equivalent of "to fuck" in colloquial Thai. Farb (1974:82) reports something similar: 'In the Nootka Indian language of Vancouver Island, British Columbia, the English word *such* so closely resembles the Nootka word meaning "cunt" that teachers find it very difficult to convince their students to utter the English word in class.' Similar reports of cross-language effects have been reported elsewhere, too (e.g., Cicero *op cit.*). Such is the power of taboo.

The Middle Dutch Perspective on Euphemism

For reasons given earlier, the authors' original intention was to discuss and compare euphemism and taboo in late twentieth century English and Middle Dutch (spoken in the Netherlands between approximately 1100 and 1550 CE) with the aim of using the comparison to gain perspective on usage in both languages. It is for that reason that Middle Dutch features very significantly in this book, even though we have been led beyond our original goal to a discussion of more theoretical matters.

We start with the MIDDLE Dutch because nothing substantial remains of Dutch at any earlier period. The only evidence of original Old Low Franconian or so-called Old Dutch is one sentence, hastily scribbled in the margins of a Latin manuscript sometime during the eleventh century. It reads:

> *Hebban olla uogala nestas bigunnan*
> *hinase hi[c e]ende thu*
> *uu[at] unbida[n uu]e nu*

> "All the birds have begun nests,
> except for me and you.
> What are we waiting for now?"

Old Dutch was clearly as euphemistic as any language today.

It is a curious assumption among certain writers today that euphemisms are of recent origin (for example, see Adler 1978:73f). Perhaps this is because of the way in which we are made to perceive languages of the past. Their verbal patterns are preserved for us now in the neat lists and regular paradigms of the modern handbook and there they remain in an almost fossilized state. As Burchfield (1985:20) says of Old English, 'it is almost as if its main reason for surviving was to supply paradigms and fine-spun sentences for grammarians and literary historians.' It is impossible from these handbooks to gain any animated impression of a living, breathing speech community existing at an earlier time. Such mummification can tell us little of the subtleties of social attitudes which might have once prevailed.

Why shouldn't these languages have been as euphemistically rich as any language today? Modern society has its deodorizing sprays and perfumes, but we should not underestimate the scent balls, the sweet-smelling vapors, and inhalations of the Middle Ages. People during this time were no more free of 'deodorizing' language than we are today. In fact, with filth, disease, and death existing on a more extensive scale, the need to hide behind the sweet-smelling euphemism was perhaps once even greater. Indeed, there is no evidence that euphemism is absent from the language of any post-Neolithic human community (cf. Frazer 1911; Griffin 1985; Burchfield 1985).

The Middle Dutch texts examined here consist of medical treatises from four-teenth and fifteenth-century Holland and are therefore written in the nonliterary style typical of this sort of technical prose or 'Fachprosa'. The medical literature of the Middle Ages is not known for its originality; in fact, original thought did not feature strongly in any contemporary works—what we would today call 'plagiarism' was an accepted practice. Most of the texts consulted contain some descriptions based on original observations, but are largely compilations of ideas contained in early Latin works. These are not translations, however; they show a language quite independent of the Latin.

These medical texts have two important advantages for a study of euphemisms. For one, they contain little, if any, literary artificiality, and reflect closely the spoken idiom of the time (so far as we are in a position to tell, see Burridge, forthcoming, Ch. 3). Certain elevated prose styles, particularly of the late middle ages, make extended use of highly elaborate and artificial expressions that go beyond the euphemism of ordinary language; and poetry is complicated by the formalisms of metre and rhyme. Not only are the medical texts free from literary ambition, but it is in precisely these types of texts that discussion of sickness, death, sex, bowel movements, menstruation, and so on, abound.

When relying on early texts, it is sometimes difficult to know what was considered appropriate and what was not; consequently, it is difficult to assess the degree of delicacy of some of the terms used. It seems clear, however, that most of the things which cause us anxiety and embarrassment today were also of concern then, and avoidance terminology was as much a part of their language as it is of ours. But, as will be shown in Chs. 3, 6, and 7, there are some interesting divergences from modern practices.

Dysphemism

Australia 4 Australians : Slopes, Wogs, Pommie Bastards, Czecks, Convicts, COMMIES—this country is just one big boiling pot of euphemism.
                (*Sic*. Men's toilet, La Trobe University, 1989)

Euphemisms are alternatives to expressions that, for one reason or another, have too many negative connotations to felicitously execute Speaker's particular communicative intention in a given context. In referring to 'a particular communicative intention in a given context,' we draw attention to the fact that Speaker chooses either to use or to not-use a euphemism in order to create a certain effect on a given occasion (e.g., there are occasions when one chooses between saying *I'm going to the loo* and *I'm going for a piss* for different effects, and others when good manners absolutely constrain the choice to the former). What motivates the choice is in many ways similar to what motivates the choice between saying *Abu Nidal is a freedom fighter* and *Abu Nidal is a terrorist*. On some occasions at least, the latter can be regarded as a 'dysphemism' because the term *terrorist* has unfavorable connotations, and is selected for that purpose. A dysphemism, then, is used for precisely the opposite reason that a euphemism is used and we define it as follows:

> A **dysphemism** is an expression with connotations that are offensive either about the denotatum or to the audience, or both, and it is substituted for a neutral or euphemistic expression for just that reason.

Dysphemisms, then, are used in talking about one's opponents, things one wishes to show disapproval of, and things one wishes to be seen to downgrade. They are therefore characteristic of political groups and cliques talking about their opponents; of feminists speaking about men; and also of larrikins and macho types speaking of women and effete behaviors.

Dysphemism employs most of the same strategies as euphemism, but there are two main differences. One is that part-for-whole (synecdochic) dysphemisms are used far more frequently than are general-for-specific ones, which is the converse of the situation with euphemisms (e.g., the use of *tits* for "breasts" is part-for-whole, as are figurative epithets like in *He's a prick*, which contrast with euphemistic counterparts showing whole-for-part substitutions like *chest* and (legal) *person*, respectively). The other difference is that the antithesis between 'hyperbole and understatement' is inappropriate. Hyperbole can be used to magnify the offense, as in *You great prick!* or *He's the rottenest bastard I ever come across.* But hyperbole is also employed when diminishing or downgrading someone, as in *You slimy little toad!* or *The nasty little creep!* or *Pea-brain!*. Bitterly sarcastic remarks from someone waiting for two counterclerks to finish their conversation and offer service sometimes take the form of dysphemistic understatements like *If you could just spare me a FÉW moments of your time . . .* It is a moot point whether that sarcastic 'FÉW' is understatement or hyperbole: there seems to be no clear distinction between them in dysphemistic discourse. Other differences between the strategies for euphemism and those for dysphemism are predictable: circumlocution is most usually dysphemistic when it manifests an unwanted jargon (see Ch. 8); the use of borrowed terms and technical jargon is only dysphemistic when intended to obfuscate or offend the audience; and so forth.

Consider some examples. NATO has a *deterrent* (euphemism) against the Russian *threat* (dysphemism). In the mid-1980s the Soviet Union claimed to have been *invited* (euphemism) into Afghanistan; the Americans claimed that the Russians were *aggressors* (dysphemism) there. Dysphemism is indicated by the term *so-called* (e.g., *the so-called democracies of the Eastern block* doesn't make a dysphemism out of 'democracies,' but it does indicate disagreement with and disapproval of the presupposition that there are such things as democracies in the Eastern block). The latter phrase, *Eastern block*, is itself dysphemistic; note the totalitarian and obstructive connotations of 'block' when contrasted with the free-among-equals connotations of *Western alliance*. (The spread of *glasnost* to central and eastern Europe and the concomitant moves toward multiparty democracies in the 'eastern block' in the latter part of 1989 is leading to the abandonment of these dysphemisms.) The acronym *GRID*, from *Gay Related Immuno-Deficiency*, once denoted the disease we now know as *AIDS*. It came to be considered dysphemistic because it associated sufferers uniquely with the gay community, thereby offending heterosexual patients such as hemophiliacs. The word *grid* also evokes negative images of the grills and gratings one associates with a prison cell. In comparisons like *I'm generous, but she's spendthrift, I'm careful, but he's mean,* or *I'm strong-minded and he's plain obstinate*, the second clause is intentionally dysphemistic by comparison with first. Compare Bertrand Russell's celebrated conjugation *I am firm, you are stubborn, he is pigheaded.*

Dysphemistic terms of insult found in personal disputes of a colloquial nature include: (1) Comparisons of people with animals conventionally ascribed certain behaviors (e.g., calling someone a *louse, mouse, bird, dove, hawk, coot, galah, chicken, bat, rat, cat, mongrel, cur, bitch, dog, fox, vixen, rabbit, sow, pig, cow, bull, ox, goat, ass/donkey, mule, snake, ape, monkey,* etc. See Ch. 5 and Leach 1964. Some of these can be used euphemistically, of course). (2) Epithets derived from tabooed bodily organs, bodily effluvia, and sexual behaviors (see Ch. 5). (3) Ascriptions of mental or physical inadequacy, such as *idiot, fuckwit, airhead, nincompoop, fool, cretin, moron, maniac,* and so on; *spastic, weakling, baldy, four-eyes,* and so on; see Ch. 5. (4) Finally there are terms of insult or disrespect, some of which invoke slurs on the target's character (e.g., *biddy, crone, hag, bag, battle-axe, codger, geezer, crank, fuddy-duddy, fuss-budget, grump, fogy, galoot,* etc.). Afrikaners often use *skepsel* "creature" when referring to Blacks and Coloreds; in Nazi German, Jews were described as *kriechend* "crawling, servile"; Nazis described the marriage of an 'Aryan' to a 'non-Aryan' as *Blutschande* "blood disgrace" or *Blutvergiftung* "blood poisoning, tetanus" (see Ch. 5 and Clyne 1987).

Like euphemism, dysphemism is not necessarily a property of the word itself, but of the way it is used. There is nothing intrinsically dysphemistic in the word *Asian,* but when they are in Australia, many people from Asia feel that being described as 'Asians' rather than more particularly as Chinese, Indian, Kampuchean, or Thai, for example, is dysphemistic. The point is more vividly demonstrated by the use of the word *liberalism* in a racist tabloid: '. . . simply a manifestation of the sickness called liberalism which is carrying Western man swiftly toward his extinction' (Strom 1984:7, quoted by Clyne 1987:38). Here 'liberalism' is a dysphemism.

It will become abundantly clear in Ch. 5 that many words suffer pejorization; it usually results from society's perception of a word's tainted denotatum contaminating the word itself. The degree of contamination perceived in the denotatum ranges on a scale that has fear, abhorrence, loathing, and contempt at one end, and nothing worse than low social esteem at the other. Earlier we showed that a number of words meaning "girl" degenerate into dysphemisms as the result of having been used as euphemisms for "prostitute," or at least "wanton woman" (which seems to be the Standard English synonym for *easy lay*). Because poor unmarried women often went into service, both *maid* and *wench* came to mean not only "girl, young woman" but, within a century, also "servant girl." From there, *wench,* typically the servant in a lower class establishment, slid down to "wanton woman"—though normally with a specifying modifier such as *light wench* or *wench of the stews.* Since the eighteenth century *wench* seems to have recovered somewhat, certainly if one compares its present day use with *harlot, whore, scrubber,* and such. It is notable that present day English *girl* can be used in all the senses *wench* has had; it will depend on context which sense is operative. If you think this proves speakers of late Middle/early Modern English were misogynistic, note that until the fourteenth century *knave* meant "boy"; then *boy* took over. The present meaning of *knave* arose during the thirteenth century, and eventually cuckooed the nondysphemistic sense. Throughout the ages countryfolk have been held in low esteem in most communities. Latin *urbanus* "townsman" gives rise to *urbane* "sophisticated, elegant, refined" versus *rusticus* "rustic" with connotations of "clownish, awkward, boorish." *Boorish* means "ill-mannered, loutish, uncouth" and derives from the noun *boor,* which existed in Old English and lives on in *neighbor*; however, that seems to have faded and been reintroduced from the Dutch *boer* "farmer" (before Afrikaners got to the Cape). English *churl* once meant "countryman of the lowest rank", and *churlish* already meant "ill-tempered, rude, ungracious" by the early Middle English period. According to the *O.E.D.*, a *villain* was "a low born or base-minded rustic" and 'in later use, an unprincipled or depraved scoundrel; a man naturally disposed to base or criminal actions, or deeply involved in the commission of disgraceful crimes'— which is the meaning today. The pejoration in all these cases seems to have been extraordinarily rapid, a century at most, if we are to believe the *O.E.D.*

We said earlier that some people will avoid using a word that even sounds similar to a taboo term. There are others, of course, who will deliberately use such words humorously, as a tease. One classic example is in the Monty Python film *Life of Brian* where Michael Palin as Pontius Pilate says with a lisp, *I have a fwiend in Wome called "Biggus Dickus"*. We know of a linguistics professor who gently taunts his graduate students by referring to the material gathered for their thesis as *thecal matter*. And there is another who while lecturing on the relationship of allophones to phonemes inadvertently said "*What the allophones of an abstract phoneme /A/ have in common, is their A-ness /'einəs/.*" Having said it, he immediately recognized the ambiguity of the focused term 'A-ness', so quick as a flash he asked, "*And what do allophones of phoneme /P/ have in common?*"

Like euphemisms, dysphemisms interact with style and therefore have the potential to produce stylistic discord; an example would be where someone at a formal dinner party publically announced *I'm off to have a piss*, rather than saying something like *excuse me for a moment*. According to our definition, euphemisms and dysphemisms are deliberate. However, they may occur inadvertently, for instance when someone commits a social gaffe as Eliza Doolittle does in Shaw's *Pygmalion* Act III.

LIZA	[*nodding to the others*] Goodbye, all.
FREDDY	[*opening the door for her*] Are you walking across the Park, Miss Doolittle? If so—
LIZA	[*perfectly elegant diction*] Walk! Not bloody likely. [*Sensation*]. I am going in a taxi. [*She goes out*]. *Pickering gasps and sits down.*
MRS EYNSFORD HILL	[*suffering from shock*] Well, I really cant get used to the new ways. (SHAW 1946:78)

Usually, the use of jargon to people not initiated to it is inadvertently dysphemistic— but it can be deliberately used to exclude Hearer, for reasons we discuss in Ch. 8. Because of their offensive nature, inadvertent dysphemisms will draw attention to themselves in a way that inadvertent euphemisms do not.

X-Phemisms and Cross-Varietal Synonymy

For convenience in discussion, we will call the set union of euphemisms and dysphemisms X-phemisms. X-phemisms (i.e., both euphemisms and dysphemisms) are cross-varietal synonyms. In other words, an X-phemism (e.g., *shit*) means the same as some other expression (e.g., *feces*), but the two are typically used in different contexts, perhaps in different varieties or dialects of the language. To be a little more technical (there is a formal definition of cross-varietal synonymy in Allan 1986a §3.6.4), cross-varietal synonyms have the same denotation but differ in connotation. In Britain or Australia one might say *I'll ring you tomorrow evening* where in the United States one would say *I'll call you tomorrow evening*. The verb *ring* in two English dialects denotes the same as the verb *call* does in a third dialect, namely "telephone". Take another example. The nouns *dandelion* and *taraxacum densleonis* are also cross-varietal synonyms: they denote exactly the same kind of plant, but because they have different connotations, they are typically used in different circumstances. People may have differing opinions about the judgment of someone who claims to like *dandelion wine*; but there is only one verdict on someone who claims to like *taraxacum densleonis wine!* Again, the same female human being may be referred to using any of the nouns *girl, woman, lady, lass, broad, chick, sheila, hen* (and a whole lot more cross-varietal synonyms); the connotations of these words differ, and there are contexts in which only one is appropriate, any other would be dysphemistic. It is quite possible to deny the applicability of one term while asserting what amounts to a preference for the appropriate connotations of its cross-varietal synonym, as in *He's not a lodger, he's a paying guest*. The difference is a difference of style, or more exactly, of jargon; and the interaction between X-phemism and jargon is something we pay close attention to in Ch. 8. Because we define euphemism and dysphemism in terms of choices between alternative expressions, we presuppose that each such expression has at least one synonym; but it is sometimes hard to figure out what the supposed alternative is, and in such cases we ask the reader to be charitable, and to bear in mind Oscar Wilde's aphorism: 'The truth is rarely pure, and never simple' (*The Importance of Being Earnest*, Act I).

Euphemistic Dysphemism and Dysphemistic Euphemism

The two phrases in this section heading may seem self-contradictory, but in fact they are not. The expletive *Shit!*, which typically expresses anger, frustration, or anguish, is ordinarily a dysphemism. The question arises about how to classify its remodelled euphemisms *Sugar!, Shoot!, Shivers!*, or *Shucks!* Our feeling is that the locution is recognized as a euphemism even though the illocutionary act might be castigated as dysphemistic, consequently we dub these euphemistic dysphemisms. This term seems equally applicable to rather flippant terms like *doodle* "penis", and some uses of rhyming slang like *jimmy-riddle* "piddle."

The following terms for menstruation are hardly euphemisms, on the other hand they are not unquestionable dysphemisms either: *have the curse, woman's complaint, be feeling that way, off the roof*, and so on. We therefore dub them dysphemistic euphemisms for some occasions (and straight dysphemisms on others, such as when a man is whingeing about the sexual unreceptiveness of his female partner). Other terms for menstruation, such as *riding the red rag* or *flying the red flag*, are either dysphemistic or, at best, dysphemistic euphemisms.

With dysphemistic euphemisms, the locution is dysphemistic, but the illocution is not.

Summing Up Chapter 1

Typical euphemisms are motivated by fear and/or distaste—both of which are driven by a desire not to offend; but, as we have seen they are also used to upgrade the denotatum, and even to amuse. Many euphemisms seem to fall into more than one of these categories at the same time: *kick the bucket* is typically a flippant downgrading of death, a taboo topic normally feared (cf. Ch. 6); on other occasions, it can function as a dysphemism. Although it is probably too strong to claim that euphemisms all have positive connotations, they do at least seek to avoid too many negative connotations. We shall see in Ch. 8 that euphemisms are also motivated by the wish to display in-group identity markers. There are many (often antithetical) sources for euphemisms: figurative imagery; circumlocution, abbreviation, omission; synecdoche and metonymy; hyperbole and understatement; use of a learned term and use of a colloquial one; using a term borrowed from another lanuage. Dysphemisms have similar sources. Dysphemisms have unpleasing connotations that are lacking in their neutral (i.e., nondysphemistic) counterparts (cf. the use of *tart* to refer just to a "girl, woman"). Like euphemisms, dysphemisms are motivated by fear and distaste, but also by hatred and contempt; and, in contrast to euphemisms, they are motivated by the desire to offensively demonstrate such feelings and to downgrade the denotatum or addressee (when deliberately used). Like euphemisms they may function as in-group identity markers and even to amuse an audience.

X-phemisms (the union set of euphemisms and dysphemisms) are members from a set of cross-varietal synonyms. They interact with style and jargon such that, on the one hand they define a style or jargon, and on the other are sometimes determined by the style or jargon adopted by the speaker.

Euphemistic dysphemisms and dysphemistic euphemisms have locutions that are at odds with their illocutionary point.

In the course of this book we shall dub some expressions 'euphemisms', some 'dysphemisms', and some 'neutral terms'. Given our definitions of euphemism and dysphemism there should be no place for 'neutral terms' since, for a given speech act, the expressions in question will be either euphemistic or dysphemistic. Such attributions, however, are without reference to a particular context of use, so we are using the middle class politeness criterion which was earlier characterized as follows: in order to be polite to a casual acquaintance of the opposite sex in a formal situation in a middle class environment, one would normally be expected to use the euphemism rather than its dispreferred counterpart(s). A dispreferred counterpart would be a dysphemism. Note that we have to judge what is probably preferred in such a context and what is probably dispreferred. It is the cases where we are undecided which way the judgment should fall that we dub 'neutral terms.'

People use euphemisms and dysphemisms in order to communicate an attitude both to Hearer and also to what is spoken about. For the most part, Speaker has some choice in what to talk about (what might be called the 'propositional content' of the utterance); this, then, is subject to considerations of euphemism and dysphemism. Furthermore, almost every language communication, particularly between fluent speakers of the same language, allows Speaker a choice of vocabulary, of syntax, and prosody; these choices, too, are subject to considerations of euphemism and dysphemism. What is it leads Speaker to use a euphemistic dysphemism like the expletive *Shoot!?* The answer is that human behavior is complex: a person may feel the inner urge to swear (see Ch. 5 for a discussion of why this may be), but at the same time not wish to appear overly coarse in their behavior. Society recognizes this problem by including conventionalized euphemistic dysphemisms within its lexicon. The act of swearing (uttering an expletive) is conventionally perceived to be dysphemistic by the middle class politeness criterion; but by using a locution that is not intrinsically dysphemistic, the dysphemistic illocutionary act can be accomplished with a euphemistic locution. If this looks like double-think, the same must be true of any form of euphemism; what it confirms is that, with matters of face, appearances are of supreme importance. Anyone who doubts this should ponder the following: if you steer your shopping trolley into someone in a supermarket, you are expected to say 'Sorry' or 'Excuse me', and you probably do. In such a slight accident do you really feel sorrow, or are you merely apologizing for appearance sake, so as to save face all round?

CHAPTER
—2—

Euphemism in Addressing and Naming

Fear-Motivated Taboos on Names

Good name in man and woman, dear my lord,
Is the immediate jewel of their souls.
Who steals my purse steals trash; 'tis something, nothing;
'Twas mine, 'tis his, and has been slave to thousands;
But he that filches from me my good name
Robs me of that which not enriches him,
And makes me poor indeed.
(Shakespeare *Othello* III.iii.155)

In the *National Geographic* Magazine Vol. 172, No. 6 (December 1987) there is an article on the African-Americans who live on the Sea Islands off the coast of South Carolina. The text stresses the African-derived customs of these Gullah-speaking people, and says, *inter alia:* 'Most black islanders have two names—one for home use and another to be told to strangers' (p. 749). What this means is that the private name (sometimes called the "true" name) is tabooed; the public name is a euphemism for it.

Personal names are taboo among some peoples on all the inhabited continents, and on many of the islands between them. It is a fear-based taboo: just as malevolent magic can be wrought with one's bodily effluvia (feces, spittle, nail parings, hair clippings, blood, etc. See Ch. 3), so can it be wrought when another person is in possession of one's true name. Frazer described the taboo in this way:

> Unable to discriminate clearly between words and things, the savage commonly fancies that the link between a name and the person or thing denominated by it is not a mere arbitrary and ideal association, but a real and substantial bond which unites the two in such a way that magic may be wrought on a man just as easily through his name as through his hair, his nails, or any other material part of his person. In fact, primitive man regards his name as a vital portion of himself and takes care of it accordingly. (Fraser 1911:318)

According to Keesing and Fifi?i (1969:159) 'The name of a person, in Kwaio (an Austronesian language spoken in Malaita) culture, is associated with the 'essence'

33

(to *ʔofungana*) of that person.' Thus in many languages a name is an inalienable possession; that is, it is assumed to be an inseparable part of the body, and this is reflected in the grammar (other properties of personal representation such as mind, spirit, soul, shadow, reflection, and so on are often treated in the same way; cf. Chappell and McGregor [forthcoming]). For instance, in the Australian language Pitjantjatjara, names are treated alike with body-parts, and unlike, for example, a boomerang: contrast (1–2) with (3).

(1) *ngayu-nya*   *ini*   (2) *ngayu-nya*   *mara*
    1:s-ACCUSATIVE   name   1:s-ACCUSATIVE   hand
    "my name"   "my hand"

(3) *ngayu-ku*   *karli*
    1:s-POSSESSIVE   boomerang
    "my boomerang".

Consequently, as Holzknecht writes of Austronesian languages (and cf. Simons 1982):

> To say a tabooed name is to assault the owner of the name, and requires sanctions to be brought against the offender. Punishment for violation of a taboo can be in the form of religious propitiation of an offended spirit, payment of goods to an offended party, exchange of goods to restore harmony between the guilty and the injured. Breaking the taboo can lead to death by murder, or suicide due to shame. An old man in Waritsian village in the Amari dialect area of Adzera told me that his father had broken a very strong name taboo in front of his father-in-law. The shame caused him to run off into the mountains where enemy groups lived; he deliberately put himself in their way and was killed.   (Holzknecht 1988:45)

Another view is the following:

> [I]t was believed that he who possessed the true name possessed the very being of god or man, and could force even a deity to obey him as a slave obeys his master.
> (Frazer 1911:389)

In ancient Egyptian mythology, Isis gained power over the sun god Ra because she persuaded him to divulge his name. In the European folktales about the evil character variously called Rumpelstiltskin (Germany, parts of England), Terry Top (Cornwall), Tom Tit Tot (Suffolk), Trit-a-Trot (Ireland), Whuppity Stoorie (Scotland), and Ricdin-Ricdon (France), the discovery of the villain's name destroyed his power. (What a contrast this is with the media-conscious twentieth century where it is thoroughly desirable to be a 'name,' and publicize it!) In some societies it seems to have been acceptable to know a personal name provided the name was never spoken. In others, the name can never be uttered by its bearer, but is freely used by others. In many Austronesian societies the names of affines and some cross-kin may not be used (see Simons 1982). In some, no two people may bear the same name: Simons (1982:195) quotes Ernest W. Lee saying of Roglai (an Austronesian language spoke in Vietnam) 'I know one example of a child's name that had to be changed because an old woman came down from the mountains to a refugee settlement and this child had the same name as hers.'

In some societies, especially in Austronesia, names of the dead are (or were until recently) taboo. John Platt told us that a Pitjantjatjara language consultant once insisted that the name of an Aborigine who had died sometime after having been audiorecorded be erased both from the recording itself and from the transcription of it. Bill and Sandra Callister report that

> People on Misima [Island, Papua New Guinea] have several names, at least three, but one of these is their "real" name, and this name is strictly tabooed by everyone in the area when they die (essentially, in their home village which may consist of 200–500 people). The penalty for breaking the taboo is to pay valuables to the offended relatives. Close relatives in the same clan are allowed to say the name of the deceased, but seldom do.   (Simons 1982:203)

In some societies, Misima among them, anyone bearing the same name as a deceased person must use another. In Tiwi (non-Pama-Nyungan isolate, Australia), the ban is even more severe than this for it extends even to those personal names that the dead person may have given to others (see Dixon 1980:28–29, 98–99). Violations of such taboos are believed to cause misfortune, sickness, and death (cf. Keesing and Fifiʔi 1969:157); often they cause offense to living descendants, too. Although native English-speaking communities do not have quite such fearsome taboos, there is a host of euphemisms for the topic of death and the dead (see Ch. 6).

It is very common among the societies of Oceania that names are, or derive from, common words. When the name is taboo, the word becomes taboo too. Thus name taboo can be extended to become word taboo (cf. Frazer 1911:349–74, Holzknecht 1988). Simons (1982:158) reports that in Santa Cruz (Austronesian, Solomon Islands), where there is a taboo against using the name of certain affines, names consist of a common word, normally with a gender marking prefix. Thus if a man's mother-in-law is called *Ikio* (*i-* "prefix to female's name," *kio* "bird") he cannot use the common word *kio* to refer to birds. Simons writes 'This means that 46% of the basic vocabulary is potentially taboo for some people on the island.' In many societies, the taboo may extend even to phonetically similar words. Euphemisms are created from circumlocution, phonological modification, or by borrowing from another language. For instance in storytelling, if a tabooed name is phonetically similar to the name of an island, it will be referred to in some round about way, and the storyteller will stop to check audience comprehension before proceeding. There is also the story of an Owa (Solomon Islands) man announcing a Bible reading of an epistle from St. Peter: the man's father-in-law was called Peter, so the name was taboo to him, consequently he said 'The lesson is from the first letter written by my father-in-law, . . .' (cf. Simons 1982:203, 206).

This has interesting linguistic consequences because the vocabularies of such languages undergo considerable and extremely rapid change, even in core items that normally resist change in other languages. 'The importance of word tabooing in Oceanic linguistics derives from the possibility that this process has significantly accelerated vocabulary differentiation between genetically related languages. . . . Such tabooing could create a spurious impression of long divergence or skew datings, or in some cases even hide genetic connections.' (Keesing and Fifiʔi

1969:155). Simons lists the following correspondences between Kwaio and Lau
(Solomon Islands):

Kwaio	Lau	Meaning
gani	*dani*	"day, daylight"
logo	*rodo*	"night, dark"
-ga	*-da*	"3p:POSS"
aga	*ada*	"look, see"
gamu	*damu*	K."chew"; L."smack lips"
guigui	*duidui*	"vinegar ant"
ugu	*udu*	"a drop of water"
age	*ade*	"do, happen"
nagama	*madama*	"moon"

There is no conditioning factor for this regular *g:d* correspondence, and Kwaio has
doublets for *gani*, *gamu*, and *guigui*, which are identical with the Lau forms. 'The
explanation for this kind of common irregular sound change is to be found in the
mechanisms of word tabooing, not in regular inheritance from the proto-language,'
Simons (1982:168) concludes.

Comparable tabooistic distortions have affected many Austronesian languages.
For instance, in many traditional Australian Aboriginal communities, any kind of
vocabulary item (including grammatical words) is proscribed if it is phonetically
similar to the name of a dead person. Thus, on the death of a man named
Ngayunya, some dialects of the Western Desert Language replaced the pronoun
ngayu "I/me" with *ngankyu*; subsequently, this term was itself tabooed and replaced
by either English *mi*, or by *ngayu* borrowed back into the language from dialects
where it had never been tabooed. This shows that the taboo on a word may cease
after some years have passed, allowing it to come back into use. This recycling is
one of the very few ways in which a tabooed item can become a euphemism. It is
more usual, however, for tabooed items to be replaced by either borrowed or newly
coined words, or to cause a semantic shift extending the meaning of a near-
synonym, thus cycling rarely used words into the basic vocabulary. Some languages
have special vocabulary items to be used in place of proscribed words; thus, in some
Kimberley languages, people whose personal names have been tabooed are addressed as *nyapurr* "no name".

Not only personal names are tabooed: in some societies even the names of
communities were not divulged to strangers.

Euphemisms for the Name of God

A *mantra* [hymn] recited with incorrect intonation and "careless", arrangement of
varṇa (letters) [reacts] like a thunderbolt and gets the reciter destroyed by God
Indra.   (*Sic.* Kachru 1984:178)

What applies to the names and naming of ordinary folk applies a fortiori to rulers
and to gods because any threat to their power endangers the entire society they

dominate. Taboos on the names of gods seek to avoid metaphysical malevolence by counteracting possible blasphemies (even, perhaps, profanities) that arouse their terrible wrath. Despite our increasing secularization in the twentieth century, there are plenty of constraints on the names of our god(s) and their acolytes. In 1989, Indian-British author Salman Rushdie has a $2½ million bounty on his head for alleged blasphemy of the Prophet Mohammed and his wives in the novel *The Satanic Verses*: the sentence of death decreed by Iranian Ayatullah Ruhollah Khomeini led to severe political repercussions in Iran, Europe, India, and Pakistan. In the Holy Communion service of the Anglican church, the Minister says 'Thou shalt not take the Name of the Lord thy God in vain: for the Lord will not hold him guiltless, that taketh his Name in vain.' (*The Book of Common Prayer* 1662/1852). This is the third of the ten commandments that God gave Moses, as reported in *Exodus* 20:7. Why should it be blasphemous to take the Lord's name in vain? Note, here, the euphemism *Lord*. Modern European constraints on the use of God's name hark back to the semitic founders of Judaism, Christianity, and Islam. In *Judges* 13:18 the angel says to Manoah 'Why asketh thou after my name, seeing it is secret?' It was blasphemous to name the god of the Jews and his cohorts, thus the Jewish god's name was written without vowels YHVH but read out as *adonai* meaning "lord"—a euphemism that has carried over into Christianity in both addressing and naming God and Jesus Christ. According to Rosten (1968), today there is a euphemism for *Adonai* used outside of formal religious service by devout Jews, *Adoshem* from the Hebrew *Ha Shem* "The Name"; very devout Jews will write *G-d* for *God*, just as they would write *Yah*, an acronym for the first two letters of YHVH. In the same tradition, the use of the inscription *IHS* in the Catholic and Anglican (Episcopalian) churches mimics the first two and either the third or final Greek letters of *Jesus*; however, these letters are also said to abbreviate the Latin *In Hoc Signo*, literally "in this sign"; and yet another explanation is that the letters abbreviate "In His Service": Are these euphemisms, folk etymologies, or plain ignorance?

Here, perhaps, we should take time out to distinguish profanity from blasphemy, although for most people, the distinction is probably arcane if not outright pedantic. To begin with, it is not always possible to distinguish particular occurrences of them. Even the *O.E.D.* defines each in terms of the other; thus the verb blaspheme is 'To utter profane or impious words, talk profanely,' and blasphemy is 'Profane speaking of God or sacred things; impious irreverence.' Profanity is 'irreverent, blasphemous, ribald; impious, irreligious, wicked.' If the *O.E.D.* sees so much overlap in meaning, small wonder that ordinary folk feel no great compunction to distinguish them from one another. And it is hardly surprising that profanity has frequently been punishable as blasphemy! The difference, such as it is, is that *blasphemy* vilifies or ridicules the deity, the deity's family, or divine mouthpieces like prophets or the priesthood, and divine scriptures. *Profanity* uses religious terms—such as the name of the deity—without blasphemous intent, but with careless irreverence. One can readily imagine that some self-righteous religious bigots will regard such irreverence as ipso facto blasphemous. So, someone who utters the expletive *Jesus Christ!* is profane, but would only be blasphemous (to a christian) if they thereby intended to vilify Jesus Christ. Unfortunately, the term *profanity* has been extended to ribald and obscene epithets; so, where necessary, we will tie profanity to blasphemy in order to clarify our meaning.

To avoid blasphemy (and, we suppose, accusations of profanity), the word *God* has been or is avoided in euphemistic expletives such as *'Od's life!*, *Zounds!*, *by gad!*, *Gog! Cock! Cod!* (all archaic) *Gosh!*, *Golly!* (earlier *Gorry!*) *Cor!*, *Gorblimey!*, *Gordonbennet!*, *Gordon'ighlanders!*, *Goodness (knows)! (Good) gracious!*, *For goodness' sake!*; these examples demonstrate various kinds of remodelling, including clippings and substitutions of phonetically similar words. In *So help me!*, *Swelp me!*, and *So save us!* there is omission of *God* (or *Lord*), and we have a feeling that *so* has not only stepped into its place, but may even be a euphemism for it. Then there are semantically related substitutions such as *Goodness!* and *Gracious!*; there are also *(Oh) Lord!*, *Lordy! Lawdy! La! Land's sake!* and *Heavens (above)!* or *Heavens to Betsy!*. *Gosh* was perhaps created from *Go-* as in *god*, + *sh* "be quiet, say no more"; the *O.E.D.* sheds no light on its etymology, nor on the etymology of *golly*. Partridge (1961) claims *golly* comes from Negro English—whatever that means; but we believe it may be a remodelling of *Good Lord(y)* / *Lawdy* / *La*—euphemistic dysphemisms which, according to Montagu (1967:225), were widely used by Victorian ladies. Doubtless they, like Ophelia in Shakespeare's *Hamlet*, were blissfully unaware of uttering oaths:

OPHELIA.   Indeed, la? without an oath, I'll make an end on't. [*Sings*]
        By Gis, and by Saint Charity,
        Alack, and fie for shame!
        Young men will do't, if they come to't;
        By Cock, they are to blame.
        Quoth she, before you tumbled me,
        You promised me to wed.
He answers
        So would I ha' done, by yonder sun,
        And thou hadst not come to my bed.
                        (Hamlet IV.v.55)

Shakespeare's magnificent wit shines through, as it does so often, and Ophelia in innocent ignorance utters the following mild oaths: 'la' "Lord," 'by Gis' "by Jesus" (see just below), 'by Saint Charity' (that there never was such a saint did not stop people swearing by him), 'fie' "[by my] faith," 'by Cock' "by God." Shakespeare weaves in several sexual innuendos, including one that plays on the meaning of *cock*—see Ch. 4 on penis as cock. In his English, the ostensible meaning of 'tumbled' was "tousled".

A rather different kind of semantic relation is exemplified in *Holy Mary!*, *Holy mother!* whence, probably, *Holy cow!* and onward to the double dysphemisms *Holy shit!*, *Holy fuck!*. *Holy Moses!* is even less potentially profane than *Holy Mary!*: it is a real euphemism—along with such as *Holy toot!*.

The names *Jesus*, *Jesus Christ*, or *Christ* are avoided in *Jis*—which very probably derives from the romanization of *IHS* and is the likely source for *Gis* as used by Ophelia. These have given way to *Jeeze!* and *Gee!* (which doubles as both a clipping from *Jesus* and being the initial of *God*); *Gee whiz!* is a remodelling of either *jeeze* or *jesus.* More adventurous remodellings with a few clippings but more substitutions are *By jingo! Jeepers creepers!, Jiminy cricket!, Christmas!, Cripes!, Crust!, Crumbs!,* and *Crikey!. Christmas!,* of course, is semantically related to *Christ.* It is likely that *By Jove!* is a clever choice of a name beginning with *J*—which is at the same time the name of the chief god in the ancient Greek pantheon. One of the best disguised remodellings of *Christ* is in *For crying out loud!* which is a euphemism for *For Christ's sake* (see Ch. 5).

Despite persisting incantations like 'Lord, have mercy upon us' in church services, it is unlikely that more than a handful of people nowadays seriously fear divine wrath to the extent that they are strongly motivated to use euphemisms to guard against it. Indeed, most people using the euphemistic dysphemisms we have been discussing are unaware of using a euphemism at all. Like euphemisms everywhere, these have become ritualized and conventional behavior. We should resist a patronizing attitude to the beliefs of so-called primitive peoples: they may have no more faith in their magic than we have in ours; for instance,

> Once when a band of !Kung Bushmen had performed their rain rituals, a small cloud appeared on the horizon, grew and darkened. Then rain fell. But the anthropologists who asked if the Bushmen reckoned the rite had produced the rain, were laughed out of court (Marshall, 1957). How naïve can we get about the beliefs of others? (Douglas 1966:58)

Ritualized superstition is revealed again in our response of *Bless you* when someone sneezes (it was to prevent the devil from entering the body momentarily emptied of its soul—notice, incidentally, the euphemistic omission of *God* as the subject of *bless you*); and in our use of *touch wood* or *knock on wood* to guard against misfortune; not to mention notions of lucky and unlucky numbers and forms of behavior. And before we leave aside euphemisms motivated by religious superstition *What the dickens . . .* for "What the devil . . ." avoids calling up the malevolent spirit of *Old Nick* a.k.a. *Old Harry, Old Bendy, Old Bogey, Old Poker, Old Roger, Old Split-Foot, the Old Gentlemen, Old Billy,* Perhaps *What in Hades?! . . .* is a euphemism for *What in hell?! . . .* (see Ch. 5). Curiously, although *What the deuce . . .* is formally analogous to *What the dickens . . .* and *What the devil . . . ,* 'deuce' here derives from the Norman French oath *Deus!* "God."

Euphemism in Styles of Naming and Addressing

Addressing/naming someone appropriately depends on the role Speaker perceives the person addressed or named—henceforth 'Hearer-or-Named'—to have adopted relative to Speaker in the situation of utterance. This role may differ in different situations. For instance, Freddie and Eddie might be on christian name terms while having lunch together before a board meeting; but when conducting official business in the boardroom where Freddie is Chairman of the Board, Eddie will probably address Freddie as 'Mr. Chairman' and name him 'the chairman' in accordance with his role. However, in an unofficial aside Eddie can quite properly revert to using 'Freddie'—even in the boardroom. This makes it clear that it is not the physical situation of Speaker and Hearer-or-Named that is relevant, but Hearer-or-Named's perceived role in that situation. High social status is not a right, but a perquisite of those who can either make or persuade other people to recognize that status.

The relative status of Speaker and Hearer-or-Named derives from two sources: their relative power, and the social distance between them. The relative power of Speaker and Hearer-or-Named is defined by social factors which obtain in the situation of utterance. For instance, the relative power of a physician and a highway patrolman is not given for every occasion, it depends on where they encounter one another: imagine how it will differ depending whether the highway patrolman is requiring a medical consultation at the doctor's office, or the doctor has been stopped on the highway for allegedly exceeding the speed limit. The social distance between Speaker and Hearer-or-Named is determined by such parameters as their comparative ages, genders, and sociocultural backgrounds (cf. Brown and Levinson, 1987). The management of social status (i.e., of power and social distance relations), involves the management of face, and consequently the management of X-phemisms. Hence the style of language Speaker uses will depend on (1) the role s/he perceives Hearer-or-Named to have adopted relative to Speaker in the current situation of utterance, or perhaps on some prior occasion; and (2) Speaker's communicative purpose on this present occasion—in particular, whether s/he intends to be dysphemistic or not. We shall examine relevant strategies for dysphemism in Ch. 5. Here we discuss euphemistic behavior: that is, Speaker's use of address/naming forms that seek to either enhance Hearer-or-Named's face or, at the least, to avoid loss of face by any party. For English, Joos (1961:11) identified the following five levels of decreasing formality, which we shall be using to assess styles of addressing and naming in different situations:

frozen > formal > consultative > casual > intimate

These five points of reference are intended to exhaust all possible manners of addressing and naming. They are guidelines, and we do not claim that it is possible to recognize firm boundaries between adjacent pairs. We invite the reader to anticipate the application of these terms to exemplary occasions while awaiting our demonstration of these during the course of the following discussion.

The harm that kings and chiefs can do ensures they are nearly always surrounded by taboos (see Frazer 1911). Most of these taboos were originally instituted to protect the stability of the community by protecting the ruler against malevolent witchcraft (exercised on the feces, spittle, nail parings, hair clippings, blood, etc.), as well as against malevolent physical and political acts. Consequently, special language is often used both when communicating with rulers and when talking about them. Perhaps as an antidote to downgrading a ruler, naming or addressing them often involves extreme euphemism, not to mention pomp and circumstance. In *Gulliver's Travels* Jonathan Swift mocked the splendiferous titles given to contemporary princes in the following address to the Emperor of Lilliput—a man slightly taller than Gulliver's middle finger was long:

> GOLBASTO MOMAREN EVLAME GURDILLO SHEFIN MULLY ULLY GUE, most Mighty
> Emperor of *Lilliput*, Delight and Terror of the Universe, whose Dominions ex-

tend five Thousand Blustrugs, (about twelve Miles in Circumference) to the
Extremities of the Globe: Monarch of all Monarchs: Taller than the Sons of Men;
whose Feet press down to the Center, and whose Head strikes against the Sun: At
whose Nod the Princes of the Earth shake their Knees; pleasant as the Spring,
comfortable as the Summer, fruitful as Autumn, dreadful as Winter. His most
sublime Majesty proposeth to the *Man-Mountain* [= Gulliver], lately arrived at
our Celestial Dominions, the following Articles . . .   (Swift 1735/1958:24)

This mode of addressing/naming exaggerates the importance of Hearer-or-Named
by magnifying their perceived or pretended higher social status. As we remarked in
Ch. 1, in bureaucratic circles it is often those in the lower ranks that get the longest
titles (e.g., *the Personal Assistant to the Secretary (Special Activities)* for "cook").
This exaggeration is a kind of euphemism. Now that we have constitutional monarchs and democratically elected presidents, the terror that our rulers once inspired
has been replaced by a notional respect, while terror has become the mark of the
petty dictator and terrorist. Yet the language used to rulers has remained much the
same, even if it is no longer so very different from the respectful deference extended
to other Hearer-or-Named persons of superior power to Speaker. At the March 1989
coronation of Prince Mangkubumi in Yogyakarta, Indonesia, the new Sultan was
given the following title: *Ngarso dalem kanjeng ratu ingkang sinuhun sri sultan
hamengku buwono adipati ingalogo ngabdurahman sayidin panoto gomo kali-
fatulluh kaping X,* which translates along the following lines: "His Exalted Majesty,
whose Honor Shines Bright, Sultan of all the World, Commander in Chief, Servant
of God, Protector of Religion, Assistant to God, the tenth."

Where Speaker is inferior to Hearer-or-Named, s/he will use unreciprocated (or
conventionally unreciprocable) deferential forms such as *Your/her Majesty, Your/his
Highness, Your Lordship, Mr. President, Madam Chairman,* and so on, and all of
which are 'frozen' or 'formal' style. These titles do not include names, but identify
roles or social positions; so, to some extent, they impersonalize. So do terms like *Sir,
Madam, this lady,* or *the gentleman,* which may be 'formal' or 'consultative' (and
much less likely, 'frozen'). Children addressing adults sometimes use the titles *Mr.*
or *Mrs.* alone, which is reminiscent of 'consultative' style. Even within that style, it
would be dysphemistic from an adult Speaker—in our dialects, at least (our dialects
are: KA, South London, England; KB, Perth, West Australia). In rather stilted
English, Hearer can be addressed in the third person; we have done it ourselves, as
the astute reader (that's you) will doubtless have noticed; occasionally one encounters similar forms in the more expensive shops (e.g., *If Madam so desires she could
have our tailor alter the waistband just a touch*).

This sort of impersonalizing manner of naming and addressing might be compared with the regular use in some languages of third person address forms to
Hearer; for example, the deferential Polish question in (1) contrasts with the
familiar version in (2)—which roughly corresponds to the 'intimate'/?'casual' styles
of English.

(1) *Co   mama robi?*
    what mother 3:s:do
    "What are you doing mother?"

(2) *Co   robisz, mama?*
    what 2:s:do, mum
    "Wotcha doin', mum?"

In the typical traditional speech situation where Speaker and Hearer are in face-to-face conversation, there is a greater psychosocial distance, and often a greater physical distance, between Speaker and a third person, than between Speaker and the second person (i.e., Hearer). This difference in relative distance is captured in the ordinals of the terms 'first', 'second', and 'third' persons. Thus THIRD person is intrinsically more distant from Speaker than SECOND person is; hence, its use to Hearer exaggerates the social distance between Speaker and Hearer. Such exaggeration is a widely used strategy for indicating deference; and it is, as we have said, euphemistic.

Another way for Speaker to indicate deference is to address or name not the individual Hearer-or-Named, but to include Hearer-or-Named among a number of people. For example in French, Speaker uses the second person plural as a deferential mode for addressing a singular Hearer in (3), the 'intimate'/?'casual' form in (4):

(3) *Vous êtes très gentille, madam.*　(4) *T'es　très gentille, maman.*
　　2:p　are very kind　madam　　　2:s:are very kind　mummy
　　"You are very kind, madam."　　　"You're very kind, mummy."

The deferential address form in German uses the third person plural form, though *Sie* "you" is orthographically marked by an initial capital. Compare:

(5) *Sie　sehen　gut　aus.*　　(6) *Du siehst gut　aus, Mutti.*
　　3:p/2:s 3:p:see good out　　　2:s　2:s:see good out, Mum
　　"They/You look good."　　　　"You look good, Mum."

Spoken Tamil shows respect by using a third person PLURAL form when naming a third person singular; for example, in a radio commentary on an unfolding event, one might hear:

(7) *motal mantiri avaanka mantikal kooṭa　　　pooranka*
　　first　minister 3:p:poss ministers together:with 3:p:going
　　　　　　his　　　　　　　　is going
　　"The prime minister is going, accompanied by his ministers"

In examples (3), (5), and (7), Speaker acts on the normal presumption that any individual is representative of a group, and derives social standing accordingly. Because there is safety in numbers, Hearer-or-Named is less vulnerable as a member of a group than if s/he were alone—any threat to Hearer-or-Named may be perceived as a threat to the whole group; thus, Speaker will pretend to greater respect for Hearer-or-Named than if Hearer-or-Named were a lone individual. We can look upon this strategy as exaggerating the relative power of Hearer-or-Named. And, as we have shown, the vehicle for this is a plural form instead of the singular form for a single Hearer-or-Named. This strategy is somewhat less impersonalizing than the use of third person in place of second person.

The strategies we have been discussing can be ranked on personalizing scale as follows:

MOST PERSONALIZED　2:s − 2:p − 3:s − 3:p　**LEAST PERSONALIZED**

But we know of no language that employs more than three of these (e.g., eighteenth-century German), so the ranking is essentially nonfunctional for communicative purposes within any one language. This suggests that, rather than conflating the two strategies for marking deference on one personalizing scale correlating with relative status, it may be more appropriate just to recognize two distinct systems motivated by the components of social status: namely, social distance and power (cf. Haiman, 1980:530). We have already done this in discussion, but we would sum it up as follows:

> **Deference to Hearer** can be marked by exaggerating the social distance between Speaker and Hearer (e.g., by using third person to Hearer).
>
> **Deference to Hearer-or-Named** can be marked by exaggerating his or her power relative to Speaker (e.g., by using plural number of a single denotatum).

In every language, Speaker may register a change in attitude toward Hearer-or-Named by changing the style of naming or addressing from that which s/he has been using in prior discourse, or which s/he normally uses (see Ch. 5 for a discussion of dysphemistic uses of this device).

We return to the use of personalized naming and addressing with the 'consultative' style in English. 'Consultative' style is used where Speaker is superior in status to, but of friendly disposition toward Hearer-or-Named; or where Speaker and Hearer-or-Named are of similar social status but there is considerable social distance between them: Speaker will address or name Hearer-or-Named using title + surname for the task (e.g., *Mr. Milquetoast*).

If Speaker is superior in status to Hearer-or-Named, s/he can choose either to maintain the status difference or choose to be less formal and show solidarity by using in-group markers that demonstrate a concern to enhance Hearer-or-Named's positive face. Where Speaker and Hearer-or-Named are of similar social status and there is little social distance between them, the informal in-group language found in 'casual' and 'intimate' styles is the regular mark of solidarity. These styles are marked by colloquialisms, slang, swearing, diminutives, contractions, ellipsis, and so on. Among adults, address forms in 'casual' style include given name or nickname, perhaps with the surname; also American *bud(dy)*, Australian and southern British *mate*, northern British *marra*, southern British *old boy* (possibly passé), and *brother* or *sister* in various American, Australian and British sociolects. These forms of address are also used in 'intimate' style, and in addition to the address forms just mentioned we find such terms as *baby, bud, daddy, darling, dear, duckie, ducks, fella(s), gorgeous, grandad, guys, handsome, honey, love (luv), lover, mac, momma, sexy, sis, sugar,* and *sweetheart*.

The reader may have noticed that we have sought to maintain *gender-neutral* language throughout this book, referring, for example to Speaker and Hearer as *s/he* and *him/her,* and the like. While it is probable that SOMEone among our readership will be offended by this, we believe it should offend the smallest number of people overall; consequently, it is our own preferred habit. In the late 1960s, the feminist movement began to make itself heard objecting to a community attitude that downgraded women by comparison with men. People in the movement perceived this depreciating attitude to be reflected in language, and sought to change at least public language so that it should become less dysphemistic to women. They held, and continue to hold, the view that revising habits of language use will change community attitudes; whether or not this belief is credible, is no concern of ours in this book. What is relevant, is that since the 1970s there have been issued a large number of guidelines for nonsexist language usage in private and public institutions, government offices, and so on. A couple of the most recent are *Style Manual for Authors, Editors and Printers* (4th edn) Australian Government Publishing Service (1988) and *Language, Gender, and Professional Writing: Theoretical Approaches and Guidelines for Nonsexist Usage*, produced by the Commission on the Status of Women in the Profession, edited by Frank and Treichler, and published by the Modern Language Association of America in 1989. The latter, like many publications with similar goals, identifies dysphemistic expressions and offers what are claimed to be neutral alternatives. (Given our definitions, they might be judged neutral locutions in what, for many users at the present time, are euphemistic illocutions. Discussing such a politically and emotionally charged quibble at this time will generate more heat than light, and we leave it to readers to make up their own minds.) Speakers and writers are advised to choose the neutral alternative from such lists as the following:

DYSPHEMISTIC LOCUTION	NEUTRAL LOCUTION
man(kind)	human beings, humanity, people
chairman	chairperson, chair
congressman	member of Congress, representative
fireman	firefighter
policeman	police officer
mailman	mail carrier
foreman	supervisor
salesman	salesperson
actress	actor
(air) stewardess	flight attendant

Obviously, the *-man* locutions are not dysphemistic when used of a male denotatum, and the neutral locution is primarily intended to name the office (job) itself so as to acknowledge that women may hold such an office. It might be thought that terms suffixed *-ess*, and others such as *lady/woman doctor*, should be acceptable to a female denotatum; but it is widely perceived that women referred to using such terms are less highly valued than their male counterparts, therefore the terms are dysphemistic and the neutral alternatives are preferred for a female denotatum. From this, together with the fact that neutral terms are preferred in nonreferential contexts when there is to be no specific mention of gender, it follows that many speakers prefer to use the neutral term all the time, even when denoting males; thus does the neutral term become truly gender neutral.

In Ch. 1 we puzzled over the fact that some terms are euphemisms, while others denoting precisely the same thing are dysphemistic: it is the connotations of each that are different; the connotations reflect the community attitude towards the term itself. Hence, *air stewardess* gives way to *female flight attendant*—for the time being, at least. It is worth mentioning that individuals differ greatly in their attitudes to the terms listed as dysphemistic; in particular we have encountered a number of women who are quite happy to be *Madam Chairman*, because they understand the word *chairman* as an idiom denoting the office of chairperson: on this view (cf. Allan 1986a §4.4 on compounds) it should no more be decomposed into "chair" and "man" than *moonshine* meaning "illegal liquor," or *fathead* meaning "idiot, fool," should be decomposed into a semantically transparent pair of morphemes. We are suggesting that women who do not find titles like *chairman* dysphemistic should not be dismissed as unreconstructed nerds, brainwashed by outdated social attitudes; as time passes, the semantic feature "male" could well be bleached out of the *-man* suffix. On the other hand, since that has not yet occurred, it is perfectly legitimate for other women to feel affronted and wrongfully excluded by such compounds.

It is also recommended in guides to nonsexist usage that address forms for women should be comparable with the address forms for men used in the same context. In the absence of knowing anything about the particular preferences of the individuals concerned, introducing a couple as *Dr. and Mrs. John Doe* MIGHT not cause offense to the lady thus named, but there is a growing number of women who have no wish to be named as if they are an appendage to their husband—which is one of the motivations for women not adopting the husband's surname. Moreover, John Doe's wife might well be due the title *Dr.* herself; and we know of several occasions when offense has been caused by an insensitive introduction of this nature. Even worse, of course, are occasions where the superior title due to a wife or other female companion is wrongly transferred to her male partner because of entrenched expectations of differing relative achievements of men and women. Because a new convention is slowly replacing the old one, Speaker (particularly, one regrets to say, the male Speaker) needs to be wary.

Being nondysphemistically p.c. regarding nonsexist language is hazardous: one must steer a way between the Scylla of sexism and the Charybdis of the cooperative maxim of manner. The following is drowned by Charybdis:

> Some of them [trainee therapists], may eventually be in a position in which they are expected to help a homosexual person who has cracked up as a result of society's abhorrence of her/him, or who has fallen foul of the law. Many therapists are genuinely concerned to help that person overcome her/his problems, without trying to effect a fundamental change in her/him. Unfortunately, quite a large number of therapists do in fact try to change the gay person, to make her/him into a heterosexual. (Birke 1980:119)

Gender differentiation between third person singular pronouns in English leads to the use of *her/him, him or her, s/he, he or she, hers/his*, and the like in nonreferential contexts. Because the third person plural—*they, them, their(s)*—is not gender differentiated, it is recommended for use wherever possible. However, the MLA Commission, referred to earlier (cf. Frank and Treichler 1989:181), decided against the use of *they* to denote singular (e.g., in contexts like *If anyone needs to know about euphemism, they should read this book from cover to cover*); this is a sadly pedantic decision in view of the fact that such usage has been common practice for many centuries, especially in colloquial speech (cf. Smith 1985:50–53). Whatever one's views on the use of *they* for singulars, it is stylistically preferable to write *As readers can judge for themselves* rather than *As the reader can judge for him or herself* because the plural version does less violence to the cooperative maxim of manner. A different cause of violence to the maxim of manner is demonstrated by Jean Aitchison in *The Articulate Mammal*: Aitchison uses an equal number of masculine and feminine nonreferential pronouns in order, she says, not to give 'the misleading impression that only male mammals are articulate' (1983:9). Thus, toward the top of p. 148 she writes: 'These are, firstly, the child's everyday needs; second, his general mental development; third, the speech of his parents'; and further down the page she writes 'In brief, the argument that the child learns language in order to help her to manipulate the world does not explain why she does not stop learning as soon as she starts obtaining what she wants.' Although recommended by the MLA Commission (*loc.cit.*), this strategy of swapping whimsically between masculine and feminine pronouns proves extremely distracting to readers, who keep thinking they have missed the antecedent noun phrase somewhere. Violations of the manner maxim are, of course, dysphemistic; and, as conscientious authors well know, one sometimes has to weigh the dysphemism of sexist language against the dysphemism of distracting style.

In most English-speaking families it is dysphemistic to address or name consanguineal kin of an ascending generation by their given names. Instead we use kin titles such as *Dad, Nan,* or *Grandpa* for lineal kin; and for lateral kin, a kin title like *Auntie,* or kin title + given name, (e.g., *Aunt Jemima*). If lineal kin of the second and higher ascending generations need to be distinguished from a collateral with the same title, kin title + surname is the usual form used (e.g., *Grandma Robinson*). The social taboo against omitting the kin title is weakest with kin from the first ascending generation who are about the same age as Speaker, particularly collateral kin and step-kin; it is strongest with kin of the second and higher ascending generations. These are, of course, asymmetric conventions: given name only is the norm when Hearer-or-Named is close kin of a descending generation; though more distant affinal kin from a descending generation may warrant title + surname. (On address forms in English, see Allan 1986a, I:16ff.; Brown and Ford 1961; Brown & Gilman 1960, 1989; Erwin-Tripp 1969; Ervin-Tripp et al. 1984.)

In many societies (e.g., among the Zia of Papua New Guinea) personal names are not used among spouses, siblings, and often not to descending generations. Instead kin titles, and the translation equivalents of terms like *person, man, woman, boy, girl,* are used. Other societies use public names, nicknames, clan names, and kin descriptions, like *mother of* X.

For contrast, there are communities where there is virtually no choice in the style of naming and addressing to express deference and attribute power or prestige. Among the religious groups of Old Order Amish and Old Order Mennonites in Pennsylvania and Ontario, Canada, everyone gives and receives first names only, regardless of relative familiarity, status, age, and sex. There exist no titles and no honorifics; this is true even for use on public occasions like a church service or during any of the rites of passage. This practice arises from the people's doctrine of humility. These people speak a dialect of German, and on the rare occasion where deference may need to be shown to an outsider (rare because English would normally be the language spoken at such times), the second person plural form can be used, but it is considered awkward and people prefer not to use any form at all (cf. Enninger et al. 1989:146–147).

Interestingly, there are no terms of endearment like 'darling' either—couples address each other by their given name or by no name at all (cf. Hostetler 1980:156). Though nicknames are important, they seem to be used purely for the purpose of identification. Among the Old Order Amish and Mennonites, first names derive only from the Old Testament, and there is a limited number of family names: for instance, for the 85,000 Amish in Pennsylvania, there are only 120 names to go around; and in the small Mennonite town of St Jacobs in Waterloo County, Ontario, there are 27 David Martins registered at the local post office! So most people have a distinguishing nickname. It is often derogatory, characterizing the bearer's behavior or appearance (e.g., *Chubby Jonas*), or perhaps referring to some memorable event (e.g., *Gravy Dan* once poured gravy instead of cream into his coffee) (cf. Hostetler 1980:241–244). However, these nicknames are normally used in intimate settings to refer to some absent person, and it is difficult for an outsider to get information about them.

The use of **IN**appropriate styles for naming and addressing is dysphemistic and likely to affront Hearer-or-Named's face (see Ch. 5). For just that reason, we regard appropriate naming and addressing as not only polite, but also as euphemistic behavior.

Taboos on Naming and Addressing Kinsfolk

Many Austronesian languages have taboos on names for kinsfolk, especially affines and, to a lesser extent, cross-consanguineal kin. The latter is generally restricted to taboos on using the names of siblings of the other sex—although this is often extended to cross-cousins and 'clan brothers and sisters' (cf. Simons 1982:177–179). It is very possible that taboos on naming or addressing cross-kin are motivated by the danger of incest: this would seem to account for the widespread taboos on social intercourse between mother-in-law and son-in-law or father-in-law and daughter-in-law, which have led to 'turn tongue' or 'mother-in-law languages' in Australia (see Frazer 1911:346f; Dixon 1972:32, 1982 Ch. 2). However, there is also widespread avoidance of normal language use in the presence of parallel-kin, where an incest taboo is inapplicable. For instance in Bauro (Solomon Islands) there is a strong taboo on cross-sibling names and, in some parts, on cross-cousins, too; however, there is also an exactly similar taboo on a boy using his elder brother's name (cf. Simons 1982:206). What alternative explanations can there be? First, we note that there are no reports of verbal strategies as extreme as the use of 'mother-in-law languages' for use between consanguineal kin of opposite sexes. Since consanguineal incest is not as rare as hens' teeth, it seems unlikely that cross-affinal incest should lead to the development of avoidance languages, when consanguineal incest does not (see Ellis 1963). Second, we note that relationships with in-laws are notoriously difficult in all societies, and conclude that the 'mother-in-law' languages and all similar taboos on naming and addressing kinsfolk are grounded in the desire to maintain social harmony in what all human communities recognize to be the most difficult area of intrahuman relationships.

Euphemisms Used for Dangerous Animals, and When Engaged in Hazardous Pursuits

In many communities euphemisms are used when mining, harvesting, hunting, and fishing in order to avoid conflict with the spirit world. People seek to avoid mining disasters, failed harvests, dangerous beasts, and getting lost at sea. In Ch. 6 we discuss euphemisms for death, which are prevalent in all communities; one motivation for such euphemisms is the fear of being haunted by spirits of the dead. Among some peoples, these fears extend to the spirits of game, particularly dangerous game. Frazer wrote

> we may with some probability assume that, just as the dread of the spirits of his enemies is the main motive for the seclusion and purification of the warrior who hopes to take or has already taken their lives, so the huntsman or fisherman who complies with similar customs is principally actuated by a fear of the spirits of the beasts, birds, or fish which he has killed or intends to kill. For the savage commonly conceives animals to be endowed with souls and intelligences like his own, and hence he naturally treats them with similar respect. Just as he attempts to appease the ghosts of the men he has slain, so he assays to propitiate the spirits of the animals he has killed.   (Frazer 1911:190)

See also Douglas (1966, Ch. 5). In many communities, therefore, a hunter will conceal his purpose, saying such things as 'I'm going to collect rattan,' 'I'm going to climb a betel-nut tree' (cf. Simons 1982:194, 206). And often he will not name the game that he is hunting, even if he sees it. Sometimes men will not address other hunters by name. It is as if their quarry can monitor their language. The same is true for fishermen.

In addition, many peoples have taboos against mentioning the names of dangerous animals, whether hunting or not. The motivation for it can be gleaned from the Ukrainian proverb *Pro vovka pomovka a vovk u khatu*, "One speaks of the wolf and it runs into the house" [lit. 'about wolf talk and wolf into house']—which is not unlike the English proverb *speak of the devil and he comes running*. In a Walmatjari story from central Australia we find the following:

> The child said, "How come people say that cattle are always goring people? They didn't gore us. Why do they say that?"
>   The child talked on persistently, "Do they say that cattle are always goring people? How do they talk about cattle? They say they always gore people but they didn't gore us this time." The man answered the child. "Don't talk like that! The cattle might hear us, and attack and kill us."   (Hudson & Richards 1978:15)

This attitude reflects a fear that the souls of animals can understand human language, and if a dangerous creature hears its proper name called, it will come running—perhaps to protect itself against evil. So wolves and bears, snakes and other venomous creatures, are often named by euphemisms. Consequently, in most

Slavic languages, for instance, the bear is called something like 'the honey eater,' and the English word *bear* derives from the euphemistic "[the] brown one"— compare *Bruno the bear* in which 'bruno' is Italian for "brown". Among the Zia, a non-Austronesian people who live in Morobe Province in southeastern Papua New Guinea, fishermen avoid talking about, and try even to avoid thinking about, dangerous creatures like sharks, rays, and saltwater crocodiles for fear of inviting attack. If such creatures are seen they are referred to using the class name *woo* "fish" instead of *bawang* "shark" or *beoto* "ray"; and the crocodile will not be called *ugama* but *emo meko* "bad man" [lit. 'man bad'] instead. Among the Zia, such dangerous creatures are feared because they are believed to be possessed by the spirits of dead humans, see earlier discussion and Ch. 6.

The Zia also invoke the names of powerful dead ancestors for a help when hunting. For instance, when throwing his spear at a quarry a Zia man might yell 'A Omguta!' hoping for the ancestor's aid in scoring a hit. This is a euphemistic use of names of the dead, and can only be successful when employed by mature men, otherwise it will bring misfortune to the hunter.

The only verbal taboos associated with hunting and fishing reported for English are those employed by Scots and—earlier this century—by Cornish fishermen (cf. Frazer 1911:394ff; Cove 1978; Knipe 1984; Knipe and Bromley 1984). Euphemisms are or were employed during fishing expeditions when speaking of churches and clergymen, women, and land animals—in particular the rabbit—plus in Scotland, anadromous fish (i.e., those which live part of their life in saltwater and part in freshwater). Cove suggests that the practice derives from a conjectured contrast between sea and land: because the sea is productive and a male province, no euphemism is necessary; but the land is nonproductive and a female province, so that speaking of it while fishing is unlucky. Knipe suggests that this language behavior is based on a contrast between the insider world of the fishermen, and an outsider world which, for superstitious reasons, will only be spoken about euphemistically. The anadromous fish, for instance, are game fish like salmon and trout, traditionally the fare of the laird not the fisherfolk. Knipe (1984) and Knipe and Bromley (1984) document the demise of the taboos as the catch has become more predictable and the occupation much safer: until recently there were dreadful hazards attendant on life at sea and great uncertainty in locating fish. Other old superstitions, too, have ceased to hold sway in the face of greater personal security and where what was once mysterious and feared has given way to rational explanation and become the predictable effect of known causes. Against this background, one can see that the topics tabooed by fishermen are things associated with the security and comforts of the land, for which a man at sea might yearn and so cause his luck to turn against him. There is some evidence for a similar constraint among Malay fishermen (cf. Simons 1982:193). Fear of comparable hazards leads the Zia fishermen of Papua, New Guinea, (among other peoples) to taboo the names of headlands and other shoreline features, particularly those where accidents have happened in the past, lest some misfortune befall the canoes.

In most societies congress with women, and thus women's sexuality, is believed to be a danger to hunters, fishers, warriors, and sportsmen; for discussion, see Brain 1979 and Ch. 3. The reason that rabbits have been singled out for taboo is probably the centuries old link (examined in Ch. 4) between coneys, bunnies, and cunnies, which ties them in with the taboo on women. Yet boats are often given women's names. Is it woman as mother rather than lover? The Zia will not allow women on hunting trips, believing that they frighten animals away; the belief is so strong that if attacked by a wild boar, cassowary, or other dangerous creature, a hunter will call out his wife's name supposing that invoking a woman's name will be enough to drive the creature away from him. This behavior is reminiscent of invoking God to frighten away the Devil and saying *Bless you!* when someone sneezes. Cynthia Reed (January 1990) tells us that the North Americans she fishes with 'still pretend to think it's bad luck to have a woman on a fishing boat. I always get blamed when nobody catches.' Some traditions will not die.

Summary of Chapter 2

In this chapter we have looked at the euphemisms prevalent in naming and address-ing. In many societies, a name, that is a "true" name, is regarded as a proper part of the name bearer; it is not just a symbol but the verbal expression of his or her personality; and to offend it is as great a blow as physical assault. It is almost the same for Anglos. When we say of Bill Hodgson's son Eric that 'Eric's a real Hodg-son,' we are speaking as if the surname itself carries the genes that make Eric 'a chip off the old block.' The same is true for phrases like *make a name for oneself, have a good name, bring one's name into disrepute, clearing one's name,* and so forth. Whatever we believe in fact, even in our society we speak as if the name carries the properties of the name bearer! And names do have some such force: this is why proper names enter the general lexicon, not only in direct reference to an original celebrated name-bearer as in the case of, for example, *He's a little Hitler* (spoken of, for instance, Eric Hodgson), but also in the case of words like *lynch, boycott, hoover, kleenex, biro,* and so on (cf. Allan 1986a §4.7.6). Because "true" names are so closely associated with their name-bearer as to be a proper part of him or her, in some societies (e.g., in Austronesia) there are strict taboos preventing two living persons from going by the same name; though it may be possible for a name to be reused after death—given to a child, for example, who has been connected in some way with the dead person (see Dixon 1980:27; Simons 1982). Furthermore, "true" names are often secret, rendering euphemistic names necessary for public naming and addressing. For English, the taboo on a "true" name is observed only for *God* or *Christ* and only inconsistently for them under two conditions: (1) when borrowing the traditional Judaic euphemism; and (2) when 'taking the Lord's name in vain' almost exclusively in dysphemistic illocutionary acts. The Devil's name is also sometimes avoided—not lest it be annoyed but lest it (s/he?) be called.

We argued that to use an appropriate style of naming and addressing is euphemis-tic behavior, and we set about sketching the various conditions under which styles of address are appropriate. We emphasized the importance of context—both situation of utterance and world spoken of, but particularly the former—in determining these matters. The style of naming and addressing depends on the role Speaker perceives Hearer-or-Named to have adopted relative to Speaker in the situation of utterance, and on Speaker's attitude toward Hearer-or-Named at the time of utterance. When Hearer-or-Named is more powerful than Speaker, or there is a great social distance between Speaker and Hearer-or-Named, the normal euphemistic behavior is for Speaker to be deferential to Hearer-or-Named and to use a style whose degree of formality is greater than, or at worst equal to, 'consultative.' Where Speaker is superior in status to Hearer-or-Named, s/he has more choice among styles: where the context of utterance allows for Speaker to drop down to a 'casual' style, this is will enhance Hearer-or-Named's face—provided Speaker is not perceived to be patronising.

We briefly discussed X-phemism and gender, reviewing the motivations for using gender-neutral terms and also the strategies recommended for avoiding terminology deemed dysphemistic toward women. Whereas gender-neutral terms for jobs and offices can readily be adopted into the language, the employment of gender-neutral pronouns poses some hazard to a clear and succinct presentation of material.

After discussing appropriate ways to name and address kinsfolk, we finished off with a brief investigation of taboos on names of dangerous animals and naming taboos observed by people undertaking hazardous pursuits such as mining, hunting, and fishing. These practices are motivated by fears comparable with those on death and disease (cf. Chs. 6 & 7), and people use similar strategies to avoid calling down misfortune upon themselves. Many occupational and recreational pursuits have their own taboos and euphemisms for similar reasons: they are to avoid bad luck. Think, for instance, of taboos in the theater: no whistling; don't say *Macbeth*; don't say *Good luck!*, that's tempting fate, so instead it's *Break a leg!*

CHAPTER
—3—

Bodily Effluvia, Sex, and Tabooed
Body-Parts

Inter faeces et urinam nascimur

Bodily Effluvia and Revoltingness Ratings (r-Ratings)

Many communities have believed that malevolent magic can be wrought on a person's shit, spittle, blood, nail-parings, hair-clippings, and other bodily effluvia; and there are taboos on these effluvia, and on associated body-parts, in order to protect an individual from such danger (cf. Frazer 1911). In Greece, from ancient times through to the twentieth century, a target's bodily effluvia were formally cursed and either buried or drowned together with a tablet on which the curse was inscribed; hundreds of such tablets have been discovered (cf. Montagu 1967:39). In some cultures, including the one in which we currently live, the taboos do not derive from fear of witchcraft; instead they are motivated by fear of pollution (cf. Meigs 1978), or perhaps just by distaste for certain bodily effluvia and the organs that vent them. Intuitively we seem to find nearly all the bodily effluvia of anyone, especially any nonintimate, revolting to all of our senses. A stranger's dirty underwear, socks, cast off condoms, and the like are so much more revolting than our own; a sentiment that is captured in the German proverb *Eigener Dreck stinkt nicht* "your own shit doesn't stink." The revulsion and shame we have for these things is not shared by animals or even children: it seems to be something we learn. Because they are not all equally revolting, we will assign to each effluvium a revoltingness rating (r-rating) determined from a questionnaire distributed among staff and students of Monash and La Trobe Universities in Melbourne, Australia (the questionnaire we used and our basic findings are presented in Appendix R at the end of this chapter). In this study subjects were asked to assign to bodily effluvia a rank on a five-point revoltingness scale as RRR, or RR, or R, or ½R or Not-R such that RRR-rated effluvia are judged worse than RR- ones, and so on. We anticipate that the r-ratings will vary from society to society, but the taboo topics are almost universal.

Bodily effluvia include: RRR-rated shit and vomit (84 percent of subjects gave these a r-rating greater than R, the central value), also sperm/semen and urine (58 percent gave these a r-rating greater than R), and menstrual blood (80 percent of men gave this a r-rating greater than R, while only 47 percent of women did so); then there are RR-rated snot and fart (70 percent of subjects gave these a r-rating greater than R), pus (67 percent gave this a r-rating greater than R), and spit (50 percent gave this a r-rating greater than R); R-rated belched breath (78 percent found this worth a r-rating greater than or equal to R, whereas other breath is judged Not-R with 71 percent giving it a r-rating less than R); sweat was rated ½R with 58 percent of subjects rating it less than R; and in descending order of revoltingness, the following were judged Not-R; nail parings (65 percent gave these less than R), breath (71 percent gave this less than R), blood from a wound (79 percent gave this less than R), hair clippings (84 percent gave this less than R), and breast milk (86 percent gave this less than R). In considering these responses it should be taken into account that the blood of a stranger is welcomed when one needs a blood transfusion; and although wet-nurses have been superseded by bottle-feeding, it would appear that the nurturant qualities of breast milk have until comparatively recently made it acceptable from a stranger—a memory that may perhaps influence judgments on its revoltingness. Breast milk is presumably acceptable as food for the neonate, but it is probably distasteful when it leaks from the nipple or stains clothing. Least revolting of all effluvia are tears, to which 94 percent gave a r-rating less than R. Tears may hold this position because unlike the other liquid effluvia—except for blood from a wound—they are not waste products; and unlike blood, tears do not stain, and their flow cannot lead to death. For William Blake, even, 'a tear is an intellectual thing' (*Jerusalem* pl. 52 1.25—there is discussion of the nature–intellect dichotomy later in this chapter). Degrees of exception exist for effluvia that result from sexual congress (a nice euphemism), from very young children, and perhaps from the sick and injured. It is probably only during sexual intercourse that certain of someone else's bodily effluvia can be enjoyed as opposed to being borne stoically when circumstances, such as looking after a young child, demand it; (the exception might be a sadistic torturer enjoying the sight of the victim bleeding to death or suffering the humiliation of some uncontrollable efflux).

Constraints on the Mentioning of Certain Body-Parts

Body parts connected with RRR-rated effluvia are often highly restricted as to mentionability, being only readily mentionable to a doctor, lover or close friend, and so on. In our survey, we discovered that shit and vomit are the most revolting effluvia, but whereas the anus is the second least freely mentionable body-part after the vagina (11 percent judge the anus freely mentionable, versus 7 percent for the vagina), the mouth—which is the discharging orifice for vomit—is judged freely mentionable by 97 percent (the other 3 percent judge it somewhat restricted as to mentionability). The reason for this discrepancy is that the primary function of the mouth is not to discharge vomit, whereas the primary function of the anus is to discharge shit and fart—another noxious effluvium. Hence the anus is tabooed, the mouth is not.

For men in particular, menstrual blood has a very high revoltingness rating, the reasons for which are discussed shortly. This might be reason enough for tabooing the vagina, but the strongest motivation for doing so is that the vagina is a sex organ. Taboos on sex organs are primordially motivated by worries about genealogy; this justifies the jealous possessiveness people exhibit over their lover's sex organs. But bodily effluvia loom large as well, because heterosexuality is defined on the male ability to inject semen into the female and the female's ability to conceive and give birth to a child. Each of these events involve bodily effluvia, and recently there has been extensive, and appropriate, use of the euphemism *exchange of bodily fluids* to describe what happens during sexual intercourse. (This particular euphemism has been dispensed with as misleading in the context of AIDS, because people understood it to include effluvia like sweat and saliva, which apparently do not transmit AIDS.) Furthermore, the external orifices of the organs used in sexual intercourse are also used in urination. Thus, sex and bodily effluvia are closely associated conceptually—which is why attitudes to the taboo terms for them are closely associated (e.g. in the euphemism *four-letter words*). Paradoxically, the very enjoyment to be had exchanging bodily fluids with one's sexual partner promotes possessiveness and jealousy, which—along with genealogical fears, and perhaps the distaste for the urinary functions of the genitalia—renders them taboo. Bound into this emotive area of vilified body-parts and *dirty deeds* represented in *dirty books* is another paradox: an awareness that because nature brings forth life from organic decay so the sex organs must be somehow associated with decay. Their effluvia and not infrequently their odor would seem to confirm that this is indeed the case; and in many different cultures, too. For instance, 'The Hua [a people from the Eastern Highlands District of Papua New Guinea] associate . . . black with the dark rotting substances alleged to be crucial in female fertility.' (Meigs 1978:317 fn.8). In the least, there is some justification for Yeats writing (in 'Crazy Jane talks with the Bishop') 'Love has pitched his mansion in/The place of excrement.' We will have more to say on this paradox later.

The taboos we have been discussing explain why the vagina, anus, and penis are restricted as to mentionability, being only freely and readily mentionable to a doctor, lover, close friend, or the like. The figures from our survey show that women are somewhat more circumspect than men in speaking of such body-parts: for instance the vagina is 'freely mentionable' to 10 percent of men, but only 5 percent of women; the anus is 'freely mentionable' to 25 percent of men, but only to 6 percent of women; and the penis is also 'freely mentionable' to 25 percent of men, but only to 8 percent of women. On the other hand, men are more polarized than women when talking of genitals, since 55 percent of men find both the penis and the vagina 'highly restricted as to mentionability'! The effect is that far more women than men find genitals 'SOMEWHAT restricted as to mentionability' (cf. Appendix R). It is because such body-parts as these are, overall, restricted as to mentionability that they are known as *private parts*—though this phrase is normally restricted to the genitals, as are its counterparts in other languages; for example, Dutch *schaamdelen* "shameful parts," Indonesian *kemaluan* "shame, embarrassment," or Latin *pudendum* "that of which one ought to be ashamed" (also used in English—whence the clipped form *pud*). They contrast with body parts like the hand, which are normally freely mentionable to anyone, anywhere. A woman's breasts and nipples (also a source of effluvia, but an unrevolting one) are subject to the sex-fear, too; and they are not freely mentionable. It is significant that we see far more nipples on public show than genitals or arseholes: they are not so strongly tabooed, because although they are subject to the sex-fear, their principle function is perceived to be the provision of milk to newborns; and, used for this purpose, they are exposed in public by some nursing mothers who would not otherwise so expose them. (The lactatory function of breasts is recognized in slang terms, such as French *roberts*—based on the tradename for a nursing bottle, *laiterie* "dairy"; and English *dairies, udders, jugs, Nörgen-Vaazes*—based on the tradename for an Australian ice-cream company.) Similarly, although the nose and mouth are the source for (R)RR-rated effluvia, their principle functions are not taboo, nor are they subject to the sex-fear: for those reasons they are not tabooed and may normally be exposed to public view (by at least some groups) in all societies.

What we need now is a historical perspective, so once more we shift our attention to the Middle Ages.

Tabooed Body Parts, Bodily Effluvia and Sex: The Middle Dutch Perspective

> Man is nothing but stinking sperm, a sack of excrement and food for worms. After
> man comes the worm, stench and horror. And thus is every man's fate.
> (St Bernard *Meditiones*, quoted in Camporesi 1988:78)

Distaste for our body and bodily processes goes back a long way. Early Christian teaching, for example, maintained that human life was only a temporary stage in what was seen as a journey toward a much better future existence. Therefore only the soul was worthy of attention, and any attempt to prolong life was deemed irreverent. The body was too base a thing to be of any importance and was certainly not the affair of the pious, some of whom went to extraordinary lengths to uphold this belief: there are even records of people canonized solely on account of the fact that they did not wash (cf. Gordon 1959:19).

Given the lack of sanitation and the generally appalling state of people's health at this time, this gloomy vision of the human body is probably not all that surprising. Foul-smelling fistulas, gangrene, weeping sores, cancers, and such brought on by the ravages of disease would have seemed evidence enough of bodily corruption. Then there was the usual stench of the cities and towns; pigs and other animals wandered streets already choked with mud and sewage, and the smell of rotting meat. The bodies of criminals were left hanging in the streets to rot, sometimes for years!

Indifference to, if not loathing for, the body is very apparent in the way the dead were buried (see Ch. 6 and Aries 1974). All attention was on the soul, nothing was done for the body. There was no attempt to individualize graves: Except for the illustrious few, they were anonymous, and the bones of the dead were continually being dug up to make way for new bodies (witness the finding of poor Yorick's skull in the grave yard scene of Shakespeare's *Hamlet*, V.i.163ff). There was no concept of burying the dead in a plot of land where they would remain forever—we have to wait until the eighteenth century before we begin to find anything like the modern veneration of burial grounds.

Bodily orifices and their emissions call for euphemisms in most societies, as is clear from their centrality in any discussion of euphemism (cf. Brain 1979 ch. 13). Middle Dutch society was no exception, although there are some intriguing differences from modern English. What is most striking is the apparently inconsistent treatment of these topics: to modern readers, expression seems to swing from poetic prudery on the one hand to something very close to indecency on the other. Medical writers typically used some fairly evasive terminology when they had to describe a patient's generative and excretory organs. With almost poetic flourish, they refer to *hemelijke/gheestelijke weghe, porten, leden* "hidden/secret paths, doors, members." On the other hand, *Wie niet pissen en mach* "People who cannot piss" strikes the modern reader as extraordinarily coarse for the title of a section on urinary disorders within a medical text.

Discussing Urological Problems in Middle Dutch

We will start with treatment of urine and associated disorders. The practice of 'uroscopy' or 'watercasting,' the inspection of urine, was regarded as the supreme diagnostic tool in medicine and was considered an honorable profession for any physician. Urine was believed to have been filtered from the four humors and was therefore seen as the key to understanding a patient's complaint. There was therefore a preoccupation with urine that seems obsessive to us now. Uroscopy was the subject of many works of art. Paintings and woodcuts of the time show the physician in close scrutiny of the patient's urine flask, carefully examining it for color, quality, density, and sediment. It was reverently handled in gorgeous flasks and carried in wicker baskets like the very best of select wines. These flasks were sometimes sent over great distances for diagnosis by physicians famous for their accurate readings. Urine was also a vital ingredient in many of the fantastic remedies used. There was certainly no need to shelter behind any elegant phraseology here and the only expressions that could perhaps be viewed as an attempt at delicacy is the use of *water* and *nat* "wetness", which appeared regularly in one text. But there is certainly nothing delicate or evasive about expressions like the following:

> *hi doet vele urine maken* "he produces much urine"
>
> *Dyabete—dat is een die vele pisset sonder wille* "diabetes—that is one who pisses a lot without will". The term *diabetes*, derived from the Greek *dia* "through" and *betes* "to go", is in origin euphemistic, and refers to the frequent urination characteristic of this disease.
>
> *Strangiuria—dats die coude pisse* "Strangiuria—that is cold piss"
>
> *dat is die passye, dat hem die lude al slapende bepissen* "that is the disease such that people piss themselves while sleeping".

Whether it was treasured in flasks or hurled out of chamberpots onto the road below, it seems there was little shyness surrounding urine and consequently little reason to be euphemistic about it.

Anal and Bowel Problems in the Middle Ages

In the Middle Ages, the practice of purging the body was central to the maintenance of good health. Although perhaps in itself not uplifting, defecation became so elevated in importance that it could be described as one of the central focuses of medicine during this time. In contrast to the open mention of pissing, however, reference to defecation was always indirect. Many of the words to do with the evacuation of bowels refer simply to "going" (cp. Ch. 4). The following are plainly euphemistic *ganc* "(a) going"; *ter cameren gaen* "to go to the room or chamber"; *bloetganc* "blood going" (a reference to the bloody stools caused by ulceration); *ende daerop sitten* "and sit on there"; *ten stoel gaen* whose English counterpart "to go to the stool" was the source for the present day *stool* "shit" in English medicalese.

It seems people in the Middle Ages viewed a foul-smelling 'camer' with the same distaste as we who live in the century of the flush toilet. (*Camer* is cognate with *camera*, both for taking photos with and as in the legal 'a hearing in camera'—in chambers, out of the public eye.) In a text devoted almost exclusively to enemas and suppositories, the lack of a single direct mention of feces indicates a topic about which the medievals felt uncomfortable; instead, they used euphemisms like *materie* "matter" and *(be)roeringhe* "movement." Writers talk picturesquely about *die weghe des lichaeme open te maken* "opening up the paths of the body," *werden die weghe des lichaems bestopt ende die porten besloten* "when the paths are blocked and the doors are locked"—these texts reveal a poetry in constipation hitherto unrecognized! They abound with decently vague references to *perse beneden* "pressure below," to *laxeren ende purgeren van verhouden humoren* "relaxing and purging the body of ill humours;" to *die lede beneden sachten* "softening the parts underneath"; to *onreynicheit van hem doen* "getting rid of impurity"; to *die ripe materi . . . uut te seynden* "sending out the ripe matter"; even to "the stomach running" (*loept die buuc te zeere*) or "blocking" (*es oec die buuc bestopt*). One of the most useful verbal escape-hatches available to writers of the time were the words *lichaem* and *lijf*. Both words had the general meaning of "body," as they still do in Modern Dutch (the difference between them need not concern us here). They could also be used to mean the "stomach" or perhaps more specifically the "bowel." For example, *teghen vasten/besloten lichaem* "for a firm/locked body" are titles heading sections on cures for constipation. Both *lichaem* and *lijf* were also used to refer to bowel movement—*den lichaem hebben*, lit. 'to have the body' meant "to defecate."

Sometimes different versions are available of the same text. In these cases, it is interesting to see that writers show definite preferences for different terms. The term *aers(gat)* "arse(hole)," for example, is conspicuously absent from most texts; one notable exception is the 'Circa Instans,' which uses it throughout. In a second version of this text, however, earthy Germanic *aers* has been replaced on every occasion by the more genteel *fundament* borrowed from Latin.

Why were the medievals euphemistic about defecation but not urination? Why is urine used in remedies in many parts of the world, whereas feces is not (or very rarely, see later discussion)? Why do urban men stand side by side with other men to urinate in public lavatories but sit in secluded privies to defecate? Is it simply a matter of modesty not having to remove clothing, or one's vulnerability in a sitting position (cp.

normal practice in women's public urinals), or is it simply that feces smells much worse than urine and is messier if you trample in it? We might well imagine that people in medieval Holland did not respond so adversely to bad smells as we do now, so great must have been the stench pervading their villages and towns. Yet many of the medical texts warn of the dangers of foul odors to health; "bad smells" theories of disease were extremely popular at this time. If bad air was impossible to avoid, then writers advised that good air be made artificially, by sprays and vapors—particularly, in summer. The following is a fairly typical recommendation:

> *Men sal vlien nevelighe lucht, dicke lucht, besloten lucht, stinckende ende ghecorrumpeerde lucht. Ende is openbaer mit experimenten dat onzuverheit der lucht den sinne plompt, der zielen sinne begripen belet, huer vonnisse ende huer oerdeel verdonkert, huer ghepens verdwaest, vriendscape minret ende si maken vele ziecheyden.*

> "One should flee from misty air, thick air, closed air, stinking and corrupting air, And it is clear from experiments that impurity of air dulls the intellect, hinders the comprehension of the soul's senses, clouds their judgment and their discernment, dulls their understanding, lessens enjoyment and they cause many illnesses."
>
> (Boec van Medicinen in Dietsche)

The smell of human feces is almost universally regarded as disgusting, perhaps as Brain (1979:133–138) suggests, because it reminds us of the smell of the putrefying corpse; so the taboo surrounding defecation may be associated with the fear of death (see Ch.6). However, the medieval obsession with purging, essentially a purification ritual, suggests another motivation: defecation is an evacuation of bad humors that presumably reside in the shit and account for its unpleasant smell. Bad humors are naturally to be avoided as a possible source of physical if not metaphysical contagion. Urine was not viewed in this way and although diuretics existed, they were not recommended for the purpose of cleansing from defilement. Presumably this is why we find the very different language treatments of these two subjects.

'The Organ of the Natures': Sexual Therapy in the Middle Ages

Without a doubt medieval life was generally one of extremes, but it is in this area of human activity in particular that the contradictions are most obvious. In the literature of the time, erotic sentiment finds expression sometimes in earthy explicit language and at others with chaste sentiment in the literature of delicate courtly love. Similar variety is also apparent in the medical texts of the period. We are told of the rewards of a virtuous and chaste existence and are warned of the consequences of excessive sexual indulgence: an unchaste existence, one text warns, will shorten one's life thirty years! At the same time, texts abound in references to oils and potions recommended to increase sexual potency and drive, and to enhance lovemaking by *meeren/stercken nature* "increasing or strengthening the natures." The word *nature* both in Middle Dutch and English generally denotes the quality or essence of something; but its scope was once much wider and encompassed both the abstract and concrete aspects of whatever it was that gave something its fundamental character. For that reason, it could serve as something like a euphemistic omnibus for writers of the time. In the Middle Dutch texts, for example, it is frequently used to denote the sexual organs of a patient.

> *als die natuur bezwaert ende bedruct is, beyde in mannen of in wiven, van de*
> *quaden humoren, soe pijnt haer die natuur altoes te reynighen.*

> "when the nature [sexual organ] is burdened and afflicted, both in men and
> women, by bad humors, then the nature tries to cleanse itself."
>
> (Boec van Medicinen in Dietsche)

The term nicely avoids direct reference to the organs themselves, or for that matter their effluvia, for these are also included in its range of reference. *Die natuer vanden mannen dat men heet sperma int latiin* "The nature of men, which one calls sperm in Latin" is viewed as their life force; and sexual intercourse, because it drains men of their nature, is considered debilitating to them (*ex debilitate virtutum*) (cp. the early English slang expression *spend* meaning "to ejaculate"). Those men "who have been too often with women" (*die te vele met vrouwen gheweest hebben*) so that "they have lost their strength" (*haer craft verloren hebben*) and "their nature has left them" (*die haer nature ontgaet*) are to be rehabilitated on a special diet. There is even poetic reference in one text to a gonorrhoea sufferer as *dat een sijn natuur ontloept* "one whose nature flows out"—a nice euphemism for a discharging dick!

On the whole, copulation is well camouflaged in these texts behind evasive circumlocutions, which mostly seem to involve general companionship rather than the act itself. Nonetheless, one text does refer to *rijden* "riding," a metaphor that appears in a number of languages (e.g., English and early Greek) (cf. Griffin 1985:36). Unlike most euphemism, however, this expression seems to draw attention rather than divert it, and it creates something like the heightened reality of the euphemistic allegories we discuss in Ch. 9. It is difficult to accept that modesty is the motive behind such an actively enthusiastic euphemism as this one. "Being or going with someone," however, was the typical euphemistic formula, as it is in English (most certainly when the someone is a prostitute). Also common was the expression "to have to do with someone". Wet nurses were warned not to have anything to do with a man (*die mamme en sal mit ghenen man te doen hebben*). One medical writer from fourteenth-century Brabant describes the victims of various penial disorders as *sotten* "fools" because they "have to do with unclean women"; he goes on to clarify his meaning in parenthesis with *als si by hem leggen* "since they lie with them" (cp. English).

Men who wanted *wel te wesen met vrouwen* "well to be with a woman" were advised to eat the ash of a mole's burnt penis and testicles. A special oil from beavers' testicles rubbed on the genitals was strongly recommended when *als een mit vrouwen niet en mach wesen* "one is not able to be with women"; this is presumably a remedy against inopportune flaccidity. As in many cultures, this affliction was thought to result from being *betouert* "bewitched" and remedies were usually available in collections of what were known as *secreten* "secrets." The word *secret* itself is interesting. In both Middle Dutch and early English it was commonly used to describe the genitals, their effluvia, and associated diseases—particularly those of women. In its usual sense of "hidden, not disclosed," it was a convenient euphemistic label to describe what was otherwise the literally unspeak-

able. The following comes from a sixteenth-century pamphlet from Antwerp. In it
a quack healer boasts:

> *Ten anderen can desen Meester helpen ende ghenesen alle secrete ghebreken van*
> *mannen ende vrouwen, hoe desolaet die selfde persoonen syn, die hier niet en*
> *dienen verhaalt te zyne om der eerbaerheyt wille.*

> "Furthermore, this Master can help and cure all secret illnesses of men and
> women, however devastated they are, which here cannot serve to be repeated for
> decency's sake." (Quoted in Hüsken 1987:103)

(This is reminiscent of the 'invisible words' of the Victorian novel; for example, the
unmentionables, indescribables, inexpressibles, viz. "underclothing, breeches"; cf.
Burchfield 1985:16.) *Secret* also once had strong associations with the occult. A
'Book of Secrets' often contained references to matters of sorcery and recipes that
aimed at somehow changing the natural order of things. For example, the following
describes the effects that a certain herbal potion will have on a woman:

> *Sij wert stahans soe luxurieus, sij volghet den man waer dat hijse hebben wille ende*
> *doet sinen wille.*

> "She will soon become so lascivious, she will follow the man wherever he wants to
> have her and do his will."

Publishers then, like advertisers today, were keenly aware of the advantages of a
titillating title; they knew that book titles which contained the word *secret* were
assured of a readership; and so it was used where it was barely, if at all, appropriate
(cf. Braekman 1987:271).

But to the more virtuous side of medieval life. Those not wishing to *bruden*
"marry" (this was a metonymic euphemism for "copulate") were advised to wash *die
lede der naturen* "the organ of the natures" in a special preparation of *agnus castus.*
The name of this herb reflects its special power *houd den menschen reyne als i. lam*
"to keep the person as pure as a lamb" by *alle hitte van luxurien vercoelt* "cooling the
heat of exuberance" and *ende hine wint nemmermeer kint* "and he will never more
want children" (recall the Catholic dogma is no copulation without procreation).
The same preparation *der moeder ende haren mont ende den inghanc vernauwet
ende maecten enghe* "reduces and narrows the mouth and entrance of the mother [=
womb]"; consequently, it was believed to be a means of ensuring chastity in women.

In addition to being named "the organ of the natures", male and female genitalia are referred to simply as *ghemacht* or *macht* literally, a "shape" or "something
made/shaped", and possibly best translated by "groin". Other synecdochic subter-
fuges encountered are *die nedersten steden beide achter ende vore* "the nether regions
both behind and in front", and simply *eynde* "end". Terms like *die manlicheit/
wijflichede* "the manhood/womanhood" are euphemistically applied, as they can be
in modern English, to the sexual organs (e.g., *sine manlijcheit ende al die steden
daer omtrent worden al vol wormen* "his manhood and all the places around were
full of worms"). One writer uses *vernoy smans/swijfs* "irritation of the man's/
woman's" where the name for the organ itself is simply omitted altogether.

On a lighter note is *tknechtekens seer* "the little boy's sore," referring to some
inexplicit affliction of the penis, possibly inflammation of the glans and/or prepuce.

Whatever the complaint is, the name certainly does not suggest a dangerous affliction. 'Nicknames' of this sort are common for diseases which are not of a serious
nature or which can be seen as mildly humiliating to the patient (cf. *the trots, the
barfs, Bali Belly* and *Montezuma's revenge*. See also Johnson & Murray 1985:153).

A Woman's Body and the Attendant Euphemisms

> A woman can be proud and stiff
> When on love intent;
> But Love has pitched his mansion in
> The place of excrement;
> For nothing can be sole or whole
> That has not been rent.
> (William B. Yeats 'Crazy Jane
> talks with the Bishop' 11.13–18)

There is no doubt that more taboos surround the body and effluvia of women than
those of men. Concomitantly, there is a branch of medicine, *gynecology*, that
pertains to the functions and diseases of womankind, but nothing comparable that
pertains to the functions and diseases of male human beings: *anthropology* is, of
course, something altogether different, and neither *andrology* nor *anerology* appear
to exist. If we identify the physiological milestones in a woman's life as menarche,
pregnancy, and menopause, they are stained with blood; and there can be little
doubt that the original cause for the very different ways in which men and women
are perceived is female physiology, which renders the female at a comparative
disadvantage to the male when she is menstruating or childbearing, during lacta-
tion, and even at later stages of child-rearing. For example, look at the following
from a Middle Dutch text:

> *Want vrouwen vele cranker sijn . . . daer om hebben si menigherhande ziechede.
> Ende sunderlinghe bi dien leden, die natuur ter drachten ghevoecht heeft ende hem
> die ziecheit in hemeliken leden aencoemt, dat si van scaemten gheen meyster
> ontdecken en dorren. Daer om heeft mi ontfarmt haer scamelheyt, dat ic hebbe
> ghemact een boec allen vrouwen mede te helpen.*

> "Because women are much weaker . . . therefore they may have many kinds of
> sicknesses. And especially in those parts which nature has added for pregnancy and
> the sickness affects them in secret parts, so that from shame they dare not reveal
> (them) to any master. Therefore I pity their shame and have prepared a book to
> help all women." (Boec van Medicinen in Dietsche)

The text from which this excerpt is taken is unusual for having a section devoted
exclusively to women's problems. It is important to remember that the authors of all
the ancient and medieval medical treatises were male, as was their intended audience. The problem of communication between female patients and male
physicians—although often assisted by female attendants—is well documented in
the ancient treatises, and a number of medical writers discussed the fact that many
women are inhibited when describing their complaints to males (cf. Lloyd 1983:58–

111 on the treatment of women in the Hippocratic treatises). There was apparently a reciprocal reserve on the part of the male physicians and writers when discussing gynecological problems—as is indicated by the fact that all ancient and medieval medical works spend most of their ink discussing male patients. If women were mentioned at all in a medical text, it was more often than not as carriers of disease, despite the fact that suggested remedies always envisaged male rather than female patients. The only texts that freely mentioned the female sex were collected works of recipes and potions, and then it was always in connection with methods of testing for virginity, pregnancy, the sex of a fetus, or sterility—arguably, those things of greatest concern to men. Curiously, sterility was viewed as a female problem, even though at this time it was the male who was believed to make the more significant contribution to reproduction (see discussion Lloyd 1983:86–111). Women seem to have been viewed as a kind of human petri dish to hold the male seed!

Particularly striking for their neglect of women are the surgical treatises of the day; for example, the detailed treatise examined here *Boeck van Surgien* by the fourteenth-century Brabant surgeon Meester Thomaes van Scellinck. In this long work, Meester Thomaes makes no reference to gynecological operations and in fact refers to women only twice; once indirectly when discussing the treatment of hermaphroditism; and a second time in connection with the removal of kidney stones. Such reticence is all the more remarkable given the importance placed at the time on the different physical characteristics of patients: texts went to great lengths to point out that different treatments were necessary depending on whether the patient was fat or thin, old or young, bilious or not (cf. Lloyd 1983:64); yet sexual differences were almost completely ignored.

The following observations by the famous thirteenth-century physician Arnold of Villanova in his book on poisons (!) suggests that we are perhaps looking in the wrong place for accounts of women's complaints: 'In this book, I propose, with God's help, to consider diseases of women, since women are poisonous creatures, I shall then treat the bites of venomous snakes' (quoted from Guthrie 1945:113). There seems to be no end to the rantings about the dangers of women! They have always been linked strongly with matters of the occult—their association with the periodicities of the moon, for example, and also with the left, or sinister, side of the body. During the Middle Ages it was believed that male embryos developed on the right side of the uterus and female embryos on the left, a belief that goes back to the infamous Aristotelian concept of women as 'natural deformities' (cf. Lloyd 1983:94–105). Women were seen as the source of magical actions and when causes had to be found for calamities like the outbreak of an epidemic or some natural disaster, any unfortunate woman whose behavior did not fit in with the community norms was likely to be accused of witchery and made a scapegoat.

Ortner (1974) has argued that the secondary status of women in society derives indirectly from their physiology, which we regard as self-evidently true. Because women and not men bear children, and consequently menstruate and lactate, women are perceived to be more closely bound by and to their bodies and body functions than are men, which renders women more like (other) animals and therefore closer to nature than are men. Men, not being physiologically bound in such ways, not only had the opportunity to become politically and economically dominant, but furthermore had the time and energy to expend on things of the mind rather than of the body; that is, to control the domain that supposedly distinguishes humans from animals. The association of women with the animal side of humans, and men with power and intellectual pursuits, quite naturally produces a cultural and social appraisal in which men are superior to women. (Not everyone agrees with this hypothesis, but we have not been persuaded that there is any other that comes close to accounting for all the relevant cultural, social, behavioral, historical, and linguistic facts that this hypothesis DOES explain. Note that we do NOT claim that it morally justifies any of them; that's another matter altogether.) Two effects of these perceived differences between men and women are relevant to our particular theme in this book: (1) the pollution taboos on women's unique physiological processes at certain times; and (2) the downgrading of a man by ascribing to him the characteristics of a woman, in contrast with the converse: a woman is not generally downgraded to a comparable degree when ascribed the characteristics of a man.

The disadvantage that women's physiology imposes on them has been exploited by men to assert social dominance and even ownership rights over them; it has led to peculiar taboos over women's procreative organs (and often over their entire bodies) that purportedly aim to protect a genealogical investment. Irrespective of one's moral evaluation of this, there is reason behind it: until the advent of in vitro fertilization, a woman invariably knew that the child she has borne is genetically her own, whereas a man can only be sure that his wife's child is genetically his if he is certain she has not had sexual intercourse with another man. In the interests of self-protection, women have generally accepted and even encouraged the taboos on their bodies and effluvia as measures toward ensuring their personal safety. These taboos have been confirmed by the dominant religions in our culture, and in many others too. Against this background, any question about a woman's sexual behavior has been seen as an offense against a desirable social and religious—and even rational—norm. The particulars of these taboos on women, as they relate to our theme, are explored below.

The Menstruation Taboo

> Oh! menstruating woman, thou'rt a fiend
> From whom all nature should be closely screened.
> (Crawley 1927/1960:77)

> menstruation [would] have been the locus for glorification had it been the experience of men
> (Spender 1984:200)

In most societies menstruating women are or have been taboo. The taboo in our Judeo–Christian tradition is written into the Bible:

> 1The LORD said to Moses and Aaron . . .
> 19 "When a woman has her regular flow of blood, the impurity of her monthly period will last seven days, and anyone who touches her will be unclean till evening.

20 "Anything she lies on during her period will be unclean, and anything she sits on will be unclean. 21 Whoever touches her bed must wash his clothes and bathe with water, and he will be unclean till evening. 22 Whoever touches anything she sits on must wash his clothes and bathe with water, and he will be unclean till evening. 23 Whether it is the bed or anything she was sitting on, when anyone touches it, he will be unclean till evening.

24 "If a man lies with her and her monthly flow touches him, he will be unclean for seven days; any bed he lies on will be unclean."   (*Leviticus* 15)

In Manhattan, the Hasidic Jews who work in the diamond district have a private bus to take them to and from Brooklyn where they live in order to avoid contamination from menstruating women on public transportation. Perhaps because men can only experience catamenia (menstrual discharge) as the effluvium of another person, a woman, menstrual blood is especially tabooed by men. In a recent survey, mostly of students and staff at Monash and La Trobe Universities in Melbourne, Australia, we confirmed that men viewed menstrual blood with far more horror than do women: 80 percent of men gave menstrual blood a r-rating greater than R, while only 47 percent of women did so; furthermore, not one single man gave it a Not-R rating, whereas 17 percent of women did so. In some societies, men will not even walk where a menstruating woman might have passed above (see Frazer 1911:250); and in many more, men are forbidden to have intercourse with a menstruating woman. Traditionally, the taboo is based on a fear of contagion because catamenia is seen as purging or purifying, a cleansing away of ill humors—hence the Middle Dutch terms *purgacy* and *reynicheit* (cf. [1] p. 65). In the Middle Ages, menstrual blood was believed to contain defiled spirits; and many of the most feared diseases, like leprosy and syphilis, were thought to be transmitted through menstruating women. Because of the danger they posed, such women were usually prevented from mingling freely in the community. In sum, menstruation is seen as a polluting discharge that weakens the woman; the pollution and weakening can be transferred to a male partner and, at worst, may lead to the man's death.

To some contemporary feminists, the menstruation taboo seems an outrageous denigration of womanhood, and which has no counterpart for men. The only peculiarly male effluvium is semen/sperm; and although male ejaculate has a RRR rating and 58 percent of all subjects gave it an r-rating greater than R (versus 55 percent for menstrual blood) women found it less revolting than did men: 56 percent of women gave semen/sperm an r-rating greater than R whereas 63 percent of men did so! It is not too difficult to figure out why menstrual blood and male ejaculate are regarded so differently. This was just as true for the ancient Israelites:

16 "When a man has an emission of semen, he must bathe his whole body with water, and he will be unclean till evening. 17 Any clothing or leather that has semen on it must be washed with water, and it will be unclean till evening. 18 When a man lies with a woman and there is an emission of semen, both must bathe with water, and they will be unclean till evening."   (*Leviticus* 15)

Note that the woman who comes in contact with semen is only unclean till evening, but the man who comes in contact with menstrual blood is unclean for seven days.

But it is naïve to dismiss men's reaction to menstrual blood as simple gynophobia, and leave it at that. The taboo often inconveniences men, as well as women; and it is most likely that a menstruating woman is set apart for a combination of some or all of the following reasons. (1) Menstruation is typically induced by hormonal changes associated with an unfertilized ovum; these changes lead to the break down of the endometrium, in which blood soaked regions of it crumble away and are discharged along with mucous and cellular debris (including the unfertilized ovum): thus, in a very real sense, catamenia is the discharge of decaying matter. (2) There is the pollution taboo of any discharge through the vagina (cf. *Leviticus* 12:1–8, 15:25–30). (3) Menstrual discharge renders the vagina comparatively messy and unpleasant smelling (and tasting)—consequently a menstruous woman is regarded as unclean; hence the German terms *Schweinerei* and *Sauerei* (roughly "piggery" and "*sowery*"). (4) Loss of blood is otherwise correlated with injury and loss of strength. (5) Unlike other blood, menstrual blood does not coagulate. (6) In many women menstruation is accompanied by cramps or similar discomfort (dysmenorrhea). (7) Some women have reduced libido during the first days of menstruation. (8) Some women suffer a brief period of premenstrual tension that makes them cranky or worse. (9) On some occasions, menstruation is counterevidence to a desired pregnancy—indeed it is reported that 'Maoris regard menstrual blood as a sort of human being *manqué*' (Douglas 1966:96). If we put all these together, we get a composite of the menstruating woman that typically puts her below her best in temper, in her sense of well-being, in sexual receptivity and sexual attractiveness, and in some contexts, she will be disparaged for having failed to achieve pregnancy; so it is hardly surprising that one synonym for menstruation is *the curse*. None of these disadvantages of catamenia is shared by male ejaculate—at worst, the latter is often perceived to lead to the temporary weakening of the man.

E Putridis Vita: "Polluting Effluvia" as Agents of Health and Well-Being

> If there were no decay, the universe would come to nothing . . . There is no
> birth without previous corruption.
> (Lorenzo Ventura 1613, quoted in Camporesi 1988:157)

There is one curious aspect to the concepts of pollution and taboo that our discussion has not yet addressed, although it has alluded to it on a number of occasions; this is the interesting interaction that exists between filth on the one hand and purity and healthiness on the other. The effluvia we have been discussing here are not simply impure, odious, and devitalizing substances—they are also believed to have important health-promoting aspects that, over the years, have been used in witchcraft, pharmacy, and folk remedies in many different societies. And it seems that the most polluting of substances, those subject to the most severe taboos, also have the greatest amuletive, inoculative, and healing powers of all.

During medieval times, for example, ashes from the bones of the dead, particularly the remains of those like executed criminals who had died a violent death, were attributed special medicinal and protective properties. Extract of the cranium was reputed to be efficacious against epilepsy; teeth and hair against gout. Virtually any bodily emission could be turned into a powerful curative agent. Camporesi (1988:154) describes some of the early 'dung-based health programmes,' using either human or animal feces, in Europe. Meigs (1978) describes the positive powers attributed to polluting substances among the Hua of New Guinea. For the Australian Aborigines, feces has always been regarded as having protective and curative powers. One traditional practice, for instance, has been to smear the faces of new born babies with excrement. Knowing this, it should come as less of a surprise that our own practice of throwing rice and confetti at newlyweds has its origins in the throwing of excrement! Blood, and in particular menstrual blood, was used in Europe to treat ailments of all description, especially serious diseases like leprosy—despite its reputation as the dangerous purveyor of these diseases (cf. Durkheim 1963:94). It is presumably homeopathic, like "the hair of the dog" (alcohol as the remedy for a hangover). We also find that symbols of blood are endowed with special powers. Coral, because of its red color, has been traditionally used in babies' teething rings. In the north of England, one custom has been to apply red flannel to bronchial chests and joints that are sore and inflamed. Bloodstone is worn as an amulet because the red spots symbolize Christ's crucifixion. The Hua, according to Meigs (1978:309f), attribute exceptional curative powers to red pandanus oil and those foods that release a reddish color when cooked.

Many more examples could be given, but the question is: How can something be perceived as both polluting and health-giving? Inoculation and homeopathy perhaps supply some of the answer. The rest, as we have already suggested, is that effluvia are perceived as the evacuation from the body of substance that is decaying, rotting, dying and which—like *night soil* (what a euphemism that is!)—will fertilize new life and growth. In short, bodily effluvia have regenerative and life-producing powers.

St Bernard was wont to say that the human body is a *sterquilinium*, a walking dung-heap, full of worms. Medical belief held that these creatures perpetually inhabited the human body, being born out of the body's own putrid substances. They were loathed, much as bacteria are today, but at the same time their association with rotting matter gave them strong curative and protective powers. Pulverized worms of all types were the prime ingredient in many Middle Dutch remedies; so were toads, which were also mistakenly believed to be born of decay. Women, needless to say, were reputedly more full of vermin than were men. In particular, the womb as it is the source of human life, had to be full of compost and other decaying fertilizer. This is reminiscent of the 'dark, smelly, explicitly rotting interiors attributed to females' bodies' by the Hua people in Meigs (1978:312). No wonder women were viewed as such dangerously polluting creatures that the taboos on them were similar to those on corpses! We wonder if, were it not for women's biological handicap resulting from their childbearing capacity, such strange physiology would have afforded them a dominating position in society—revered as divine, and not disparaged as evil and dirty.

But what has all this to do with contemporary Western society? Medieval notions of physiology and their medical doctrines were largely built upon imagination and superstition—ours is the age of science and technology. But the old ideas and prejudices that we have been describing 'have given birth to ways of doing things which have survived and to which we have become attached', to quote Durkheim (1963:113) talking about the incest taboo, but the remark is relevant here. It is an ancient routine which we follow in toilet training children (and house-pets) and teaching that bodily effluvia are 'dirty.' And we will continue to follow it because of our deep attachment to and our apparent need for ritual; especially when the ritual has practical advantages: there is, after all, something to the medieval theory that effluvia are connected with disease.

To sum up: from decay and death comes life and growth. Bodily effluvia are both contaminating and life-giving. Taken either way, they are fearful substances, and this probably explains why it is they are able to repulse and attract at the same time. For us in the twentieth century, the fear has changed to distaste, but it is enough to ensure that the linguistic sanctions endure. We may no longer impute supernatural powers to our bodily effluvia, but they remain a powerful source of verbal taboo.

Summary of Chapter 3

In this chapter we have examined the reasons for taboos on mentioning certain body-parts and their functions; in the next chapter we shall discuss the English lexicon for circumventing or defying such taboos. There are two motivations for the taboos: procreation and the discharge of waste products from within the body. The taboo on procreation applies to sex organs and all varieties of sex acts. Primary examples of the taboo on human waste products are shit, vomit, piss, semen (as a waste product) and catamenia; fart, snot, pus, and spit are little better regarded. The bodily organs whose principal function is to discharge such revolting and tabooed effluvia are for that very reason tabooed along with them. Pollution by association extends from the effluvia not only to the body parts, but also to underclothing, to lavatories, even to people with an occupational association with lavatories: which is why Harijans are taboo to Brahmins within the Hindu caste system.

We went on to consider the medieval Dutch perspective. Particularly striking here was the almost oxymoronic treatment these subjects received in the Middle Dutch texts. This was most evident with the excretions. The richly euphemistic language surrounding feces was most certainly motivated by strong pollution taboos; feces was believed to contain dangerous defilement and regular purging was recommended for the maintenance of good health. There was no corresponding taboo on urine, and mention of it was always straightforward and usually quite graphic. Urine was considered the physician's prognostic tool; it gave indications of the temperament and harmony of the humors, and therefore held the secret to good health. This accounts for its special position during this time, which meant there was no need for the kind of euphemism we use today.

We also sought to explain why there are more taboos on the body and effluvia of a woman than on the body and effluvia of a man. We ascribe the difference ultimately to the unique physiological processes deriving from her ability to bear children. The exploitation of woman's comparative physiological disadvantage has probably contributed to, if not resulted in, her generally inferior social and economic status; but this is of little direct relevance to the particular concerns of Chapter 3. We argued that catamenia is more obviously a waste product than semen, and therefore is regarded less favorably; it follows that the external organ through which it discharges suffers the same fate (see Ch.4). We also argued that the stronger taboo on the vagina than on the penis is partly motivated by a genealogical worry: until very recently, a father had to take his paternity on trust and circumstantial evidence; a mother always knew a child was hers (even if she didn't know who the father was). The greater the taboo on the woman's genitals and sexual activity, the stronger is the circumstantial evidence for paternity. One additional thought: some part of the explanation for the greater use among women of perfumes and deodorants is to counter the stronger taboo on the effluvia of women than of men: Was there ever a male-targeted counterpart to vaginal deodorant?

Finally, we sought to resolve the curious paradox between bodily effluvia perceived as polluting and these same substances viewed as health-giving. We argued that dangerous substances can be used as inoculants or homeopathic remedies against the very dangers that exposure to them can bring down upon one. But in addition, there is a very widespread perception among human beings that because bodily effluvia are organic waste products, and organic waste products in nature are seen to bring forth—or at least provide a nursery for—new life, then bodily effluvia must also have generative or regenerative powers. All this, despite remaining subject to taboo under most circumstances.

Appendix R, Revoltingness Ratings

The Questionnaire

Many communities have believed that malevolent magic can be wrought on a person's spittle, nails, hair or blood, etc. and exercise taboos to protect an individual from such danger. Taboos concerning bodily effluvia in current English do not synchronically result from fears of witchcraft, but intuitively we seem to find nearly all the bodily effluvia of anyone, especially any non-intimate, revolting to all of our senses. They are not all equally revolting, and we can assign them r-ratings (revoltingness ratings) such that RRR-rated ones seem worse than RR- ones, etc. The r-ratings will vary from society to society; but the taboo topics are almost universal.

Please identify yourself according to the following criteria:

Sex: Male / Female   Age: under 25 / 25-45 / over 45

Language & cultural background if not Anglo-Australian:

Assess the following SUBSTANCES **as produced by an adult stranger** using a scale of **RRR for the most revolting, RR for less revolting, R for even less revolting, etc.** Ring the appropriate response.

belched breath	RRR	RR	R	½R	Not-R
breath	RRR	RR	R	½R	Not-R
menstrual blood	RRR	RR	R	½R	Not-R
blood from a wound (blood not exuded from the anus, penis or vagina)					
	RRR	RR	R	½R	Not-R
breast milk	RRR	RR	R	½R	Not-R
shit	RRR	RR	R	½R	Not-R
fart	RRR	RR	R	½R	Not-R
hair clippings	RRR	RR	R	½R	Not-R
nail parings	RRR	RR	R	½R	Not-R
spit	RRR	RR	R	½R	Not-R
pus	RRR	RR	R	½R	Not-R
semen/sperm	RRR	RR	R	½R	Not-R
skin parings	RRR	RR	R	½R	Not-R
snot	RRR	RR	R	½R	Not-R
sweat	RRR	RR	R	½R	Not-R
tears	RRR	RR	R	½R	Not-R
urine	RRR	RR	R	½R	Not-R
vomit	RRR	RR	R	½R	Not-R

Any additions or comments?

The Questionnaire - continued

In this test we assume three degrees of constraint on mentioning body parts, e.g. imagining that the body part is sore/painful. We are **not** concerned with the names for the body parts, but with the body parts themselves.

FM　　(freely mentionable - to anyone, anywhere)

SR　　(somewhat restricted - not freely mentionable)

HR　　(highly restricted mention - only readily mentionable to a doctor,
　　　　　　　lover or close friend, etc.)

Please assess the following body-part terms according to the above constraints on mentionability, by ringing the appropriate response.

ankle	HR	SR	FM
anus	HR	SR	FM
buttocks	HR	SR	FM
ear	HR	SR	FM
foot/feet	HR	SR	FM
head	HR	SR	FM
man's chest	HR	SR	FM
mouth	HR	SR	FM
nose	HR	SR	FM
penis	HR	SR	FM
thigh	HR	SR	FM
tongue	HR	SR	FM
vagina	HR	SR	FM
woman's breasts	HR	SR	FM

Any additions or comments?

THANK YOU FOR YOUR ASSISTANCE.　KA & KB　　　　　　June 22, 1988

Table 3.1. Percentages and [Raw Scores]*

	RRR	RR	R	½R	Not-R	Omitted
belched breath	20.00 [13]	29.23 [19]	30.77 [20]	18.46 [12]	1.54 [1]	[1]
breath	3.03 [2]	10.60 [7]	12.12 [8]	19.70 [13]	54.55 [36]	
menstrual blood	28.79 [19]	18.18 [12]	15.15 [10]	21.21 [14]	16.67 [11]	
blood from a wound	1.54 [1]	9.23 [6]	12.31 [8]	13.85 [9]	63.08 [41]	[1]
breast milk	3.08 [2]	1.54 [1]	9.23 [6]	10.77 [7]	75.38 [49]	[1]
shit	60.60 [40]	19.70 [13]	12.12 [8]	6.06 [4]	1.52 [1]	
fart	27.70 [18]	43.08 [28]	16.92 [11]	12.31 [8]		[1]
hair clippings	1.52 [1]	1.52 [1]	13.64 [9]	15.15 [10]	68.18 [45]	
nail parings	4.55 [3]	7.58 [5]	21.21 [14]	28.79 [19]	37.88 [25]	
spit	23.08 [15]	36.92 [24]	26.15 [17]	12.31 [8]	1.54 [1]	[1]
pus	30.30 [20]	36.36 [24]	19.70 [13]	10.60 [7]	3.03 [2]	
semen/sperm	35.94 [23]	20.31 [13]	17.19 [11]	12.50 [8]	15.08 [9]	[2]
skin parings	7.94 [5]	23.81 [15]	22.22 [14]	20.63 [13]	25.40 [16]	[3]
snot	33.33 [22]	31.82 [21]	30.30 [20]	4.55 [3]		
sweat	6.06 [4]	10.60 [7]	24.24 [16]	42.42 [28]	16.67 [11]	
tears			6.06 [4]	1.52 [1]	92.42 [61]	
urine	30.30 [20]	24.24 [16]	21.21 [14]	10.60 [7]	13.64 [9]	
vomit	62.12 [41]	18.18 [12]	9.09 [6]	6.06 [4]	4.55 [3]	

	HR	SR	FM	Omitted
anus	41.54 [27]	52.31 [34]	6.15 [4]	[1]
buttocks	1.54 [1]	44.62 [29]	53.85 [35]	[1]
man's chest		9.23 [6]	90.77 [59]	[1]
penis	46.15 [30]	46.15 [30]	7.69 [5]	[1]
thigh		19.70 [13]	80.30 [53]	
tongue		12.12 [8]	87.88 [58]	
vagina	48.48 [32]	46.97 [31]	4.55 [3]	
woman's breasts	9.09 [6]	59.09 [39]	31.82 [21]	

*Females: N = 66

Table 3.2. Percentages and [Raw Scores]*

	RRR	RR	R	½R	Not-R	Omitted
belched breath	10.00 [2]	40.00 [8]	40.00 [8]	10.00 [2]		
breath	5.00 [1]	15.00 [3]	20.00 [4]	30.00 [6]	30.00 [6]	
menstrual blood	60.00 [12]	20.00 [4]	5.00 [1]	15.00 [3]		
blood from a wound	10.53 [2]		5.26 [1]	36.84 [7]	47.37 [9]	[1]
breast milk			15.00 [3]	15.00 [3]	70.00 [14]	
shit	60.00 [12]	35.00 [7]	5.00 [1]			
fart	30.00 [6]	35.00 [7]	30.00 [6]	5.00 [1]		
hair clippings		15.00 [3]		20.00 [4]	65.00 [13]	
nail parings	5.00 [1]	15.00 [3]	20.00 [4]	25.00 [5]	35.00 [7]	
spit	25.00 [5]	25.00 [5]	40.00 [8]	5.00 [1]	5.00 [1]	
pus	30.00 [6]	40.00 [8]	25.00 [5]	5.00 [1]		
semen/sperm	57.89 [11]	5.26 [1]	21.05 [4]	10.53 [2]	5.26 [1]	[1]
skin parings	15.00 [3]	35.00 [7]	15.00 [3]	25.00 [5]	10.00 [2]	
snot	35.00 [7]	45.00 [9]	20.00 [4]			
sweat	15.79 [3]	10.53 [2]	21.05 [4]	36.84 [7]	15.79 [3]	[1]
tears			5.00 [1]	5.00 [1]	90.00 [18]	
urine	25.00 [5]	45.00 [9]	10.00 [2]	15.00 [3]	5.00 [1]	
vomit	60.00 [12]	35.00 [7]	5.00 [1]			

	HR	SR	FM	Omitted
anus	30.00 [6]	45.00 [9]	25.00 [5]	
buttocks		40.00 [8]	60.00 [12]	
man's chest		5.00 [1]	95.00 [19]	
penis	55.00 [11]	20.00 [4]	25.00 [5]	
thigh		25.00 [5]	75.00 [15]	
tongue		10.00 [2]	90.00 [18]	
vagina	55.00 [11]	30.00 [6]	15.00 [3]	
woman's breasts	5.00 [1]	60.00 [12]	35.00 [7]	

*Males: $N = 20$

Table 3.3. Percentages and [Raw Scores]*

	RRR	RR	R	½R	Not-R	Omitted
belched breath	17.65 [15]	31.77 [27]	32.94 [28]	16.47 [14]	1.18 [1]	[1]
breath	3.49 [3]	11.63 [10]	13.95 [12]	22.09 [19]	48.84 [42]	
menstrual blood	36.05 [31]	18.60 [16]	12.79 [11]	19.77 [17]	12.79 [11]	
blood from a wound	3.57 [3]	7.14 [6]	10.71 [9]	19.05 [16]	59.52 [50]	[2]
breast milk	2.35 [2]	1.18 [1]	10.59 [9]	11.76 [10]	74.12 [63]	[1]
shit	60.47 [52]	23.26 [20]	10.47 [9]	4.65 [4]	1.16 [1]	
fart	28.24 [24]	41.18 [35]	20.00 [17]	10.59 [9]		[1]
hair clippings	1.16 [1]	4.65 [4]	10.47 [9]	16.28 [14]	67.44 [58]	
nail parings	4.65 [4]	9.30 [8]	20.93 [18]	27.90 [24]	37.21 [32]	
spit	23.53 [20]	34.12 [29]	29.41 [25]	10.59 [9]	2.35 [2]	[1]
pus	30.23 [26]	37.21 [32]	20.93 [18]	9.30 [8]	2.33 [2]	
semen/sperm	40.96 [34]	16.87 [14]	18.70 [15]	12.05 [10]	12.05 [10]	[3]
skin parings	9.64 [8]	26.51 [22]	20.48 [17]	21.69 [18]	21.69 [18]	[3]
snot	33.72 [29]	34.88 [30]	27.90 [24]	3.49 [3]		
sweat	8.24 [7]	10.59 [9]	23.53 [20]	41.18 [35]	16.47 [14]	[1]
tears			5.81 [5]	2.33 [2]	91.86 [79]	
urine	29.07 [25]	29.07 [25]	18.60 [16]	11.63 [10]	11.63 [10]	
vomit	61.63 [53]	22.09 [19]	8.14 [7]	4.65 [4]	3.49 [3]	

	HR	SR	FM	Omitted
anus	38.82 [33]	50.59 [43]	10.59 [9]	[1]
buttocks	1.18 [1]	43.53 [37]	55.30 [47]	[1]
man's chest		8.24 [7]	91.76 [78]	[1]
penis	48.24 [41]	40.00 [34]	11.76 [10]	[1]
thigh		20.93 [18]	79.07 [68]	
tongue		11.63 [10]	88.37 [76]	
vagina	50.00 [43]	43.02 [37]	6.98 [6]	
woman's breasts	8.14 [7]	59.30 [51]	32.56 [28]	

*All subjects: N = 86

Table 3.4. Summary Findings on Revoltingness Ratings*

shit, vomit	RRR,	84% > R	(61% > RR)
urine, semen/sperm	RRR,	58% > R	
menstrual blood	RRR,	55% > R	(36% > RR)
(MEN		80% > R, 0% Not-R)	
(WOMEN		47% > R, 17% Not-R)	
fart, snot	RR,	70% > R	(72% < RRR, 11% < R)
pus	RR,	67% > R	
spit	RR,	56% > R	
belched breath	R,	78% ≥ R	(51% < RR, 49% ≥ RR)
skin parings	R,	64% ≤ R	
sweat	½R,	58% < R	(19% ≥ RR)
nail parings	Not-R,	65% < R	
breath	Not-R,	71% < R	
blood from a wound	Not-R,	79% < R	
hair clippings	Not-R,	84% < R	
breast milk	Not-R,	86% < R	
tears	Not-R,	94% < R	(0% > R)

*All subjects: N = 86. The ranking scale is RRR, RR, R, ½R, Not-R, hence 'R' is the midpoint. In the table '>' means "greater than", '≥' means "greater than or equal to"; '<' means "less than", and '≤' means "less than or equal to".

CHAPTER
—4—

The Lexicon for Bodily Effluvia, Sex, and Tabooed Body-Parts

Feminine moisture end it now and stay fresher all day
(Advertisement for Vagisil™ Feminine Itching Powder)

[Antonia is objecting that Nanna is using too many euphemisms when comparing the life styles of a nun, a wife, and a courtesan] *Io te lo ho voluto dire, ed emmivi coordato: parla alla libera, e di "cu', ca', po' e fo' ", che non sarai intesa se non dalla Sapienza Capranica con contesto tuo "cordone nello annello", "guglia nel coliseo", "porro nello orto", "chiavistello ne l'uscio", "chiave nella serratura", "pestello nel mortaio", "rossignuolo nel nido", "piantono nel fosso", "sgonfiatoio nella animella", "stocco nella guaina"; e così "il piuolo", "il pastorale", "la pastinaca", "la monina", "la cotale", "il cotale", "le mele", "le carte della messale", "quel fatto", "il verbigrazia", "quella cosa", "quella faccenda", "quella novella", "il manico", "la frecata", "la varuta", "la radice" e la manda che ti sia non vo' dire in gola, poi che vuoi andare su le punte dei zoccoli; ora dì sì al sì e no al no: se non, tientelo.*

"I meant to tell you and then I forgot: speak plainly and say 'ass' [*cul(o)*], 'prick' [*cazzo*], 'cunt' [*potta*] and 'fuck' [*fottere*] if you want anyone other than scholars from the Sapienza Capranica to understand you. You with your 'rope in the ring', 'obelisk in the Colosseum', 'leek in the garden', 'key in the lock', 'bolt in the door', 'pestle in the mortar', 'nightingale in the nest', 'tree in the ditch', 'syringe in the flap-valve', 'sword in the scabbard'; not to mention 'the stake', 'the crozier', 'the parsnip', 'the little monkey', 'her thing', 'his thing', 'the apples', 'leaves of the missal', 'the fact', 'the thanks-be-to-God', 'that thing', 'that affair', 'that big news', 'the handle', 'the arrow', 'the carrot', 'the root' and the shit that you don't want to say straight out, so you go tiptoeing round it. So say yes when you mean yes and no when you mean no; otherwise, keep it to yourself."

(Pietro Aretino 'Ragionamento della Nanna e della Antonia' 1534; quoted in Frantz 1989:74f)

75

Introduction

In this chapter we examine the vocabulary for bodily effluvia, sex, and tabooed body-parts, which surely manifests significantly more synonymy than one finds in other parts of the English lexicon. Many of the terms show a keen inventiveness that ranges from the delightfully poetic to the disgustedly crass. We use the term 'inventiveness' deliberately and, for the sake of any reader who is a linguist, would contrast it with Chomsky's notion of 'creativity,' which is little more than a synonym for 'the ability to apply grammatical rules recursively' (cf. Chomsky 1966:3–31). In classifying the vocabulary of defecation and urination, but more especially that pertaining to sex, the lexical evidence suggests to us that every avenue the imagination could reasonably take has been ventured upon; yet we cannot doubt that human minds will find new ones. Nonetheless, we have attempted to group vocabulary items into nondiscrete classes. Many such classes are productive in the sense that they can be expanded by new members; however, they are not predictive in the sense that they define the entire set of classes possible. Most, if not all, of the classes are themselves what a rational being would expect to find, given the nature of the denotata and what we know about the human imagination; but perhaps that is to say no more than that there are no surprises for the imaginative analyst.

In addition to cataloguing the literal vocabulary, we investigate the bases for some of the figurative uses of language in this domain. Once these are established, we ask *Why do some figurative terms become conventionally established and widely used, while others fall by the wayside?* Close examination of the etymologies of taboo terms will often not clearly identify a unique starter for the present-day meaning of the taboo term, and it becomes apparent that extensive usage of such a term derives from the coalescence of a demonstrable network of associated influences on its development and maintenance, each of these reinforcing the others.

Bad Breath, Running Nose, Belching, Vomiting, and Body Odor

Our community places great value on *personal freshness*. This euphemism is motivated by considerations of health and social acceptability: the healthy person is supposedly free of the odorous and unsightly debris of bodily effluvia, disease, and bodily neglect. *Personal freshness* results from the absence of obnoxious bodily effluvia: the smell of sweat, halitosis, and the like can be kept at bay with toothpaste, mouthwashes, soaps, deodorants, perfumes and *personal hygiene products* like kleenex, toilet tissue, tampons, and sanitary napkins.

It is potentially dysphemistic to accuse someone *You've got bad breath*; euphemistic hints make reference to teeth-cleaning, catarrh, or diet: *Have you cleaned your teeth today?; Do you use a mouthwash?; Do you have catarrh?; Did you have garlic for dinner last night?;* and so on.

There seems to be no euphemistic way of referring to nasal effluvium from an adult, except to ask if s/he has a cold or perhaps offering a tissue. Expressions like *Your nose is running!; Wipe your nose* or *Shall I wipe your nose?* are intrinsically neutral expressions, though they could certainly be used in dysphemistic speech acts, to an adult for instance. The intonation and tone of voice used could be critical here. Intrinsically dysphemistic terms for nasal effluvium are *snot* for the liquid form and *bogey* for the solid form.

Belching is said to be used in some communities to mark a proper appreciation of a fulfilling meal; but not in Anglo-societies, where belching in public is frowned upon. Strangely, though, when an individual inadvertently belches in public, it often leads them to giggle; and some people, young men particularly, flout the usual convention by belching loudly in public and thus seem to create a great deal of amusement among their friends of both sexes. This form of behavior is taken up in comedy shows, where belching in public is a source of low comedy. To a lesser extent, farting is treated the same way (cf. Reynolds Price *The Tongues of Angels* 1990:26 and earlier Chaucer *The Miller's Tale* 1386, D'Urfey 1719 Vol.3:332f, Clemens 1968 [1876]); though no other effluvium is. The colloquial term *belch* and the formal *eructation* are neutral, and co-exist with the euphemisms *burp, heartburn*; and *my food is repeating on me* (there is also the smart-alecky excuse *Better an empty house than a bad tenant*); but there seem to be no intrinsic dysphemisms for belching.

Vomiting is a neutral term, for what is described in the following:

> In mid laugh, as if it were a natural extension of the joke, Gesh [*who was drunk and stoned on mixture of cocaine and Quaaludes*] suddenly doubled over to evacuate the contents of his stomach. I listened to his wet heaving gasps—the dying throes of a hero with a sword twisted in his gut—and began to feel queasy myself. "The smug son of a bitch," he choked, and then gagged again. When he was finished, he straightened up, wiped his mouth with the back of his sleeve and said, "What next?"   (T. Coraghessan Boyle *Budding Prospects* 1984:154)

Feel queasy/unwell is a state that often precedes vomiting. *Retching* describes the muscular actions and contractions and the noises thereof—in fact all the motions of vomiting without the flood of the effluvium itself. Other euphemisms for *vomiting,* in addition to *evacuate the contents of one's stomach,* are *be sick* and *throw up; gag, puke, spew (up), barf, honk,* and so on, edge progressively toward the dysphemistic.

Finally, to *perspiration.* For us, this is a neutral term rather than a euphemism, but other people find it euphemistic. Euphemisms include *be hot, hot and sticky,* and *a little flushed.* There is the mildly dysphemistic *sweat* and *be dripping, soaked.* The normal effect of sweat is *body odor,* euphemistically abbreviated to *b.o.* There is no intrinsic dysphemism to describe this condition, the closest to it is a comparison such as *sweating like a pig;* and one can speak of *reeking/stinking of sweat.*

There are effluvia we have not remarked on, such as spittle and the mucous which clogs the eyes, because there is little to say about them. We began this discussion with *personal freshness*: this is not something that results from wiping one's nose, or any other noxious effluvium, such as the vestiges of vomit (see the preceding quote), on one's sleeve; nor from remaining unwashed or wearing unwashed clothes. The contrary of personal freshness is to be found in phrases like *smell/reek/stink like a dero/ wino* or simply *I/you stink.* The unpleasantness of an assault on the olfactory organs is reflected in a variety of dysphemistic uses for the verb *stink,* inspired by the obnoxious nature of several kinds of bodily effluvia (see Ch. 5).

The Lexicon of Urination and Defecation

> If you sprinkle
> When you tinkle,
> Kindly wipe the floor.
>> (Notice in a men's toilet in MASDAR House,
>> Finchampstead, England)

This salutary verse prompted one reader to observe:

> I know a more developed version of this rhyme, which I have seen several times as graffiti in various NY area ladies rooms:
>
>> "If you sprinkle when you tinkle,
>> Please be neat and wipe the seat."
>
> In this form it addresses a pressing social issue of great concern to all women who use public facilities and as such deserves a wider audience.   (CR, June 1990)

Who could disagree?

The lexicon of urination and defecation divides into three categories: terms for the place where it is done, vocabulary for the acts performed, and names for the effluvia.

The Place

ROOM AND RECEPTACLE: It is in *the littlest room* that one finds *the throne*: the latter term is hyperbolic, but the former is quite literally true for most private houses. A more general-for-specific example occurs in the invitation to a guest: *Can I show you the geography of the house?* There are the old terms *bench, stool*—whence the meaning transfer to *stool* "turd". The *can* was originally used of what were later called *bedpans* in hospitals, and it got shifted to "a lavatory" (cf. Read 1977:40). The slightly dysphemistic *dunny* is presumably correlated with *dung* or *dun*. (The *O.E.D.* makes no etymological correlation between *dun* "brown" and *dung* "excrement," but there is little doubt that ordinary folk associate the two; for example, consider expressions like *shit-colored, brown hole, brown-nose*; see Ch. 5.) There is a collection of one-for-one substitutions: *head, latrine, kazi, john, jakes, bog, crapper*, and so forth. The first four are neutral terms; the last four tend to the dysphemistic. We speculate that *kazi* originated in East Africa, probably among soldiers: it is the Kiswahili word for "work" and, we think, referred to the work involved in digging latrines. We do not concur with McDonald's (1988) suggestion that it derives from Italian *casa*. *Crapper* is curious: it ought to be neutral if it is named after Thomas Crapper who patented a valveless 'water waste preventer' which guaranteed 'a certain flush with an easy pull' in 1882 (which led him to become plumber to the Queen; cf. Palmer 1973:106; according to Wright (1960) the first patent for a flush toilet was taken out in 1775); but *crapper* is dysphemistic, presumably because of the verb and noun *crap* "shit," which we say later (following the *O.E.D.*) derives from a Dutch verb. Either Crapper's name was remarkably coincidental, or the stressed vowel in the old word *croppen* (see immediately below) got changed to 'a' as a result of Crapper's name. The archaic euphemism *necessary house*, like other synonyms compounded on *-house* reflects the bygone era when loos were all outside, hence *outhouse* (cf. French *cabinet*, Italian *gabinetto*) and the somewhat dysphemistic *croppen ken* ('ken' = "house", and it may be a source for *can*) equivalent to our *craphouse*. The latter, like *shithouse* is normally dysphemistic.

Given our definitions of euphemism and dysphemism in Ch. 1 there should be no place for 'neutral terms' since, for a given speech act, the lexical items in question will be either euphemistic or dysphemistic. But here, we are classifying lexical items and phrases without making reference to a particular context of their use; to do so we have to employ 'the middle class politeness criterion'. When we are undecided whether, according to that criterion, these expressions would be preferred or dispreferred, we dub them 'neutral terms.'

Now back to the lexicon for lavatory.

A PLACE WHERE THERE IS PRIVACY: *privy*, *ladies* and *gents*; *women's* and *men's* for segregated public loos. The latter pattern has given rise to variations like *boys* and *girls*; *guys* and *gals*; and in places like seafood restaurants *buoys* and *gulls*; and in a dinner theater called the 'Olde West' *stallions* and *mares*.

A PLACE FOR EASEMENT: *comfort station*, *restroom*, *facilities*.

A PLACE WHERE THERE IS WATER: *water closet*, *w.c.*, *lavatory*, *bathroom*, *washroom*, ? *toilet*. The common British and Australian euphemism *loo* is associated—rightly or not—with *Waterloo*, though both French *l'eau* "water" and *lieu* "place" are other supposed sources—perhaps mutually reinforcing ones. We think the most likely source is *bourdalou*, and since this term does not appear even in the revised *O.E.D.* (1989), we quote the following from *The Penguin Dictionary of Decorative Arts*:

> Bourdalou(e) A C18 type of oval slipper-shaped urinal or chamber-pot intended mainly for the use of ladies when travelling, sometimes said to have been carried concealed in a muff[! see discussion latter]. It is supposedly named after the famous French Jesuit preacher Louis Bourdaloue (1632–1704) whose sermons at Versailles were so popular that his congregations assembled hours in advance. [Presumably the reader is to infer that some members of the congregation needed to privily relieve themselves before the end. What a nice example of euphemism!] But the earliest surviving examples date from c.1710 (made at DELFT) and the word—as *bourdalou* (m.)—is not recorded in print before it was defined in the *Dictionnaire de Trévoux* in 1771. Such chamber-pots were made throughout Europe, including England, and even in China and Japan (for export) . . . [they] are often very delicately decorated and are sometimes mistaken for sauce-boats.
> (Fleming and Honour 1977:110)

Doubtless the foreclipping of *bourdalou* to *loo* was influenced by the associations with water that we have identified already. McDonald (1988:84) offers the following etymology, which we believe is incorrect: Lady Louise Hamilton 'was staying at the vice-regal lodge in Dublin around 1870, [when] the name card on her bedroom door was transferred to the lavatory door. The card read simply 'Lou' and the joke was immortalized by the other guests who persisted in referring to the lavatory as *lou*.' The story is very possibly true; but we suspect that Lady Louise's name card was transferred to the toilet door because the toilet was already called a *loo*. Certainly apocryphal is the tale that in English hotels the number on the toilet door is normally 100 = *loo* (if this were true, we would almost certainly find something like *the century* meaning "toilet," and we don't). *Thunderbox* might have been inspired by the noise of toilet flushing or the noises made prior—or very probably, both. While on noises made: *long drop* for a "bush dunny" with a long drop into the cesspit (well, maybe this term is inspired by visual rather than aural experience).

Lavatory paper is perhaps a less refined expression than *toilet paper*, which is further upgraded to *toilet tissue*. In America the euphemism predictably went to *bathroom tissue*; but less predictably has been further disguised by becoming *bath tissue*—an expression which most Brits, Ozzies and Kiwis find mystifying when it is first encountered. Is it conceivable that a lavatory may one day be called a *bath*?

The Acts Performed

> All pets must be taken *off of the property* for their "pottie" needs.
> (Los Altos Village [Apartments] Newsletter, Tucson, Arizona, March 1990)

One of the principal metaphors used is GOING AWAY FROM ONE'S CURRENT LOCATION—terminology that English shares with many other languages, even perhaps Proto Indo-European: English *shit* derives from the Indo-European root **skei-* "cut, split, separate." The *O.E.D.* relates English *crap* to Dutch *krappen* "pluck off, cut off, separate." It could be that the present day meaning of *crap* (and the various words derived from it) has a variety of sources: Dutch *krappen*, eighteenth-century English *croppen*, and Thomas *Crapper*. Such a variety of possible sources can be identified for a number of words we shall be discussing in this chapter. Euphemisms: *Mummy, I need to go-o-o!*; *Please teacher, may I be excused?*; *Please excuse me for a minute*; *I'll meet you in the lobby*; *I'll have to make a pit stop* (a nice idea, notice the connexion between *pit* and *toilet*; but in a literal pitstop the vehicle takes on fuel, it doesn't unload it!). GO: *to the restroom, loo, down the hall, round the corner* (for predictable reasons this is the favoured euphemism among the ladies of the Linguistics Department at Monash), *for a short walk, to answer a call of nature, to spend a penny, to powder one's nose* (of women). Cf.

> Back in the Long Gallery some of the women went upstairs to "powder their noses." Lady Montdore was scornful.
> "I go in the morning," she said, "and that is that. I don't have to be let out like a dog at intervals."   (N. Mitford 1979:294)

In West African English, where appropriate, *I go for bush* (out of the living area and into the 'bush'). Also, *The dog needs to got out*; *The cat needs to be put outside*; and so on. Neutral terms (?): *Go to the toilet/lavatory/lav/lavvie* (the last two are particularly British); *Go to the john* (American). Dysphemisms: *Go for a piss, leak, shit*. The abbreviation in *go for a pee* is a euphemistic dysphemism.

HAVING or TAKING *a whizz, leak,* or *squat* are euphemistic, or perhaps euphemistic dysphemisms. *Having a bowel movement* is neutral. *To have the shits* and *to have/take a piss, crap* or *shit* are dysphemistic.

DOING (occasionally MAKING) *a weewee, number one, pooey, poop, doo-doo, job(bie), number two, one's business, a beanie, pooey smell*. Most of these are nursery terms, and the following was addressed to a child: *If you don't hang your coat up, the cat will make a puddle in it.* There is also *make a mess*, which is a general-for-specific euphemism, often used with animals (see Ch. 5 for additional comment).

EASEMENT: *relieve oneself, dump a load* (people talk of diapers being *loaded*); *pass gas, let off/let loose/cut a fart.*

GROOMING: *freshen up, wash one's hands, powder one's nose* (females only); the archaic expression *to be at one's toilet* was the source for *toilet*—the name of the room.

ONOMATOPOEIC: *piss; tinkle; fart; plop-plop.*

MISCELLANEOUS. An adult person who is unable to control their urinary or defecatory processes (*the call of nature*) is said *to be incontinent* or *to suffer incontinence*; advertisements for adult diapers refer to *bladder control*; these are indubitably euphemistic. *Wet/soil the bed* are neutral. Terms not included earlier: *urinate, defecate, break wind* are neutral terms rather than euphemisms. *Be natural* and *anal emission of gas* for "fart" are attempts at euphemism. *Point percy at the porcelain, shake hands with the unemployed, siphon the python* (all Australian and applicable to males) are joky. *Fart* is a dysphemism. A really objectionable dysphemism is *strangle a darkie* for "defecate": the white racial overtones, and the suggestion of a dark, secret act, add vicious force to the clever figures in both 'strangle' and 'darkie'.

The Effluvia

Many of the terms for effluvia have already been introduced, but there are a few others. Euphemisms include *road apples, cow patty, buffalo/cow/camel chips, bird lime, doggie doo/dirt.* The following are neutral terms: *urine; faeces, stool, excrement* (from Latin *ex-crementum* "what grows out from [the body]"), *manure, scat, dander, droppings, cowclap, [dog's] mess; flatulence.* Perhaps *gas* is also a neutral term, but we suspect that like *raspberry* and *bottom cough* it is euphemistic. Dysphemisms include *turd* and *fart.*

The Lexicon for Menstruation

In Ch. 3 we examined perceptions of menstruation in searching for the bases for the strong taboos upon it. The perceptions reported there are borne out by the lexicon for menstruation. We can identify six nondiscrete categories of X-phemisms for menstruation (cf. Farb 1974:82; Joffe 1948).

1. ILLNESS or DISCOMFORT. Euphemisms: *be sick, unwell, indisposed, under the weather; have stomachache, the cramps* (perhaps a dysphemistic euphemism), and the like. The following are hardly euphemisms, on the other hand they are not unquestionable dysphemisms either; they are rather what we have called 'dysphemistic euphemisms': *have the woman's complaint; be bleeding, bloody, be feeling that way, off the roof,* and so on. They might be dysphemistic when a man is whingeing about the sexual unreceptiveness of his female partner (cf. *closed for repairs*).

2. ANTIPATHY. Generally dysphemistic expressions or at best dysphemistic

euphemisms. *The jinz, the pain, the foe, the plague, the curse, the drip, cramps, be off the roof,* and so on. Some of these fall into other categories, too.

3. PERIODICITY. Generally euphemistic. The currently most neutral and widely used euphemism seems to be *It's (I've got/I'm having) my period.* There are also *It's that time (of the month); It's my time; It's the wrong time of the month; I've got the monthlies; It looks like a wet weekend.* Possibly dysphemistic is *It's blood week.* Latin *mensis* "month" (plural *menses*) is the source for the *menses* of medical jargon, and the basis for Latin *menstruare* whence English *menstruate.* We also find *I've got my regular,* and in Irish English the bestial metaphor *be in season* (with its connotation of the woman being close to nature).

4. A VISITOR. Generally euphemistic. *To have a visitor; little sister's here; the girl friend; my buddy/friend(s)/foe's here; Fred's here; George has come; I've got Fred/ George/Jack; little Willie/Aunt Susie is here/Kit has come* (for the last see later discussion). There seems to be no accounting for why these names are chosen, though only a handful are used. It is not so surprising that some 'visitors' have female names for this uniquely female event; male names are somewhat more mysterious, especially since they are more common. There are two possible explanations: one is that a visitor to the vagina under other circumstances would most likely be male, so this one is male by analogy; the other explanation, which we owe to Sarah Castle (tongue in cheek?), is that a period is a pain in the proverbial—just like a man is. This category is often combined with category (5).

5. The Color Red. Mostly euphemistic. *My aunt/grandma's here from Red Creek/Red Bank/Reading.* And, because of the military's pre-twentieth-century predilection for red coats, *the cavalry's here, entertaining the general,* and especially the French *j'ai mon anglais.* This kind of metaphor recently (mid-1988) turned up in a television advertisement for Fleur tampons shown in Melbourne, Australia: two women's voices discuss the discovery of the neat little flower decorated box (see Category 6) among the contents of a handbag; one identifies it as Fleur tampons and describes the superiority of these over other tampons; then, as the shot changes to an elderly man in hunting red and riding hat carrying a tray of drinks, the second voice says 'Oh good, the cavalry's arrived.' The color red is also invoked in *bloody Mary, the Red Sea's in, it's a red letter day.* This category is sometimes linked with category 6, particularly in men's speech (e.g., in *riding the red flag, flying the red flag,* etc., which are dysphemistic—or at best, dysphemistic euphemisms).

6. REFERRING TO SANITARY PROTECTION: 'Sanitary protection' is itself a euphemism with links to category (1); and note the implication of cleanliness and health implicit in *sanitary;* there is a similar implication from 'hygiene' in the phrase *feminine hygiene.* Those are euphemisms, but most members of category 6 are dysphemistic or dysphemistic euphemisms: *on the rag/jamrag; wearing the rag; riding the cotton bicycle; the hammock is swinging;* and so on. Reference to *the rag* shows the language lags a generation behind contemporary means for sanitary protection. There is also the tasteless joke about an imaginary western 'Blood on the Saddle' starring the Kotex Kid. Other dysphemisms or dysphemistic euphemisms are *covering the waterfront; wearing the manhole cover; plugged up; closed for repairs;* and so on, which indicate sexual unavailability. Euphemisms are in short supply for articles of *feminine hygiene* (like *brassiere* this may not be an intrinsic

euphemism, since there is no real alternative): *tampon* (largely displaced by the euphemistic extension of the trade name *tampax* in many dialects) and (*sanitary*) *pad/napkin* are neutral, the omission of 'sanitary' creating a euphemism which, along with *panty-liner* is about as euphemistic as is possible—though the latter are not intended for the full flood of menstruation, but rather for 'spotting' and other leaks. In the days of *sanitary towels* there was the euphemistic abbreviations *STs*. In Australian and British English it is possible that the use of *napkin* for "sanitary napkin" is leading to the demise of the term used in the sense "table napkin", which is replaced by *serviette* in Australia, Britain, and New Zealand; this is not the case in American where the trade name *kotex* has generalized to function as a euphemism (cp. *tampax*). In her book *Lily on the Dustbin* Keesing offers some Australian euphemisms, but we don't know the extent of their usage:

> Mary, Kathy, Bill and Ron leave work together to walk to the bus stop. Kathy says to Mary 'You lot go on. Don't miss the bus, but I've just remembered I'm out of white bread, and George is calling tonight.' Mary understands that Kathy wants to dash into a chemist shop to buy sanitary napkins. She may choose 'sliced sandwich' (thin size) or 'sliced toast' (thick), or maybe she used 'fruit cocktails' (tampons)—a cylindrical toffee wrapped in white paper looks very similar.
>
> (Keesing 1982:31)

There are some quite poetical expressions for menstruation to be found in Joffe (1948:185), including *her cherry is in sherry, the gal's at the stockyards, snatch box decorated with red roses*. She even records a retired prostitute from the Barbary Coast (California) using *I've got my flowers*—an old expression that has almost died out except perhaps in Irish English. Yet perhaps it is not archaic even in British English; or if it is, it may linger on in folk memory. In her article 'An urge deeper than the skin,' Ruby Wallace described her reasons for having her left shoulder blade tattooed:

> From the age of 15 onwards my periods stopped [because she was severely anorexic . . . W]hen my periods restarted, so many years later [. . . they] were a sign of health, of recovery, and to me they were a sign of womanhood that I welcomed and had in fact longed for.
>   When I felt secure that I was fully "well" and had learned to cherish my female body and femaleness, I wanted to celebrate. I performed what amounted to a menstrual rite. I had myself tattooed.
>
> (*The Age*, Melbourne, September 17, 1988, Saturday Extra p. 5)

And the tattoo? Flowers: scarlet blossoms! We cannot know whether this is folk memory, a symbol for spring, or pure accident. Even the possibility that it is either of the former, demonstrates the powerful association between *flowers* and menstruation.

The expression *flowers* or *monthly flowers* and its translation equivalents in other European languages may have a variety of sources. The most likely one is the plant growth metaphor of *seed* for "ovum [and semen]" (also *sap* for "semen"), *flower* or *bloom* "menstruation (i.e., prima facie evidence of fertility"), *fruit* "[one's] children"; (cf. Grose 1811 'FRUITFUL VINE. A woman's private parts; i.e., that has *flowers* every month, and bears fruit in nine months). This metaphor is found in the Middle Dutch medical texts we examined where the most frequent term for menstruation is *bloeme* "flowers," along with the verb *bloyen* "bloom"; (in case we had any doubt about the meaning, in one text every mention of a remedy to bring on *bloeme* had prompted one reader to resolutely supply in the margin the Latin *provocat menstrua* or *producit menstrua*). The word *dracht*, which usually denotes a yield of fruit or the fruit itself, was also used to mean "fetus," and more often "pregnancy." The child was referred to as *vrucht* "fruit." For example,

> *want ghelijc dat die boeme sonder bloemen geen vrucht en winnen, alsoe werden die vrouwen beroeft van haerre dracht, als si sonder purgacy sijn.*

> "for just as trees without flowers will bear no fruit, so too will women be bereft of their pregnancy (literally, 'yield'), if they are without [this] purging."
> (*Boec van Medicinen in Dietsche*)

It has been suggested (e.g., by Neaman and Silver 1983:57) that *flowers* derives by either remodelling or folk etymology from Latin *fluere* "flow" via French *flueurs* (a medical term for "discharge" from the sixteenth century; cf. Dauzat 1938) that was remodelled as, or perhaps in England misinterpreted as, *fleurs* "flowers." Enright (1985:10) suggests its origin is *flow* (cf. the archaic *a woman in her courses*); and *flow* is used currently among Canadian female students (if not elsewhere, too), along with *flushing* (cf. Aman and Sardo 1983). Middle Dutch *vloet* "flow" was an occasional alternative to *bloeme*. While such etymological speculations are suggestive, they will not account for the archaic German euphemism *Blumen* or Dutch *bloeme* unless these are presumed to be loan translations of English *flowers*—which is hardly likely. Thus, we would contend that the plant growth metaphor provides the most likely source for archaic *flowers* meaning "menstruation." Yet we shall present plenty of evidence in the course of this book that X-phemistic popular terms for taboo topics often have mongrel origins; that is, they seem often to have multiple sources, each reinforcing the others. For instance, if *flowers* "menstruation" was indeed based on a plant growth metaphor, it may well have been reinforced by an association with the notion of 'flow.'

We should remember that virginity was (quaintly) held to be the flower of maidenhood, and it was lost on *deflowering* (so much nicer a term that *popping her cherry*); in *Memoirs of a Woman of Pleasure*, Fanny Hill writes of her 'virgin flower' and of another girl's 'richest flower' being 'cropped' (Cleland 1975:77, 143) and gives graphic descriptions of the bloody result. In the thirteenth-century French allegory *Le Roman de la Rose*, the 'rose' is the maiden's hymen. Thus there is a well-established figurative link between flowers and blood issuing from the vaginal orifice, whether catamenic or the result of a ruptured hymen. The fragility of virginity is captured in figures like *she has cracked her pitcher/pipkin/tea-cup*, and so on (cf. Grose 1811).

Yet another category of euphemism for menstruation is revealed in the following Middle Dutch recommendation for the daily dose of a special preparation:

> *es goet . . . den vrouwen haer dinch wel te doen comene*

> "is good . . . to cause women's thing to come."

A euphemism like *a woman's thing* can be used for any number of delicate topics, leaving the context to determine which.

Menstruation is a chronic recurrence in the first-hand experience of most women and the second-hand experience of many men; and although there are occasions on which its onset is welcome, the vocabulary for it reveals that menstruation is usually an unpleasing experience. All common experiential aspects of menstruation turn up in the lexicon: it is viewed like an illness; as a troublesome source of discomfort; there is reference to its periodicity; to the flow or drip and the means and effects of stanching it; to the blood and the color of blood. The most picturesque euphemism for menstruation is the unfortunately archaic metaphor of blooms and flowers—which is perhaps a coalescence of influences from the color red, the notion of flow, and centrally the plant growth metaphor. This metaphor is the most positive in the vocabulary for menstruation; it may have declined with the social attitude that nowadays places less overall emphasis on the ability to conceive than on preventing conception.

Of Gynecology and Obstetrics

There is probably an oral thesaurus for gynecological problems awaiting the folk lexicographer. Here is an annotated quote from Keesing (1982); our annotations are italicized.

> One woman mentioned in a letter that during her childhood the overheard phrase 'She's had her tonsils out' invariably referred to any major [*gynecological*] operation. A woman might have 'dropped a few pegs' [?], 'had a re-bore' [*womb scrape or dilation and curettage, normally euphemized to D & C*] or need treatment 'in the pipes' [*fallopian tubes, cp. to have tube trouble*]. 'Mummy's going in [to hospital] for a little op', might indicate a simple curettage or be euphemism for abortion; men more often spoke of a woman 'having the boilers scraped down'. Other synonyms [*for abortion*] include 'slip a joey', 'crack an egg', 'need a scrape' (which can be a curette recommended for other reasons), or 'have appendicitis'. This last example is equivocal Sheilaspeak [*women's language*]; with the co-operation of a sympathetic 'medico' it can be used to excuse absence from work for unaccountable illness.   (Keesing 1982:36)

Euphemisms for pregnancy include: *be expecting (a happy event)*, *in the family way*, *in the (pudding) club*; *have a bun in the oven*; *be in trouble* (cp. *get a girl in trouble*); *be in a delicate/an interesting condition*, *be with child* (these last two seem a trifle archaic). An expression similar to the last of these is found in Middle Dutch: *met kinde gaen* "go with child." In Modern English we also have the dysphemisms *be knocked up*, *banged up*, *up the spout*, *up the duff*, *up the stump*, *up a gum tree*, *up the creek*, *up shit creek*—notice the ubiquitous 'up'; the only exception to it seems to be in the phrase *to get caught out* (after taking a risk, presumably). Why does the preposition or particle *up* occur in nearly all these expressions? In general, there are connotations of being somewhere no one wants to be, in an unpleasant or unwelcome state, and helpless to do anything about it—indeed, *up the creek*, *up shit creek* are used in this sense for other conditions than pregnancy. These connotations probably arise from association between expressions such as the previously listed dysphemisms and others like *filled up, tied up, bunged up, broken up, beaten up, stuffed up,* or *fucked up.*

Many societies have taboos on birthing women and on birth fluids (cf. Meigs 1978); but English euphemisms virtually ignore birth, though note the following from Thackeray's *Vanity Fair:*

> In the early spring of 1816, *Galignani's Journal* contained the following announce-ment in an interesting corner of the paper: 'On the 26th of March—the Lady of Lieutenant-Colonel Crawley, of the Life Guards Green—of a son and heir.'
>
> (Thackeray 1848/1962:412)

Note the avoidance of *gave birth* or *was delivered of* and any explicit mention of the child being *born* in that *interesting corner* of the paper. Today, we speak quite readily of a woman *going into labor* (cp. Middle Dutch *in arbeit gaen*), which recognizes the woman's effort. Middle Dutch has *ene vrouwe in kindel bedde leghet* "a woman lies in the child-bed", reminiscent of the archaic *be in child-bed* of English.

As we mentioned earlier, a general feature of early medical works was the absence of any reference to gynecological and obstetric ills, indicating a strong taboo. One exception, in the Middle Dutch texts at least, was the description of the medieval practice of fumigation from beneath (i.e., 'suffumigation,' Middle Dutch *suffumigacion*). The usual euphemism for this was *onderroeken* "to smoke from underneath," because the patient sat over beneficial vapors produced by burning herbs. The health-giving vapors penetrated the woman's body through *die hemelijke porte der naturen* "the hidden/secret door of the nature" (see Ch. 3). This was thought to be particularly good for the *hemelijken ziecheden* "secret illnesses" of women.

The Lexicon for Sex Acts

As one might expect, most of the lexicon for sex acts describes the act, the action, or the state of the body or bodies.

Preliminaries. *She's a good sort* is a somewhat outdated synonym for *She's a goer* or *She's hot stuff.* The counterparts for men include *He's a bit of a jack the lad, He's a skirtchaser, He's a linen lifter, He's got wandering hands, He's a groper,* or *He's a stud/a swordsman.*

Euphemisms for foreplay and its effects include *touch, tickle, handle, fondle, feel (up), rub,* and *massage,* which have the effect of getting one's partner (*sexually*) *aroused/excited.* Corresponding dysphemisms include: *finger, diddle* (from Old En-glish *dyderian* "cheat, deceive"), *fishfinger, ball,* and the archaic *firkytoodle;* also *have the hots for someone* (cp. *be in heat* of animals); *get horny; dribble; have wet knickers; get a horn, fat, boner, hard (on); get stiff, big,* etc.

There are no true euphemisms for oral sex, the most neutral terms are *fellatio* and *cunnilingus,* whose colloquial counterparts are *give head; sit on someone's face; get/go down on; eat (out); give/do a blow job; soixante-neuf/sixty-nine* ('Wine me, dine me, sixty-nine me'); *suck off; tongue;* and so on.

Orgasm (the most neutral term) is perceived as the climax to a journey (cf. *Are you there?*), and the alternatives *climax* (euphemism), *coming* (in Japanese, one "goes", *Iku yo!*), *seminal discharge, come* or *cum*—the latter being nouns for *the vaginal secretion of a sexually excited woman* as well as *a man's ejaculate* or *seminal fluid* (neutral) his *seed* (euphemism) or (the dysphemisms) *spunk, spoof* (the latter is recent Australian), and *gis(su)m/giz(zu)m* (said to be derived from Yiddish and propagated by Philip Roth's *Portnoy's Complaint* 1967). A man may also *spit white, get his rocks off, spend himself* (see the discussion of the 'Camera Song' in Ch. 9), *drop his load* and *shoot*; the latter links the action with the weapon metaphor (see discussion later on). Hyperboles like *die* or *explode* are also heard to describe the effect of orgasm.

> [*Two Boston-Irish policemen*]
> "For did not the great Sigmund Freud in 1912 conclude a symposium on masturbation with the statement: 'the subject of onanism is inexhaustible'?"
> "And will it not take time to work out our Church dogma that masturbation will render the Catholic lad blind, hairy-palmed, insane, doomed, and with the leg bones bent like an orphan with the rickets?"
> (Shem *The House of God* 1978:395)

As the 'habit that dared not speak its name' (Rusbridger 1986:42), mention of masturbation has always been veiled and indirect. In addition to descriptions like (*the crime of*) *self-pollution* and *the solitary/secret vice*, one label that became popular during the eighteenth and nineteenth centuries was *onania/onanism* (giving *onanist* and *onanistic*), a term deriving from the proper name Onan, brother of Er (*Genesis* 38:9–10). It is not clear why 'the secret vice' came to be associated with Onan. According to *Genesis*, when Onan lay with his brother's widow, he chose to 'spill [his seed] onto the ground'; this could describe *coitus interruptus*, in which case the vice of masturbation has been wrongly attributed to Onan.

Accounts of masturbation early this century describe at length the dangerous consequences of the act, but are at the same time quite remarkable for never explicitly naming it! In his history of the sex manual, Rusbridger (1986:41–43) quotes some early twentieth century euphemisms: 'this dangerous shoal, this evil, this secret vice' (Dr. Emma Drake 1901); 'self-defilement' and 'the stormy physical trial [facing adolescents]' (Honourable E. Lyttelton 1900); 'self-abuse . . . the scourge of the human race . . . a vice . . . a perversion . . . an evil . . . this insidious disease . . . an influence that seems to spring from the Prince of Darkness' (Hall, Professor of Psychology and Pedagogy, Clark University, 1911); 'a hidden evil, like a hidden ulcer' (the anonymous author of *For Men Only* 1925). This last writer later explains his reticence to name the vice directly by saying that 'it is not easy to describe or discuss any of these secret habits and vices without giving offence, indeed it is so extremely difficult that some may regard it as impossible.' A more recent booklet *Eight Protein Passions with Care* (issued by the Protein Wisdom group) warns how an overdose of protein during one's adolescent years could make it 'very hard to be well behaved with a sexual friend, and to be headstrong in one's lonely bed!' It concludes 'Do not let passion defeat you, ALONE, nor with a sexual friend' (p. 2, emphasis in original).

Current euphemisms (?) for masturbation include *self-abuse, play/fiddle/twiddle with oneself*, and *self-pleasuring*; *touching oneself* is said to be the preferred expression at a Roman Catholic confessional. There is *frig* (same root as *friction*); *rodwalloping*—one of many words compounded on stem verb with the sense "strike with the hand"; *wank (off)* (*wank* is etymologically related to *wench*, curiously enough; the Germanic verb meant something like "shake, waver" hence perhaps the misogynistic "be inconsistent like a woman is"), *whack off, get off, come off, beat off, toss off* (Grose glosses this as 'Manual pollution'—a wonderful euphemism); *bring off someone* (oneself or a partner) also *diddle* (a female—Partridge 1955 reckons that *dildo* derives from *diddle-o*; McDonald 1988 favors Italian *diletto* "darling" as the source); *beat the beaver/bishop, five finger exercise, finger fucking, make love to one's fist, pull one's pud*. In this last phrase the term *pud* would seem to be an abbreviation of the Latin *pudendum*, but Wentworth and Flexner (1975) surmise that it derives from *pudding head*—a mental abnormality which is just one of the many abnormalities said to result from masturbation. We believe Wentworth and Flexner to be wrong on this occasion. *Jerk/jack/jay/j- off* is normally only applicable to men. Note the common occurrence of the particle *off* in phrasal verbs of masturbating: this particle captures the release from pent up earthy desire that motivates masturbation; it also associates the act with the image of a rocket shooting off into space—an image perhaps more appropriate to men than to women. It is notable that the name *Jack* seems to have many dysphemistic associations. In addition to this one, there its in use in *I've got Jack* to mean "I've got my period"; its use in the past to mean "lavatory" (still around in the forms *jakes* and *john*); *I'm jack of it* means "I'm fed up with it"; and Grose (1811) lists *Jack Adams* "a fool," *Jack at a pinch* "a poor hackney parson," *Jack in a box* "a sharper, or cheat," *Jack in an office* "an insolent fellow in authority," *Jack Ketch* "the hangman." *Jack* is often an alias for *John* and Grose contains the following: 'JOHNNY BUM. A he or jack ass: so called by a lady that affected to be extremely polite and modest, who would not say Jack because it was vulgar, nor ass because it was indecent.' (Cp. Gwendolen's views on the inferiority of the name *Jack* in Oscar Wilde's *The Importance of Being Ernest*.) It is clear that *Jack* has a long history of dysphemistic association.

Euphemisms for methods of *birth control* or *family planning* (themselves nice euphemisms) are many and varied. During the nineteenth century, writers on this subject could not bring themselves to name outright any of the devices they were describing; this is perhaps not surprising given that, at the time, contraceptive advice by physicians usually met with severe criticism from colleagues and even legal action. So, for example, a well-known English physician of the time, G.H. Darwin, in his dissertation on the subject refers to 'certain practices' that are 'much more common in America, and on the continent of Europe, but not altogether unknown among ourselves, by which it is sought to limit the number of children brought into a family, or to entirely prevent conception.' He goes on to aver that he will refrain from any further 'repulsive detail' (cf. Rusbridger 1986:186). Other writers referred variously to *such practices/methods, rubber mechanisms* and even in some instances in *Satanic devices*. Today, even though discussion is much more frank, contraception remains a fertile breeding ground for euphemism. For example, contraception is 'natural' if *the rhythm method* is used; but *contraceptives, contraceptive devices* are *mechanical methods*. Common general-for-specific euphemisms for these are *protectives* and *prophylactics* (from the Greek *pro-* meaning "before in time" and *phylaxis* meaning "watching or guarding against"). Prophylaxis in general refers to a medicine or measure used as a precaution against disease; hence the question *Are you taking any precautions?* is, in appropriate contexts, interpreted as denoting contraception.

The euphemisms for more specific types of prophylaxis are mostly descriptive of the configuration or material used (e.g., *pill, (contraceptive) cream/gel, sheath, diaphragm, protective, coil, rubber,* etc.); also *rubber, johnny* or just *johnny.* Then there is the popular metonymic practice of describing them as simply items of clothing and the like. For example, we have *caps* and *hoods* (French, Dutch, and German), *diving suits, shower caps, raincoats* (according to folk myth men complain that wearing a condom while having sex is 'like taking a shower in a raincoat'), *riding jackets* and *gloves* (cf. 'No glove, no love' Irving 1979); *mouse's sleeping bag,* and so on.

Birth control generally used to be regarded as a depraved and evil practice harmful to both one's health and one's mental state; and some people take that view today. And, as we saw in Darwin's description above, foreigners are to be blamed as the wellspring of this degeneracy—particularly the French. This then is the source of expressions like *French methods/devices/safes,* or simply *Frenchies* (whence *frannies* or *frangers*), all of which are current today. The French of course retaliate with terms like *capote anglaise/allemande* "English/German hood" and *redingote anglaise* "English overcoat" both of which denote condoms whereas the English terms *French/Dutch cap* denote a *diaphragm* (almost another euphemism). All these nicely illustrate the common dysphemistic practice of blaming vice and immorality on outsiders (cp. terms like *French gout/measles/pox* for "syphilis" given in Ch. 7). The expression *French letter* (sometimes *Italian* and *Spanish letter*) is curious. Most writers on the subject (e.g. Paros 1984:213) seem of the opinion that the expression draws its name from the obvious association of letters and envelopes (but then why not *French envelopes!*). It is more likely that its origin lies in a now obsolete verb *let* as in *without let or hinderance* and the tennis expression *let ball* (where the ball catches the net) from the Old English *lettan* meaning "to hinder". (After regular sound change, this verb became homonymous with the modern English verb *let*— originally, *laetan*—which has the completely opposite meaning of "to permit/ allow"; obviously a language cannot tolerate homonymous antonyms, and the verb with the sense of "hinder" largely disappeared.) Nevertheless, for a time, a *letter* meant "someone or something which hindered." Since one important use of a condom is to prevent the spread of venereal infection from those who would once have been described as *frenchified,* Old English *lettan* may well be the source for the 'letter' in *french letter.* And of course French letters also 'hindered the arrangements of providence' (i.e., the conjugation of sperm with ovum) in the words of a 1682 Papal Bull issued by Pope Leo XII against the use of such devices.

The history of the term *pill* in the sense of "oral contraceptive" exemplifies euphemistic narrowing. Two decades ago, the term would have required an appropriate modifying phrase (e.g., *birth control/contraceptive pill*). Recently, however, the phrase *the pill* has all signs of narrowing to this sense alone: its definite article gives it comparable identifiability to *the Pope.* The taboo context is undoubtedly the cause of this rapid meaning shift and it now seems that advertisers favor terms like *tablet* and *capsule* over *pill*.

Have you got protection? Since the onset of (what is perceived as) the AIDS epidemic, the term *condom* (earlier *cundum*) has all but lost its tabooed status. In earlier times it frequently appeared in print euphemistically abbreviated to *c-d-m*. Now, however, the term *condom* is widely used in the media, both in reporting and in advertising, and condoms are occasionally even to be seen on television (though not usually located on the organ for which they are manufactured). Like advertisements for tampons and sanitary napkins, advertisements for condoms normally employ visual euphemism. The etymology of the word *condom* is mysterious. Grose (1811) attributes the term variously to (1) a Colonel Cundum, the alleged inventor of the device, (2) a false scabbard over a sword, (3) an oilskin case for holding the colors of a regiment. Paros (1984:212) attributes it to a Doctor Condom (or Conton), a seventeenth-century physician at the court of Charles II. He allegedly invented the device to halt the king's growing number of illegitimate offspring. As Paros admits, however, there is some doubt as to the authenticity of this information. Other imaginative etymologists argue that the name derives from a blend of *cunnus* or *cunt* and *dum(b)*. If it is at all comparable with some of the other etymologies we shall examine, that of *condom* may well be a potpourri of all these and more!

The generalization of brand names is a common practice and it can be the cause of cross-cultural embarrassment. For example, in Britain *durex* is a brand name for condoms that has developed into a common noun meaning "condom," and is thus comparable with *hoover* meaning "vacuum cleaner." It also happens to be the name of a widely available brand of sticky tape in Australia, which causes considerable confusion for an Australian visitor to Britain who innocently requests what they think is sticky tape and baffles the Briton who understands them to be asking for condoms. The same type of confusion arises with the term *rubber,* an American euphemism for "condom" but which in Australian and British English means "eraser." A common North American brand of condom is 4-X that, when spoken is pronounced /foreks/; this happens to be the name for a popular brand of Australian beer XXXX; some Americans are shocked and others highly amused when they hear one advertising slogan for this beer. 'I feel a XXXX coming on.'

Sterilization is the ultimate in contraception; this is the neutral term. *She's had her tubes tied* is euphemistic; we haven't heard *he's had his tubes tied* for "vasectomy" but *I've had the chop* and *I've had my tubes chopped* are attested in this sense. These are euphemistic, but mainly different in style from the more clinical and socially distant *I've had a vasectomy* or *I've been sterilized*.

The euphemisms *sexual intercourse, sexual congress, copulation,* and *coition* all take note of the dyadic nature of copulation—that *it takes (at least) two to tango.* The lexicon for copulation can be classified according to at least the following general categories, each of which focuses on a different aspect in the perception of the act. Some terms appear in more than one category. The widely employed general-for-specific euphemism for copulation is of course *it*; for instance,

> Hardly had I arrived in the house when I was lugged off to their secret meeting place, the Hons' cupboard, to be asked what IT was like.

"Linda says it's not all it's cracked up to be," said Jassy, "and we don't wonder when we think of Tony."

"But Louisa says, once you get used to it, it's utter utter utter blissikins," said Victoria, "and we do wonder, when we think of John."

"What's wrong with poor Tony and John?"

"Dull and old. Go on then Fanny—tell."

I said I agreed with Louisa, but refused to enter into details.

(N. Mitford *Love in a Cold Climate* 1979:372)

1. DYADIC

your daughter and the Moor are now making the beast with two backs.

(Shakespeare *Othello* I.i.114)

Despite this quote from Shakespeare, most members of the dyadic category are marked by the structure 'NOUN PHRASE PREDICATE *with* NOUN PHRASE'. *Have sexual intercourse with, copulate with, couple with, fuck with, screw with, score with, have one's way with, have coitus with, have union with, have sex with, have a good time with, have fun with, play around with, have it away with, have intercourse/(sexual) relations/an affair with, be intimate with, get together with, go with, be seeing, be with, lie with, go to bed with, match with, mingle with, knot with, sleep together, sleep with, take a roll with, get down with,* and so on. And there's another.

Louisa said to me, her eyes big as saucers: "He rushes into her room before tea and lives with her." Louisa always describes the act of love as living with.

(N. Mitford *The Pursuit of Love* 1979:219)

Most, if not all, of the dyadic examples fall into other categories too.

2. HAVING. To have: *a lover, a fuck, a screw, a bang, a rattle,* sex (perhaps the growing use of the euphemism *gender* in place of *sex* is partly motivated by the fact that 'sex' in *have sex* means "copulate"), *a piece of ass, a bit of tail, a beef injection, a naughty, a good time, a cuddle, a jump with, fun (with), have someone away, have it away with, have an affair with, carnal knowledge of, (sexual) relations/intercourse with,* and so on.

3. GETTING. To get: *some, lucky, laid, on (top of), in(to), up, (down) to it, under, over, down with, into the pants of, into bed with, a leg over, one's oats, one's rocks off,* some *nookie* (perhaps from the vagina as *nook,* but more likely from the obsolete noun *nug* "love(r)"), and so on. There is also *grant sexual favors to* (and *withdraw* them).

4. ACHIEVING. *Make it, make out, get to it, score, have one's way with, go all the way, have someone away, possess,* and so on.

5. HAVING FUN. *Have fun, make whoopee, have a good time, play around, fool around, slap 'n' tickle, hankypanky, jigging, jigjogging, up-tails-all, tumble, roll in the hay,* and so on.

6. CLOSENESS. *Love, make love with, be intimate with, be with, cuddle, belly slapping, exchange flesh/bodily fluids, (the biblical) know,* and so on.

7. BED. *Bed, go to bed with, get someone into the cot, lie with, get between the*

sheets with, share a pillow with, sleep with, hit the hay, bury face upwards, and so on.

8. FOOD. The verb *pork* "copulate" may be associated with the food metaphor for genitalia (see next section); hence *sinking the pork sword/the sausage, beef injection.* A common euphemism for "copulate" in many African languages is *eat;* which is also used occasionally in English both for copulation and oral sex; this links with the notion of the vagina as mouth (see the next section). Food may also be the figure in American *Cut yourself a piece of poontang,* which echoes the cutting of pie or cake. (*Poontang* is said to have come via Louisiana Creole from French *putain* "whore".) There is also the notion of the vagina as fish or fishpond (see later) and so *go fishing, groping for trout,* and so on (see Partridge 1955 under *diver* and *groping*).

9. ACTION

> The wren goes to't, and the small gilded fly
> Does lecher in my sight.
> Let copulation thrive;
> > (Shakespeare *King Lear* IV.vi.116–18)

> Quoth she what is this so stiff and warm,
> Sing trolly, lolly, lolly, lo;
> 'Tis Ball my Nag he will do you no harm,
> Ho, ho, won't he so, won't he so, won't he so.
> But what is this hangs under his Chin,
> Sing trolly, etc.
> 'Tis the Bag he puts his Provender in,
> Ho, ho, it is so, etc.
> Quoth he what is this? Quoth she 'tis a Well
> Sing trolly, etc.
> Where Ball your Nag may drink his fill,
> Ho, ho, may he so, etc.
> But what if my Nag should chance to slip in
> Sing trolly, etc.
> Then catch hold of the Grass that grows on the brim
> Ho, ho, must I so, etc.
> But what if the Grass should chance to fail,
> Sing trolly, etc.
> Shove him in by the Head, pull him out by the Tail,
> Ho, ho, must I so, etc.
> > (from 'The Trooper watering his
> > Nagg', D'Urfey 1719 Vol.5:13f)

Euphemistic or neutral terms—*copulate, couple, have sexual intercourse with, do it/ the deed/the do, make it, make love, on the job, get to it, get at it, get going, plough, fall to it.* Dysphemistic terms not mentioned later include *ball, lay, cork, grind, knee tremble, knock off, scrog, do the nasty, whitewash her/your guts, sluice.*

Terms of attack: the archaic *firk* and hence *fuck* (see 10 below), *bang, boff, bonk, hit, knock off/up, lay siege to, pluck, rape, root, scale, screw, strike, struggle with, thump,* and so on. All dysphemistic.

Riding: *cover, mount, ride, tup,* and so on. Since these terms, and *serve,* are frequently used of copulating animals, they tend to be dysphemistic.

Rolling about: *hump, roll (in the hay), shag, horizontal jogging/folk-dancing, doggy-dancing, belly slapping,* and so on.

Various terms for copulation invoke penetration. In sex education and law courts the expressions *[sexually] penetrate [a woman]* and *insert the penis* are to be taken literally, but they imply at least the onset of sexual intercourse. The more colloquial (and erotic) expressions are, depending on Speaker's point of view, *go* or *come in/into/inside, slip in,* and so on. There is a line in the film *Ruthless People* where the Danny de Vito character says to his mistress 'I was hoping we'd have enough time for a little poke in the whiskers' (is *pork* a remodeling of *poke* or is it associated with food? See 8 above). Other penetration terms are *shaft, prod, root, get (it) up, dip your wick* (the counterpart for the female partner is presumably *spread/open your legs*), *sinking the pork sword/the sausage, beef injection,* and other figures.

10. FUCK

> Fairweill with chestetie
> Fra wenchis fall to chucking,
> Their followis thingis three
> To gar thame ga in gucking:
> Brasing, graping, and plucking,
> Thir foure, the suth to sane,
> Enforsis thame to fucking.

> "Farewell to chastity
> when girls take to fondling,
> three things follow
> which cause them to go fooling around;
> embracing, groping, plucking [or ploughing].
> These four, the truth to tell,
> will get them to fucking."
>
> (Alexander Scott 'Ane ballat maid to the
> derisioun and scorne of wantoun women',
> mid-sixteenth-century, quoted in Montagu
> 1967:308; translation is ours)

It is sometimes said that *fuck* derives from German *ficken* "oscillate, move quickly back and forth, fuck"; if it does, then it must surely have been under influence from English *fig* and its fifteenth- to late-nineteenth-century alternative form *fico*. *To give the fig* or *fico* is 'A contemptuous gesture which consisted in thrusting the thumb between two of the closed fingers or into the mouth' (*O.E.D.*)—this gesture is clearly iconic for "fuck"; and note that *I don't give a fig* means "I don't give a fuck"— the expression which has supplanted it. In addition, *fig* is one of the many words for "cunt" (cp. *fica* 'Fíca, . . . Also used for a woman's quaint' Florio 1611). The euphemistic dysphemism *frig* (fully euphemized to *freak*) may be a remodelling of *fig*, although it is also connected with *friction*. The difficulty is, of course, that the vowel of *fuck* is quite different from that of *fig*.

The earliest recorded uses of *fuck* are in Scots (cf. Read 1934:238). In 1503 onetime Franciscan Friar William Dunbar published 'Ane brash of wowing' ["A bout of wooing"] which ends:

He clappit fast, he kist, and chukkit,
As with the glaikis he wer ouirgane;
Yit be his feirris he wald have fukkit;
Ye brek my hart, my bony ane!

"He quickly fondled, kissed, and chucked her under the chin,
As with sexual desire he was overcome;
Yet by his actions and gestures he would have fucked;
You break my heart, my beautiful one!"

(Quoted in Montagu 1967:308; translation ours)

The first time the word appeared in an English text was in 1598, in John Florio's *Dictionarie of the Italian and English Tongues* (the following are from the 1611 edition):

Fóttere, fótto, fottéi, fottúto, *to iape, to fucke, to sard, to swive*

Fottifterio, *a bawdy or occupying house. Also the mistery of fucking*

Fottitúra, *a iaping, a fucking, a sarding*

Fottúto, *iaped, fucked, swived, sarded*

English *fuck* does not obviously stem from the Greek *phyteyou*, a euphemism from "planting, seeding," which is cognate with Latin *futuere* the parent of Italian *fottore*, French *foutre*, and so on. Read (1934:268) suggests, inter alia, that it derives from the Latin root *pug-*, which appears in the noun *puga -ae* "buttocks, rump" and the verb *pugno -are* "fight, strain or strive against, struggle with," and in view of the various English synonyms for copulation listed previously under (9), this seems highly probable. It is interesting then to survey the entries in *O.E.D.* (1989) under **pug** and **puck**. The former, as a noun, is a dialect term for "lamb, hare, squirrel, ferret, fox" (*O.E.D.* 1989, 8), all of which are used, in certain contexts, with sexual connotation. The same noun is a variant of *puck* (cf. *O.E.D.* 1989, 5 under **puck**), meaning "evil or malicious being, devil." This could perhaps account for the dysphemism of *fuck* that we will discuss in Ch. 5; witness the following quote from Le Fanu's 1864 *Uncle Silas* II.vi.88 (*O.E.D.* 1989, 2); 'And why the puck don't you let her out?' Here, 'why the puck' can be understood either as "why the devil" or "why the fuck." Furthermore, there is a dialect verb *puck* that means "hit, strike, or butt," and the *O.E.D.* quotes from an 1870 Irish story 'The ram and the cow pucked her with their horns.' Striking with a horn is, in some contexts, quintessential fucking. What all this suggests is that *pug* and *puck* may have contributed to the development of *fuck*; though we doubt it is the only source.

According to Partridge (1955) and Shipley (1977), the most likely origin for *fuck* is *firk* "strike, stir up, tease, cheat, copulate"; hence *firkytoodle* "foreplay." *I'll firk him/her* has very much the same polysemy as modern *I'll do him/her* (cf. Shakespeare's *Henry V* IV.iv.29f: 'PISTOL: Master Fer! I'll fer him, and firk him, and ferret him.' The penis was commonly known as *ferret*, see later.) This gives credence to Partridge's suggestion that *firk*—and thence *fuck*—derives from Latin *facere* "do," in which case there has been contamination of what was once a euphemism. Modern *screw* also has a sense "cheat"; and *knock off* also has a sense akin to that, namely "steal": both are synonyms of *fuck*. Although Shakespeare uses *firk* he does not use *fuck*; however, he does apparently euphemize it to *focative* in the following:

EVANS	What is the focative case, William?
PAGE	O—*vocativo*, O.
EVANS	Remember, William: focative is *caret*.
QUICKLY	And that's a good root.
EVANS	'Oman, forbear. (*Merry Wives* IV.i.42–47)

Case and *O* are both euphemisms for "vagina"; a *caret* is wedge-shaped, a shape traditionally associated with the vagina, and there is also a play here on *carrot* "penis"; 'that's a good root' has the same meaning in modern Australian slang that Shakespeare intended here. 'Oman' is simply a play on the vocative "O man"; recall Alice's "O mouse" is Lewis Carroll's *Alice's Adventures in Wonderland*. (On Shakespeare's bawdy see Ch.9 and especially Partridge 1955.)

It is possible that the vowel in *fuck* is a variant pronunciation of the vowel in *firk* (cf. /θʌsti/ as a pronunciation of *thirsty* in some British dialects.) Furthermore there are, or have been, many phonologically similar doublets of which one member (often the more colloquial) has a short lax vowel whereas its pair has a long tense vowel; for instance,

ass–arse, bass–barse, bubby–baby, buss–burse "purse", *bust–burst, cuss–curse, gal–girl, goss–gorse, hoss–horse, mot–mort* "girl or wench", *sassy–saucy, tit–teat, whids–words, wud–world*   (Sources: Grose 1811, *O.E.D.*, Shipley 1977)

The glosses for *mot–mort* are from Grose (1811), but it seems probable that they are closer to "piece of arse, bit of cunt", since *mot* (also spelled *motte*) is a euphemism for "cunt" by borrowing the French *mot* "word." Later in this chapter we shall argue that *puss–purse* follow this pattern, and were once synonymous; the same may once have held for *dick–dirk*, too, though this is much more doubtful (see 13 in the next section). The hypothesized pairing *fuck–firk* is consistent with all the attested pairs. It is notable that one remodelling of *fuck* is *fork* (perhaps influenced by *fornicate*, and cp. *pork*), which is another alternation between a short lax vowel and a long tense vowel. It is possible that *shag* may be correlated with *shake* for this same reason (cf. *O.E.D.*).

We conclude that *fuck* has become established in its current meaning under influence from one or more of the following, any of which may have had an influence on another: the borrowing of the German verb *ficken*; the borrowing of Italian *fico* and the loan translation *fig*, particularly in expressions like *give the fico/ fig* and *not give a fig*; the English verb *firk*. The most likely scenario for the vowel change is ?/fi(r)k/ > /fɪ(r)k/ > /fəək/ > /fʌk/ (i.e., a gradual centralizing, laxing, and lowering of the vowel quality with a shortening at some stage; cp. the doublets *curse–cuss* and *girl–gal*). The final push from /fəək/ to fʌk/ may well have been under the influence of *puck* and *pug*.

11. INCEST and PEDOPHILIA. Sex educators and social workers use *touch* to children as a euphemism for *incest, paedophilia, pederasty, interfering with a child*, or *child molestation*. A 1990 poster in New York subways advertising an Incest

Hotline reads: 'When someone touches you in a way that feels bad, that's incest.'
Next to this is a list of potential touchers: Father, Mother, Uncle, Brother, etc.
Euphemistic expressions like *child molestation* or *the sexual abuse of children*,
where the nouns derived from the intrinsically dysphemistic verbs *molest* and *abuse*,
characterize society's abhorrence for the adult perpetrator. There is no conventional
euphemism for incest between consenting adults.

The Lexicon for Tabooed Body-Parts

MERCUTIO   Why, is not this better now than groaning for love? Now art thou
           sociable, now art thou Romeo; now art thou what thou art, by art
           as well as by nature: for this drivelling love is like a great natural,
           that runs lolling up and down to hide his bauble in a hole.

BENVOLIO   Stop there, stop there.

MERCUTIO   Thou desirest me to stop in my tale against the hair?

BENVOLIO   Thou wouldst else have made thy tale large.

MERCUTIO   O, thou art deceived; I would have made it short; for I was come
           to the whole depth of my tale; and meant, indeed, to occupy the
           argument no longer.

ROMEO   Here's goodly gear!
           Enter Nurse and her man Peter
           A sail! A sail!

MERCUTIO   Two. Two. A shirt and a smock.

NURSE   Peter!

PETER   Anon.

NURSE   My fan, Peter.

MERCUTIO   Good Peter, to hide her face; for her fan's the fairer face.

NURSE   God ye good morrow, gentlemen.

MERCUTIO   God ye good e'en, fair gentlewoman.

NURSE   Is it good e'en?

MERCUTIO   'Tis no less. I tell ye; for the bawdy hand of the dial
           is now upon the prick of noon.

NURSE   Out upon you. What a man are you?

                                  (Shakespeare *Romeo and Juliet* II.iv.90–117)

The degree of synonymy in the vocabulary for genitalia and copulation has no
parallel elsewhere in the English lexicon—except in the terms for "whore". There
are reportedly more than 1,200 terms for "vagina" and more than 1,000 for "penis
(with or without testicles)," 800 for "copulation" and 2,000 for "whore" (cf. Farmer
and Henley 1890–1904; Fryer 1963; Healey 1980). Every imaginable aspect of the
appearance, location, functions, and effects of the genitalia have been drawn upon
as bases for metaphor from the gross *fuckhole* to the mundane *meat and two veg* to
the highly imaginative (if precious) *the miraculous pitcher, that holds water with the
mouth downwards*. Consider some classes of synonyms for a few familiar body-parts.

   1. MAXIMALLY-GENERAL-FOR-SOMETHING-SPECIFIC: *thing, what's-
it, thingummy, et cetera*, and so on. *Member* is not maximally general, but it is
another instance of a general-for-specific euphemism.

2. GENERAL-FOR-SPECIFIC: The legal term *person* "penis" cf. Vagrancy Act, State of Victoria (Australia), §7, 1(c). *Feminine itching* for "vaginal irritation".

3. SYNECDOCHIC LOCATION: *belly, low country* (particularly of a woman), *nether regions, down there, backside, bottom, crutch, loins, tail* (genitalia of either sex; cf. *Have a bit of tail, Uptails-all*, etc.).

4. PART-FOR-WHOLE. *Tit* for "breast" although it principally denotes the tabooed part of the breast, namely, the nipple. The seventeenth- to eighteenth-century dysphemistic use of *tit* for "woman" probably has no influence on the current use of this word; its dysphemistic sense is retained in the insult *You tit!* which has more or less the same sense as *You stupid cunt!*

5. TERMS AMBIGUOUS BETWEEN BACKSIDE AND VAGINA

> "Tha'rt real, tha art! Tha'rt real, even a bit of a bitch. Here tha shits an' here tha' pisses: an' I lay my on 'em both an' like thee for it. I like thee for it. Tha's got a proper, woman's arse, proud of itself. It's none ashamed of itself, this isna'."
> (Mellors to Lady Chatterley in Lawrence *Lady Chatterley's Lover* 1961:232)

The anus and vagina are both saliently holes, whereas a penis is not. *Front bum*; children's usage of *bottom* or *botty*; adult's *tail* and *scut*; American *ass* and *tush*; British *prat* and *fanny* (the latter is a euphemism for "ass" in American). The ambiguity presumably arises because (1) they share a similar location on the lower trunk; (2) they are both tabooed; (3) both saliently contain orifices and passages that expel waste products from the body; (4) those passages are used, respectively, in anal and straight sexual intercourse.

6. METAPHORS BASED ON APPEARANCE.

FLACCID PENIS as *tail* (the term *penis* is itself Latin for "tail," a euphemism for *mentula*); *tassel, bauble, putz* (this word is much less dysphemistic in English than in Yiddish, when it derives, ultimately from the verb *putsn* "adorn"; like Yinglish *shmuck*, and *prick* in BRITISH English, *putz* has come to mean "fool", see Ch. 5), *dangle, pencil, worm, schlong* (from Yiddish *schlang* "snake"), *doodle* (from *cock* via fore and end-clipped *cockadoodledoo*), *hairy banana* (cp. Bahasa Malaysia *pisang* "banana", said to be the prevalent euphemism used among women), *noodle, dill, gherkin, wally, wire, wiener* (and other terms for sausages). One of the more delightful metaphors is *the one-eyed trouser snake* (attributed to Barry Humphries). There is the bird metaphor, discussed shortly, that gives rise to terms like *cock* and *pecker*. Then there are the bell-clapper metaphors: *ding, dink, dong, donger* (cf. the Australian proverb *Dead as a dead dingo's donger*). There are probably two sources for this metaphor. One, the vagina is seen as a bell, activated by the penis-clapper (the clitoris is sometimes called a *bell*, though this *love button* looks more like a bell-clapper; maybe it gets its name from the noisy effect from activating it). Two the silhouette of *a man's tackle* (what Americans might call his *basket*) is not unlike that of a bell, and the ungirded (flaccid) penis bouncing against the scrotum of a walking or running man is visually similar to the bell-clapper at work. The basically ono-matopoeic expression *tinkle* "urinate" is probably influenced by the bell-clapper metaphor.

ERECT PENIS as a weapon: *weapon, sword*, (cp. *vagina* as Latin for "sheath"), *pistol, gun, rod, lance, bill, pike, dart, chopper, bugle, prick*, and so on (see Ch. 5 on the etymology of *prick*). Some variants on the *prick* image are *needle, pin, thistle, hook, horn*, and in times past *pen*. These give rise to the metaphor of the *tailor* as the male partner and *stitching* for "copulating" (cf. Grose 1811 and numerous folk songs). Other metaphors include *machine, instrument, tool, hammer, poker, pipe, knob, pole, shaft, staff, stand, oar, bone(r), hard*, and *stiff*. There is also the once very common *yard*.

BOLLOCKS or *ballocks* from "little balls" (cf. *butt–buttocks, bull–bullock*). A link with *bullock* (see Clemens 1968:23) is feasible, given the long European tradition of bull-worship in which the bull, and a fortiori a bull's testicles, figured as a symbol of virility; but we know of no textual evidence to support this. Metaphors include:*balls, billiards, nuts, stones, rocks, marbles*, and *pills*. *Goolies* probably derives from Hindustani *goli* "pebble, ball, bullet" (cf. McDonald 1988). The old established *cods* (whence *codpiece*) is from Old English *codd* "bag".

VAGINA as *a ring, circle, O, do(ugh)nut, wheel*. Shortly after extravagantly sexual Zsa Zsa Gabor was convicted for slapping a cop who had arrested for her a traffic violation, the following caption appeared under a still photo from a commercial she was filming:

> ZSA ZSA'S GAME: The flamboyant Hungarian actress re-enacts her cop encoun-
> ter in her new 'Wheel of Fortune' commercial for New York's WCBS-TV. In one
> version, Gabor coos, ". . . take away my driver's license. But darling, don't touch
> my wheel." (*USA Today* Monday, October 23, 1989:2D)

The vagina is also known as the *slit, slot, crack* (hence *cracksman* for "penis," and the dysphemistic *Bit of crackling* for "woman"), *breach, gash, (everlasting) wound* (note the link with blood), *cleft, furrow, valley, oyster; boat* (the clitoris is *the man in the rowboat*); *dial; twat* (possibly associated with *two–twa* because of the silhouette of the *labia majora*—but whence the final excrescent -*t*, can it be linked to *dyad*, or is it purely accidental? See Antilla 1972:72 and Hock 1986:124); *triangle, Y, pie; gate; hole, tunnel, den, box, (genitive) case; hat* (cf. Grose 1811; 'because frequently felt'). *Well, bottle*, and *pond* all mix configuration with function in their imagery. The vagina is seen as a *mouth*, with lips and tongue (clitoris)—hence, *nether-lips*; like the mouth it salivates and drinks, and in one graphic description: may *flash an upright grin*. Such euphemistic metaphors, like others for tabooed body parts, liken it to a nontaboo part. Terms like *bite, snatch, vice, snapper, clam*, and *oyster* extend the metaphor by suggesting a mouth ready to snap up a penis (cp. Japanese *hamaguri*, after a species of mussel); *vice* is doubtless morally inspired, too. Note that the last three are also fishy—a fishy odor is commonly attributed to this organ (see later); we therefore find terms like *fish(tail)* and *ling* for "vagina."

ANUS (Latin for "ring") as *ring* (cf. *Up your ring!*), *hole, brown-eye, arsehole, shithole, blurter*, and so on. This last term reminds us that *bum* probably derives from *boom* "fart" rather than being a remodelling of *bottom*.

BREASTS as *knockers, bouncers, bulbs, balloons, bazoomas* (a remodelling of *bosoms*), *globes, headlights, melons, montezumas, mounds, molehills; a pair, a set; lungs*. It is not clear whether the term *norts* is based on *noughts* in order to capture the outline of breasts such as depicted in rough two-dimensional sketches; Australian *norks* is perhaps partly derived from *norts* by deliberate remodelling or mishearing, but more likely from the tradename *Norco Cooperative Ltd:* it was a New South Wales butter company whose advertisements once featured a cow with an exceptionally large udder.

NIPPLES as *(rose)buds, tits.*
PUBIC HAIR ON THE *MONS VENERIS.*

> One of the services offered by Dr [Takeo] Sudo's new clinic [in Japan] is artificial pubic hair for people suffering an inadequacy. He tells the story of one of his earliest clients, a young woman for whom he made a pubic toupee. While bathing with some friends at her local bath-house, her hairpiece floated gently to the surface of the water, causing her near-mortal embarrassment and prompting the doctor to remodel his product.
>
> (*The Age*, Melbourne, Sat. 12 September 1987)

The correlation of the female pudend with furry animals may result from the fact that—in contrast to many men—on most women, pubic hair is the only substantial patch of body hair. Its salience is clearly demonstrated in art, for instance in René Magritte's abstract painting 'Trois femmes' of 1922; and it is nicely recognized in the following verse:

> The cunt is a mythical creature
> All matted and covered with hair
> It looks like the face of my teacher
> And smells like the arse of a bear.
> (Collected by Lyn Menon, from
> a wall within a female toilet at the
> Royal Melbourne Institute of
> Technology)

The grass on fanny's hill is both visually and tactually salient, not to mention erotic. A graffito from Pompeii (destroyed in 79 C.E.) notes the fact: FUTUITUR CUNNUS [PIL]OSSUS MULTO MELIUS [QU]AM GLABER "a hairy cunt is much better to fuck than a hairless one" (cf. Read 1977:22). Not for nothing was pubic hair airbrushed out of soft-porn photographs until the 1960s. The significance of pubic hair on the *mons veneris* accounts for the several furry animal terms *pussy, beaver, bun[ny],* and, of course, *hare* (homophonous with *hair*) that are synonyms for "vagina" (see the next section). There are also: *the hairy chequebook* (see (9), upcoming, for an explanation for 'chequebook'); *beard, bearded clam, hair pie, hairburger, furburger* (which partake of the food metaphor, upcoming), *hairyfordshire, bird's nest, mossy mound, park, grass* ('On her belly there is a sign, "Keep off the grass, the hole is mine."'), *ling* ("heather" not the fish), *furry mound, muff* (one source for this word may be a remodelling of *mouth*—which is part of the food/eating metaphor for sex); *velvet, thatch, thicket, bush, brush, fuzz, pelt, pubes,* and so on. Not surprisingly, one term for penis is *hairsplitter* and copulation *a poke in the whiskers.*

7. FOOD. There is a food/eating metaphor in the semantic field of sex. A good looking member of the opposite sex is a *dish* perhaps a *tasty dish;* a virile man is described by the non-food animal names *bull, stallion,* or *ram,* but the woman is a *bird, chick(en), rabbit,* or *lamb.* The sexual organs and perhaps the body that goes with them are *meat:* we find *beef* for a "penis" (Grose 1811 has it for "vagina" too,

and earlier it meant "whore"); *hairburger* and *furburger* for "vagina"; and in contemporary Australian *meaty bites* for "testicles" (after the brandname of dog biscuits which are bone-shaped; that is, they have a shaft between what looks like a pair of testicles at each end). The vagina is known as a *plum* probably because of a plum's color, shape, and juice. *Jelly-roll* should be included here; we leave the motivations for this figure to the reader's imagination. There are also the various fishy terms for vagina, see p.114. The *blue-veined junket pumper* is a rather picturesque, even poetic, term for a penis. The noun *crumpet*, which according to Partridge (1970) meant "buttered bun" (for the relevance of this, see later discussion), is not so much used of the body-part per se as in the phrase *bit of crumpet*, which denotes a woman whose sole value to a man is as a sexual partner. The verb *pork* "copulate" may be associated with the food metaphor for genitalia. We have already remarked that a common euphemism for "copulate" in many African languages is *eat*, which is also used occasionally in English (though primarily for oral sex); this links with the notion of the vagina as mouth, see (6) above. Other food-associated terms include *love-juice* and *jam* for "catamenia."

The food metaphor was common in Attic Greece, particularly for female genitalia. Henderson (1975) identifies it being described as meat cooked on a phallic spit, or as the oven (cf. 'she is blackened down below [pubic hair] because the council used to put their pots [*phalli*] there for cooking before the war' p.143). They are also spoken of as licking-bowls, and bowls of soup, sauces, and juices. Henderson also mentions:

> boiled sausages (the labia one supposes) . . . the slices of fish-meat (*temáchē*) which are "fanned into flame" at the banquet . . . various foods for munching . . . *parothídes*, hors-d'oeuvre, refers to love-play that leads up to intercourse . . .

and various kinds of cakes and pastries; (cf. Henderson 1975:144f).

8. FUNCTIONS. Euphemisms for the genitalia of either sex are: *equipment*, *waterworks*. Dysphemisms: *pisser*, *pisshole* (whence *pizzle*).

PENIS as *joystick*, *fuckstick*, *creamstick*, *cracksman*, *cunny-burrow ferret*, *tool*, *love muscle*, *pee shooter* (a double whammy), *pistol* or *gun*. The neutral term *testicles* derives from a function based Latin euphemism *testiculi* "little witnesses": whether they are witness to a man's virility or not, they did function in affirming oaths during biblical times, cf. *Genesis* 24:9 'And the servant put his hand under the thigh of Abraham his master, and swore to him concerning that matter.' See also *I Chronicles* (29:24), *Genesis* (47:29), *Ezekiel* (17:18). A similar practice is reported (in Grey 1841 vol.2:342) amongst West Australian Aborigines. Note that this practice is apparently the very opposite of *having someone by the balls* (or *the short and curlies*), and there is some speculation that it symbolizes an oath not just to the man, but also to his seed.

VAGINA as *pissflaps*, *lovebox*, *fuckhole*, *cockchafer*, *cockpit*, *milkpail*, or *bottle*.
ANUS as *shithole* or *fartflap*.

BREASTS as *jugs*, *dairies*, *udders*, *dugs*, *mammaries*, *Nörgen-Vaazes* (based on the trade name of an Australian ice-cream company). Current *boobs* seems to have replaced older *bubbies*—perhaps transferred from their being used to suckle bubbies or babies.

9. VAGINA AS SOURCE FOR OR STORE OF WEALTH

no money, no coney   (Massinger 1622, *The Virgin Martyr* II.i.17)

Denied male prerogatives, [women] seek to manipulate their surroundings to maximize their situations within male-defined parameters. What are generally considered vanity, promiscuity, or piousness can be explained as female reactions to protect their most desired and saleable assets in the male sphere. [. . .] An analysis of [Kenyan novelist Grace] Ogot's work reveals the dependence of women upon men for basic security. A woman's major, and in many cases only, asset is her body, which can be bartered for this security. Until women are recognized as legitimate productive members of society, they will continue to rely on manipulation and sexual exploitation for survival.   (J. O'Barr 1984:225)

An ABC television program called 'Nobody's Children' (Melbourne, Australia, May 17, 1989) included an interview with an eighteen-year-old street-kid who claimed that in order to travel, girls like her either had to walk, or else accept a lift and 'pull out the hairy chequebook [*pause*] screw.' This figure for the vagina is very old; it has also been called *breadwinner, commodity, cornucopia, jewel, honeypot, snatch*(?); *favours, exchequer, treasury, purse, kitty, money-box, money* (Grose 1811 glosses this 'A girl's private parts, commonly applied to little children: as, Take care, Miss, or you will shew your money'—in those days people didn't wear underpants). Why should a vagina be seen as a store for or source of wealth, when male genitalia are not; (however, see (11) just below)? One place we can look for the answer is prostitution: there is abundant evidence that a woman can sell her body for heterosexual sex, but with few exceptions, a man cannot. It is principally her vagina that is a prostitute's source of income. Outside of prostitution, when women had no opportunity for economic independence, they were normally dependent on a man for the fruits of income. It is, at least in part, these circumstances that encouraged a woman to make her body as attractive as possible, and led her to spend a larger amount of time, money, and effort doing so than a man typically spent on his physical appearance. Without intending to decry any attempted improvements in character, education, and other capabilities, it is clear that women have typically been socialized into marketing their bodies to men in exchange for economic security. Although this oversimplifies the facts, it is apparent that although a woman quite properly advertises her intellectual and domestic abilities, her worthiness as a house manager, companion, and mother, she also, and to a greater extent than a man, typically advertises herself as a sex object; and for this reason society perceives her vagina as one source for or store of her intrinsic wealth.

10. SEXUAL ESSENCE: *manhood, womanhood, his/her sex* and the like.

11. GENEALOGICAL IMPORTANCE: *the crown jewels, the family jewels, the family treasure,* and so on. These mostly apply to male genitalia. For instance, Yinglish *shmuck*, originally "jewel" in Yiddish, thence "penis," though now principally denoting "fool" (see Ch.5).

12. VITAL: *arbor vitae, the vital organ.* Again, these mostly apply to male genitalia. And in the past the use of *nature* denoted the genitalia of either sex (cf. Ch.3).

13. REMODELLING. We have heard the abbreviation C (pronounced /sii/)

from *cunt*—a euphemistic dysphemism. Shakespeare made great sport with *constable* and *country*. The latter appears in a ditty favored among teenage males some years ago, and probably still going the rounds:

> A sod- a sod- a soldier I will be;
> Fuck you- fuck you- for curiousity;
> To piss- to piss- two pistols on my knee;
> To fight for the old cunt- fight for the old cunt- fight for the old country.

There are clippings like *pud* from *pudend*, *doodle* from *cockadoodledoo*—a remodelling of *cock*. And there is rhyming slang, some of which is end-clipped: for example, *groan and grunt* "cunt," *fife and drum* "bum," *dickory-dock* "cock" (if clipped to *dick* this coincides with the earlier (?) term *dick*, see 14 following; *cobblers* [*awls*] "balls" *khyber* [*pass*] "arse," *bristols* [*cities*] "titties, breasts" *berk*[*ley hunt*] "cunt" (figurative epithet only, see Ch. 5). Other examples of remodelling are '*arris* from *arse*, and *bunny* from *cunny* (see later discussion). The terms *nubbies, norks, nungs, nungers, numnums* for "breasts" suggest a possible phonesthetic import of the initial *n-* (god knows why, but possibly influenced by the onset to *nipple*): it looks at least possible that *nubbies* is a remodelling of *bubbies*, *nungs* of *lungs*. It is likely that *dink(ie)* "penis" is a remodelling from the *ding* of the bell metaphor (see 6 earlier). Some other examples of remodelling are suggested in the next section.

14. PROPER NAMES (especially *dick* and *fanny*). In addition to names like *Mr. Sausage* based on common nouns denoting genitalia, a couple of proper names beginning with 'p' are used for "penis": *percy, peter*. There are also *willie* (Will 'e or won't 'e rise to the task?), *roger* (a long-time term for fucking), *john thomas*, and so on. *Dick* presumably thrives because of the rhyme with *prick*. But it is reported (Healey 1980) to derive from *Derrick*: this was the surname of a Tyburn hangman ca. 1600; it became a euphemism for "hangman," and later transferred to the name for a lifting device. Despite the fact that the latter is not pronounced /dɪk/, the association of penis with lifting device seems quite probable. A possible source for this pronunciation is the following: *derrick* just might have been blended with *dirk* in this context, building on the penis as weapon metaphor. The link with *dick* would then lie with the hypothesized doublet *dirk–dick* in which the member with the short lax vowel is the more colloquial. An additional influence may have been remodelling from *derrick* or *dirk* under the influence of *prick* and perhaps German *dick* (see Ch. 5).

The name *fanny* is used in Britain to denote either the "arse" or a "vagina." The first sense predominates in American, where it is as widely acceptable a euphemism as, say, *bottom* or *bum* in other dialects. Wentworth and Flexner (1975) define it as 'the human rump', and thought they don't say so, it possibly derives via some remodelling from *fund*, an alternative to *fundament*—colloquially *funniment*; note that *funny* would be ambiguous, and *fanny* may be dissimilated from this under influence from the other (British) sense of *fanny*. Comparing *fanny* with American *ass* (see later discussion), one would assume that it originally meant "arse" and was extended to mean "vagina"; but the evidence for British English is that its earliest meaning was "vagina." What is the origin of *fanny*? One source is possibly *fan*: the name *fan* (in the eighteenth to nineteenth centuries sometimes *fanner* or *fanny*) was originally given to "a winnowing basket," and extended to 'things resembling a winnowing fan in shape' (*O.E.D.*). Among the things it was extended to, was a roundish object attached to a post at which knights could practice tilting with their lances (cf. Chaucer 'Manciple's Prologue' 1.42); the symbolism speaks for itself. In addition, fans were for inflaming fires: perhaps, metaphorically, the fires of love. This, no doubt, explains why Shakespeare uses *fan* to denote the female pudend in *Romeo and Juliet* (II.iv.109)

> NURSE   My fan, Peter.
> MERCUTIO   Good Peter, to hide her face; for her fan's the fairer face.

—see the full quote at the beginning of this section (p.96).

A possible second source, or rather series of sources, is *fauny, fawney,* and perhaps *fawn*. The principle objection to this hypothesis must be that the stressed syllable of *fanny* has an unrounded short low front vowel, whereas all these have a rounded long mid back vowel. There is, however, a strong possibility of semantic association. *Fauns* are mythological creatures of the wilderness, Pan's people, noted for their lustful desires (cf. Philip Massinger, 1631, *The Emperor of the East*, III.iii 'The poets' dreams of lustful fauns and satyrs'): thus *fauny* means "lusty, lustful"— an appropriate attribute for a fanny. The noun *fawney*, which was current in the eighteenth–nineteenth centuries, if not earlier, meant "ring" (from Irish *fáinne*, pronounced /fonʸ/, "finger-ring"); it was likely that, along with other words with this meaning, *fawney* was a euphemism for "vagina" (i.e., for what in our dialects is "fanny"). Add in the metaphor of *fawn* "young deer" for "young woman," mix in some influence from *fawn on* "lavish caresses on" (*O.E.D.*), and we have strong semantic associations between *fawney* and *fanny.*

Last, the bawdy meaning of *fanny* may have been boosted by the celebrated exploits of Fanny Hill, the heroine of John Cleland's *Memoirs of a Woman of Pleasure* (1748–1749).

Whether all, none, or any one of these possible sources truly identifies the emergence of *fanny* "vagina," we may never know.

The common use of quite a variety of proper names for a penis in contrast with the few used for a vagina may derive from the belief that a penis leads a life of its own to a much greater extent than its female counterpart.

15. USING BORROWED TERMS.

High-tech toilets turn Tokyo into a "Bottom Heaven"

> . . . .No longer are Japanese WCs the seatless holes in the ground of stereotypical Western dread. Consider, for example, the Washlet, a technological marvel that takes the guess-work out of cleaning up. A kind of toilet-bowl-cum-BIDET, the Washlet sprays one's DERRIERE with a water jet, then dries it with a blast of warm air. For added comfort, the seat is heated.
>   (*Time Australia*, November 28, 1988:68, our small caps)

Among Thais fluent in English, including doctors addressing well-educated patients, the English words *penis* and *vagina* are preferred to Thai terms. Many languages, English among them, use words borrowed from other languages to function as euphemisms. Recall Robert Graves' (1972:10) story of the hospitalized soldier who had been shot through the buttocks at Loos during World War I. When a lady visitor to the wards asked where he had been wounded, he replied: 'I'm so sorry, ma'am, I don't know: I never learned Latin.' Nowadays under these circumstances a New Yorker would be able to borrow from Yiddish and refer to his *tush!* The use of *anus* instead of *arsehole*, *penis* instead of *prick*, *vagina* instead of *cunt*, *labia* instead of *lips* (of the vagina [i.e., vulva]), and so forth, is the accepted practice when using Standard English (see Ch. 1). And, according to our criteria all these are euphemisms! Note that we don't resort to Latin to speak of breasts and nipples: this must be because they are not so taboo, for reasons given in Ch. 3.

Body-Parts and Animals

Why is it that the genitalia of both sexes are likened unto animals? Leach (1964) offers no solution to this question; his most relevant comment is 'Close animals may also serve as near obscene euphemisms for unmentionable parts of the human anatomy' (pp. 47–50). It would appear that the only possible answer to such a question is mundane: people perceive a similarity, and that perception somehow extends beyond one individual to become conventionalized into the concept or set of concepts evoked by the name of an animal. In the semantic field of genitalia there is an enormous amount of whole and partial synonymy: recall that there are reportedly 1,200 terms for "vagina" and more than 1,000 for "penis" (with or without testicles). At least some of these are nonce terms, and one can be sure that additional nonce terms are often created; but terms like *cock* and *pussy*, among others, are not. What we shall do here is first establish a probable basis for the similarity perceived between the animal and the body-part, and then suggest that plausible grounds for subsequent conventionalization of this perception into extensive nonliteral use of the animal name derives from the coalescence of a network of mutually reinforcing associations.

Penis as Cock: One Instance of the Bird as a Sexual Metaphor

> *Passer, deliciae meae puellae,*
> *quicum ludere, quem in sinum tenere,*
> *cui primum digitum dare appetenti*
> *et acris solet incitare morsus,*
> *cum desiderio meo nitenti*
> *carum nescio quid lubet iocari*
> *et solaciolum sui doloris,*
> *credo, ut tum grauis acquiescat ardor:*
> *tecum ludere sicut ipsa possem*
> *et tristis animi leuare curas!*

> Sparrow, my Lesbia's darling pet,
> Her playmate whom she loves to let
> Perch on her bosom and then tease
> With tantalising fingertips,
> (For my bright beauty seems to get
> A kind of pleasure from these games,

Even relief, this being her way,
I think, of damping down the flames
Of passion), I wish I could play
Silly games with you, too, to ease
My worries and my miseries.
(Catullus *Carmina II*, tr. by
Michie 1969)

e il giocar di mano con la bagatelle è meno difficile a imprare che non è lo
accarezzare lo uccello sì che ancora che non voglia si rizzi in piedi.

"And juggling with little balls is less difficult to master than how to caress the bird
[= "prick"] so that, even if desire is lacking, it gets erect."
(Pietro Aretino 1534, quoted in Frantz 1989:64)

There is a bird metaphor in English, as there is in some other languages, and it has
sexual overtones. Two exceptions to this are Australian *galah* and (the similar in
meaning) American *turkey* as in *You turkey!* The terms *bird* and *chick* can mean
"girl, woman," which is sometimes regarded as dysphemistic. *Hen*, commonly used
as a form address in Scots and occasionally in other dialects too, is not normally
dysphemistic; nor are *sparrow, dove, pigeon,* and (?archaic) *tit(mouse)*; indeed, used
with diminutives these are terms of affection. Shakespeare uses *hen* for wife or lover
and *cock* for husband or lover (cf. *The Taming of the Shrew* II.i.223–5; *Henry IV Pt I*
III.iii.54f). Used literally, *cock* is the generic term for a "male bird"; however, its
salient literal meaning is "the male of *gallus domesticus*"; and this cock, by repute,
rules the roost—whence *rooster*. This used to be what a man was expected to do (a
view that probably had some effect on the use of *Cock* as a euphemistic remodelling
of *God* in fifteenth–sixteenth-century English). Taking this perspective, for a man to
be *cock of the walk* demonstrates the very essence of manhood. From a different
point of view, the criterion of manhood—or rather maleness—is having a penis.
These two perspectives create a clear associative link between *cock* "gallus" and *cock*
"penis." A comparable metaphorical extension exists in Hindustani where *murga* is
ambiguous between "rooster" and "penis," and furthermore *murgi* is ambiguous
between "hen" and "cunt"—an ambiguity that does not, so far as we know, occur in
any dialect of English. However, there are dialects of English in which *cock* means
"cunt"; that is, in the southern United States, where the male organ is typically
called a *dick*. The use of *cock* for "vagina" may originate from Anglo-Saxon *cocer,*
cocor, cocur, Middle English *cocker* "quiver (case for arrows)," also "boots; leggings"
(cf. Baird 1981:223; *O.E.D.*).

   The original meaning of cock was, as it remains, a "male bird, in particular, a
rooster." There is, however, controversy over whether the almost archaic sense
"spout, tap, faucet" (still current in *stop-cock*) preceded or succeeded the sense
"penis." The obvious figurative connection between these last two senses raises the
question *Which of them is closer to the avian original?* and concomitantly *Which*
was the model for which? On rational grounds one would expect the direction of
meaning extension to go: MALE BIRD > MALE SEXUAL ORGAN > SPOUT, TAP. However, the textual evidence in the *O.E.D.* (12, 20) and Fryer (1963:38) suggests that
"spout, tap" is older by more than a century; and Partridge (1955, 1961) (with whom

we hesitate to quarrel) was firmly of the opinion that *cock* "spout, tap" was the direct literal model for *cock* "penis." But we will argue that the chronology matches rational expectation rather than the apparent historical record (i.e., that the direction of meaning extension was in fact MALE BIRD > MALE SEXUAL ORGAN > SPOUT, TAP).

What evidence do we have that the sense "male bird" gave rise directly to the sense "male sexual organ?" (1) First of all there is the evidence already cited of the association between the essence of manhood being defined as 'behaving as *cock of the walk*,' and the criterion of maleness being defined on 'having a penis.' (2) The second piece of evidence most likely reflects an association between cocks and manhood: the flesh and blood of a cock was believed to be a strong therapeutic and restorative agent and was recommended in Middle Dutch medical texts and in English recipes through until at least the seventeenth century, and probably much later (cf. 'A popular symbol of fertility and male virility, of "the resurrection of the flesh," insomuch as it is associated with the Easter liturgy, the cock, with its flesh and bones dismantled, minced, macerated, distilled, transformed by boiling and bain-maries into a revolting and smelly gruel, into an enigmatic elixir of life . . . a magical regenerative and revitalizing liquid') (Camporesi 1988:177). (3) Third, there is the folk myth that men often find themselves sexually aroused at cockcrow, whereas their womenfolk are more readily sexually aroused in the evening. Because the penis rises with the cock/rooster, an association is established between the two. (4) Fourth, the rooster is a randy creature that struts around with a neck that moves not unlike an erect penis on a walking man, whence one source for the idiom *keep your pecker up*. (5) *Pecker* is a frequent euphemism for "penis" (cf. 'PECKER The penis. A favorite word with many, inasmuch as the connotation is so innocuous') (Read 1977:65). (6) A *pecker* was also a narrow hoe used for digging holes when seeding; perhaps it got the name from the birdlike action involved in its use, and the association with seed. Certainly, {peckers, seeds, birds} form a natural set that intersects with another natural set whose members are {peckers, penises, seed, holes}. (7) Farmer and Henley (1890–1904) list *beak* as one of the synonyms for "penis," which is consistent with the bird as a sexual metaphor. (8) The similarity of the profiled penis-with-testicles to the outline of a bird gives rise to an image found in many languages, causing Barolsky (1978:107) to write of 'the iconographical tradition of the bird as phallus.' For instance, in Bahasa Malaysia and Bahasa Indonesia, the euphemism *burung* "bird" is used, particularly by women. The usual euphemism in Mandarin is *maque* "bird," but in the northwestern dialect they also use *jiba* and in the northeast *jizi* both of which mean "chicken." When addressing children, the diminutives *xiaoji* or *jiji* "little chicken" are used. In Sichuanese *choochoo* "bird" and also *yaya* "duck" are to be found, particularly when addressing little boys. In Italian there is *uccello* "bird" and *passerotto* "sparrow," the latter deriving from the Latin euphemism *passer* "sparrow." And, while citing Indo-European languages, it is notable that the German verb *vögeln* "fuck" is apparently based on Vogel = "bird"; and see Jongh 1968/9. We can add a further English example to all these. It occurs within a comic Mike Harding song 'Strangeways Hotel' (Rubber Records, 1975) purportedly about Strangeways Prison in Britain; the track was recorded live in the mid-1970s:

There were these two lads in a cell together, and one said to the other, "What are you in here for, Ron?"

"Robbin' a wagon. What are you in for Cedric?"

"Waggin' me robin." [*Audience laughter. Pause.*] He said, "Well, I only tried it once and as soon as I opened me raincoat she said 'No thanks. I smoke me own.'"

The raucous audience-response to *waggin' me robin*, BEFORE the contextual clarification that follows, proves that they well understood this manifestation of the bird metaphor. (9) One of the terms for the female pudend is *bird's nest*. The penis is the bird to enter this nest. For instance, Mark Twain's imaginary conversation (in mock Elizabethan English) between Queen Elizabeth I, Lady Helen (aged 15), Lady Alice (aged 70), and Francis Beaumont (aged 16):

LADY HELEN	Please your highness grace, mine old nurse hath told me there are more ways of serving God than by locking the thighs together; yet am I willing to serve him yt way too, sith your highness grace hath set ye ensample.
YE QUEENE	God's wowndes, a good answer childe.
LADY ALICE	Mayhap 'twill weaken when ye hair sprouts below ye navel.
LADY HELEN	Nay, it sprouted two yeres syne. I can scarce more than cover it with my hand now.
YE QUEENE	Hear ye that, my little Beaumont? Have ye not a little birde about ye that stirs at hearing tell of so sweete a neste?

(Clemens 1968:21f)

(10) Finally, there is the dysphemistic reference to a flaccid penis as a *worm*, and worms are food for birds; birds like worms; women like . . . and so on. On its own, this last piece of circumstantial evidence would certainly be inadmissible, but it just might have some validity given the nine arguments that precede it, all of which evidence an association between "(male) bird" and "male sexual organ." And there is more.

The ambiguity of *cock* in English is found elsewhere in Indo-European. Baird (1981) argues convincingly that Latin *gallus* had a meaning "penis" from classical times, through Vulgar Latin, and that this meaning was maintained in Italian and Spanish. Interestingly, though, Latin and its daughters do not generally associate *gallus* with taps. Baird also shows that French *coq* has many sexual associations, citing words like *coquille* ("shell, ornament in the shape of a bird's beak, vagina"), *coqueter* ("copulate with a 'chick' "[! The translation is Baird's]), *coquer* (Lyons dialect) "kiss or embrace as the cock does hens," *coquelier* ("run after young girls"), *coquine* ("prostitute, male homosexual"), *coquard* ("ridiculous old beau"), *coquardeau* ("male flirt"), and *coqueluche* ("ladies man"). There are also German *Hahn*, which means or has meant all of "rooster, penis, spout/tap"; and Swedish *kuk* and Danish *kok* meaning both "rooster, penis," but not "tap." And finally there is the early fifteenth-century English poem "I haue a gentil cook," which is contemporaneous with or slightly earlier than 1481, the earliest recorded use of *cock* in the sense "tap."

[1] I haue a gentil cook, crowyt me the day;
   He doth me rysen erly my matyins for to say.
[2] I haue a gentil cook, comyn he is of gret;
   His comb is of reed corel, his tayil is of get.

> [3] His leggis ben of asour so gentil and so smale,
> His sporis arn of syluer qwyt into the wortewale.
> [4] His eynyn arn of cristal lokyn al in aumbyr,
> And euery nyht he perchit hym in myn ladyis chaumbyr.
>
> (Silverstein 1971:129)

This delightful lyric can, of course, be interpreted quite literally as a poem about a pet rooster; but there is no doubt that it allows for a lewd interpretation, too. A very loose rendition is as follows:

> "[1] I have a fine cock that wakes me early every morning. [2] He comes of good stock. His [?uncircumcised] tip is coral red and his root is buried in black hair. [3] He has fine blue veins running up the side of him, like legs. When erect, he's milky white underneath where he joins my scrotum. [4] His eye discharges spunk the colour of crystal, but more often amber-coloured piss. And every night he enters my lady's quim."

This gross rendition destroys the subtlety, wit, and beauty of the original. The interpertation of stanzas [3] and [4] is admittedly far-fetched, and we justify it as follows. This bawdy poem cloaks its bawdiness behind a source domain: the description of a rooster. The lyric is completely consistent with this description, and like other bawdy puns it relies on one or two salient aspects of the source domain to evoke the target domain, in this case a penis. We liken the process to the construction of two parallel images: the image of a rooster and the parallel image of a penis with the attributes of the rooster transferred to those attributes of a penis that are most consistent with some coherent image of it being constructed in the interpreter's fertile mind. All understanding is a constructive process on the part of an audience (cf. Black 1984, Lakoff 1987, inter alia); and with puns and metaphors, it is doubly constructive. Two people hearing the same literal statement may well put different constructions upon it (that phrase is telling!); two people interpreting the same metaphor have far more scope to create different constructions. When interpreting a poem like this one, the two different constructions can both be right.

The poem 'I haue a gentil cook' offers very strong circumstantial evidence that, in Engish, *cock* "penis" is at least as old as *cock* "tap." Given that *cock* in the "tap" sense was likely to have been written down soon after it became current, whereas the taboo sense was probably current long before it was written down, the taboo sense is very probably older and part of the Indo-European heritage of English. Ashley Montagu writes

> that *God damn*, that favorite expression of the Englishman, though it is known to have been in common use at the beginning of the sixteenth century [for which Montagu has shown conclusive evidence] is not recorded in an English work until the end of that century. This clearly shows how much older are many of the oaths with which we have dealt and shall have to deal than the date of their first written or printed appearance.   (Montagu 1967:124)

There can be no doubt that taboo words are regularly much older than their first recorded use.

Whatever the historical truth, the reason that Americans and many Australians use *rooster* for "gallus domesticus" where the British still use *cock* is exactly because speakers readily correlate *cock* "bird" and *cock* "penis" today. As we remarked in Ch.1, nontaboo homonyms are often abandoned in this way—even though *cock* seems to have had a long innings (for those unfamiliar with it, this is a metaphor from the game of cricket).

Penis as Ferret, Weasel, or Rat

With the exception of *cock*, animal names for the "penis" are truly dysphemistic: it has long been called *rat* and was until recently known as a *ferret* (ferrets are much less common domestic animals than they once were). In Britain and Australia, especially, the ferret was put into rabbit holes to catch rabbits; and, as we shall see later, rabbit burrows and rabbits (cunnies) have long been euphemisms for the vagina: the symbolism of *ferret* is obvious. In American, the counterpart is *weasel*, and we note the dysphemistic verb *to weasel*.

We take it that the rat is known for running in sewers and other holes and tunnels, and this is an explanation for the use of *rat* for "penis"; but why a rat rather than a mouse or a mole? *Rat* is probably preferred to *mole* because a rat has a long tail and moves quickly: it is a standing jibe that copulating men move too quickly for their female partner. *Rat* is probably preferred to *mouse* for two reasons. The first is because it is more hateful and more greatly feared than a mouse. Second, and more significantly, there is the association with the epithet 'rat' as in *He's a rat*, which is represented in the *Wham, bam, thank you, ma'am* attitude widely attributed to men by women.

Why is the Vagina Called a Pussy?

We have already remarked on the salience of pubic hair on a woman's body, and speculated that it motivates furry animal nicknames; but this still leaves the specific question *Why is the vagina called a pussy?* unanswered. Folk wisdom likens men's behavior to dogs' (running in packs, chasing cats) and women's behavior to cats' (whimsical, picky, and they scratch, etc.). Certainly *cat* has long been a dysphemism for "woman." Grose 1811 lists not only *cat* but also, consecutively, 'TIB. A young lass. TIBBY. A cat. *Tib* has long meant a young woman of humble origins, and according to Partridge (1955:205) a *tib-cat* is a 'female cat—the opposite to a *tom*cat.' Perhaps this folk association, plus the fact that both cats and pussies like being stroked by approved persons (and there is perhaps a fishy connection, too; see the upcoming discussion) constitutes all the explanation there is. It may have helped that common French terms for "vagina" are *le chat* "the cat" and more especially *la chatte* "the female cat" (this term is avoided in modern French), and this would have been widely known. But there is a more specific source: the earliest recorded variant for "vagina" is not *pussy* but *puss*, which is most likely a colloquialization of the frequently used euphemism (?dysphemistic euphemism) *purse*. For instance,

> Garp nodded. The next day he brought a bottle of wine; the hospital was very
> relaxed about liquor and visitors; perhaps this was one of the luxuries one paid for.

'Even if I got out,' Charlotte [the whore] said, 'what could I do? They cut my purse
out.' She tried to drink some wine, then fell asleep. Garp asked a nurse's aide to
explain what Charlotte meant by her 'purse', though he thought he knew. The
nurse's aid was Garp's age, nineteen or maybe younger, and she blushed and
looked away from him when she translated the slang.
  A purse was a prostitute's word for her vagina.
  'Thank you,' Garp said.

(Irving *The World According to Garp* 1979:158f)

We have already reviewed evidence that the vagina has been seen as a source for or
store of wealth, and remarked on the existence of pairs of phonologically similar
doublets in which the member with the short lax vowel is the more colloquial (cf.
hoss–horse, bust–burst, cuss–curse, buss–burse "purse", etc.). To this list, we believe should be added *puss–purse*. We would speculate that the pronunciation /pʊs/
was preferred to /pʌs/ (cf. /kʌs/) in order to dissimilate *puss* from *pus* and make the
association with cats. This sense of *pussy* is possibly reinforced by a belief that *pussy*
is end-clipped rhyming slang *pussycat = twat*; but the use of *puss*, at least, predates
rhyming slang by about a century; and *twat* is pronounced /twot/ by many people.
(For the authors /twot/ is the pronunciation for the insult, e.g., in *You stupid twat!*;
whereas in literal use, e.g., in *her twat*, it is /twæt/.)

There is an interesting link between *pussy* and *kitty* and *purse*. Like a purse, a
kitty is something one puts money into; *kitty* is also a diminutive for *kitten*, and
Partridge (1970) glosses one sense of it as 'The female pudend: mostly [used by] card
players: late C. 19–20. Suggested by the synonymous *pussy*.' Partridge overlooks the
use of *puss, kitty,* and *kitten* in Thomas D'Urfey's song 'Puss in a corner' (1719 vol.
2:80–82); and he also ignores the sixteenth–seventeenth century phrase *Kit has lost
her key* meaning "menstruate," though he does list the synonymous nineteenth–
twentieth century *Kit has come* (see earlier discussion): in the earlier idiom, 'kit' may
be interpreted as "vagina." Moreover, *kit, kitty,* and *kittock* were all used in the
fifteenth–eighteenth centuries as general terms for young women (they derive from
Kate), and like many of their synonyms, they were euphemisms for 'a woman of
loose character' (*O.E.D.*). In other words, these terms denote a sex object: nothing
more nor less than, to put it crudely, pussy.

In the Middle Ages there were numerous recorded instances of women in
convents affecting feline behavior, believing themselves to have metamorphosed
into cats. Their male counterparts, cloistered men, fantasized themselves as other
animals, especially the cock! (See Ch. 7 on explanations for such psychotic disorders
in the middle ages.) Apparently, this type of monomania is particularly prevalent
among groups of people committed to celibacy. It is surely no coincidence that the
suppression of normal sexual behavior should lead to delusions manifesting sexual
icons. While the psychological and psychiatric assessment of these abberations is no
concern of ours, it is possible that the study of these sorts of fantasies would help to
shed some light on our understanding of *cocks* and *pussies* and of this little studied
field of the lexicon.

The foregoing discussion has uncovered an associative network linking the following modules: women and cats; young women and kittens, kits, or kitties; *kit* and

vagina; kitties and purses; *purse* and *puss*; puss or pussy and cat or kitten; *pussy* and *vagina*. The phrase *his kit and caboodle* can euphemistically denote male genitalia, and there may be some mutual influence in this network from the association of *kit* with genitalia of both sexes. In addition we observe that cats, kitties, or pussies are furry things, and it is notable that the furry patch of pubic hair on a woman is the most salient visual indication of her pudend. Remark, too, that the opening of a purse resembles a vulva (vagina); consequently Japanese *isoginchaku* "sea purse, *actinia malacoterem*, a kind of sea anemone," also has the senses "round coin purse (which when squeezed, opens the slit)" and "vagina" (cf. Solt 1982:78); furthermore, purses hold money, and a number of euphemisms for the vagina recognize it as a source or store of wealth. The associated modules we have identified coalesce in a mutually reinforcing way to create a lexical network that motivates the use of *pussy* to mean "vagina."

Vagina as Beaver

The female pudend is also known as a *beaver*: a beaver was a "full-beard", which invokes the salience of pubic hair on women; and *beaver* exemplifies another extension from this to the full kit and caboodle. For instance,

> The picture Garp looked at in the dream was considered among the highest in the rankings of pornographic pictures. Among pictures of naked women, there were names for how much you could see. If you could see the pubic hair, but not the sex parts, that was called a bush shot—or just a bush. If you could see the sex parts, which were sometimes partially hidden by the hair, that was a beaver; a beaver was better than just a bush; a beaver was the whole thing: the hair and the parts. If the parts were *open*, that was called a split beaver. And if the whole thing *glistened*, that was the best of all, in the world of pornography: that was a wet, split beaver. The wetness implied that the woman was not only naked and exposed and open, but she was *ready*.
> (Irving *The World According to Garp* 1979:318, emphasis in original)

In addition, a *beaver* was once a kind of hat; so called because it was made from felted beaver fur. A hat is a concave object into which a man puts his head; and the *glans penis* (note the Latin) is often referred to as its head. Hence, we find in Grose (1811): 'HAT. Old hat; a woman's privities: because frequently felt.' It is not clear whether there is any direct connection, but it is nonetheless worth mentioning that, in medieval Europe, beaver oil was regarded as an aphrodisiac.

In Ch. 1 we mentioned the hilarity that may arise when someone inadvertently brings to mind a taboo topic in circumstances to which it is very inappropriate; this happens because of the salience of taboo interpretations over other interpretations of an expression. Here is a true story exemplifying this:

> The highest award in boy scouting is, or was in the sixties, The Silver Beaver. It was the cause of endless (suppressed) merriment when Grandfather received this coveted award.

You can see why, can't you?

Cunnies, Bunnies, Hares, and Cunts

Coney (rhyming with *honey*) was the word for "rabbit" until the late nineteenth century, when it dropped out of use because of the taboo homonym meaning "cunt" (a synonym was *cunny-burrow*, hence the picturesque term for a penis as the *cunny-burrow ferret*; cf. Farmer and Henley 1890–1904). In Latin, "rabbit" is *cuniculus*, and its burrow *cuniculum*; end-clip either and you are left with *cuni[e]* (spelled variously as *coney, cony, conny, conye, conie, connie, conni, cuny, cunny, cunnie* in Robert Greene's 1591 book *A Notable Discovery of Coosnage*, cf. Baugh and Cable 1978:208). There may be influence, too, from Latin *cuneus* "wedge or triangular shaped"—like the vagina. *Cunny* "cunt" derives from Latin *cunnus* (probably as a euphemism), which is retained in modern *cunnilingus*; there may also have been some input from French *con*, also derived from Latin *cunnus* and also used for the bawdy-part from (at least) the fourteenth century (cf. Boch and von Wartburg 1975, Picoche 1979).

What about *bunny?* Playboy's 'bunnies' and 'Bunny Club' followed a long tradition going back beyond eighteenth-century London's 'Cunny House' (cf. Leach 1964:50). Though the evidence is unclear, it may well be that *bunny* was a euphemistic remodelling of *cunny*; it was a term for rabbits, rabbit tails, bony lumps on animals (reminiscent of the *mons veneris*) as well as an affectionate name for a woman. On the other hand, *rabbit* itself is usually a term of abuse when ascribed to a woman; for instance,

> Lady Patricia had loved Boy for several years before he had finally proposed to her, and [she] had indeed quite lost hope. And then how short-lived was her happiness, barely six months before she had found him in bed with a kitchen maid.
> "Boy never went out for the big stuff," I once heard Mrs. Chaddesley Corbett say. "He only liked bowling over the rabbits, and now, of course, he's a joke."
> (N. Mitford 1979:347)

There can be little doubt that *bunny* was also a euphemism for "cunt": Grose (1811) has 'BUN. A common name for a rabbit, also for the monosyllable' [= "cunt"]—the term is still used today (cf. Aman and Sardo 1982:25). (According to Partridge 1970, 'crumpet' as in *a bit of crumpet*, i.e., what Americans would denote by *a piece of ass*, is a euphemism that derives from *crumpet* with the sense "buttered BUN.".) *Bun* was also the name for the tail of a hare (associations: hare–hair–pubes–cunt; tail–cunt) (and cp. *do one's hair in a bun*). *Bun* was also used for "squirrel," while *squirrel* was one term for a prostitute; *bunter* was another (cf. Grose). There is even a link between rabbits, hares, and cats (pussies): Grose notes that *ma(u)lkin* or *mawkin* is a "cat or awkward woman" and in Scotland "a hare"; Partridge, in his 1970 supplement (p. 1351), lists "rabbit" as one meaning for *pussy* in Australian, though we can find no Australian who has heard this usage; the *O.E.D.* tells us *puss* meant, *inter alia*, "hare." There is evidently an interesting set of connections between cunnies and bunnies and hares and pussies.

To return to *bunny*, it was also a dialect term for "an opening or ravine in a cliff"—which is suggestive. If the initial 'b' was indeed some sort of euphemistic remodelling device in the case of *bunny*, consider (not the baseball term, but rather the nautical term) *bunt* "a cavity, pouch or bagging part of a sail or net; the funnel of an eel-trap": is that also remodelled? For reasons discussed in Ch. 2, rabbits were one of the few land animals that used to be tabooed by fishermen, along with talk of women. There can be no doubt that there has been a very strong folk association between bunnies and /kʌnɪz/ and cunts: what is responsible for it is the fantasy of these frisky furry creatures fornicating fecundly in their holes while hares hop happily around them.

Digression: A Short History of Cunt, of Queynte, and of Quim

The origin of the noun *cunt* is uncertain; but it is cognate with Old Norse *kunta*, Old Frisian and Middle Low German *kunte*, and Middle Dutch *conte*. Partridge (1984) suggests that it may be historically related to *cow*, and this is not impossible. The *O.E.D.* (1989) has *cu* as a form for "cow" until the fourteenth century and *ku* as an alternate from the thirteenth; the plural was *cun*, though more commonly *kyn* and *kine*; in one sense, *cow* is 'Applied to a coarse or degraded woman. Also, loosely, any woman, used esp. as a coarse form of address.' The term *cunt* is undoubtedly a worse epithet than *cow*, but the semantic link between the two is as obvious as the phonetic link between their early forms. *Cunt* was already a well-established word in Early Middle English—so it presumably existed in Old English. Strangely enough, it turns up in medieval place names: for instance, there were a number of *Gropecuntlanes* in the thirteenth century—a name suggesting a disreputable lovers' or harlots' banging-ground (cf. Fryer 1963:46; *O.E.D.* 1989). It also appeared in people's names; for example, McDonald (1988:36) names *Simon Sitbithecunte* (1167), *John Fillecunt* (1246), and *Bele Wydecunthe* (1328)—among others; these are the real life counterparts to *Biggus Dickus* in Monty Python's film *Life of Brian*. However, such names almost certainly seem even worse today than they sounded in the Middle Ages because *cunt* does not appear to have been intrinsically dysphemistic then. Thus it was used in *Lanfranc's Science of Cirurgie* (ca. 1400) where *vagina* would be required today, cf. 'In wymmen the necke of the bladdre is schort, & is maad fast to the cunte' (quoted in *O.E.D.* 1989).

In some English dialects, between the thirteenth and the late nineteenth century, this bawdy-part term was homophonous with the adjective *quaint*—which, from the thirteenth–sixteenth centuries was also spelled *queynte* (among other ways); in fact Chaucer used just this spelling in various works (see *O.E.D.*). Chaucer's famous use of *queynte* meaning "cunt,"

> And prively he [Nicholas] caughte hire [Alison] by the queynte
>   ('The Miller's Tale', 1.3276 of *The Canterbury Tales*)

uses exactly the same spelling. Florio (1611) in his Italian–English *Dictionarie* used the spelling we use today: 'Fíca, . . . *Also used for a woman's quaint.*' In those far off days, the adjective *quaint* meant much the same as it does today but, if anything, it was more laudatory. If there is a phonological link between *cunt* and *quaint*, it may lie with labialized onset to Old English *cwene*, Middle English *que(y)ne*, Modern English *quean(e)* a dysphemistic term for a "woman" that came to mean "whore" (the nineteenth-century sense "male homosexual" is irrelevant to this discussion). There is also the labialized onset to *quim* "cunt," which may also have been influenced by *queane*, but whose more likely source was *queme* (sometimes spelled *quim* from the seventeenth century) whose nontaboo sense was the utterly appropriate "something pleasurable, snug, intimate." This seems more likely than Grose's dysphemistic suggestion that *quim* comes from Spanish *quemar* "burn." Finally, it is not impossible that Welsh *cym* "cleft, valley," and its cognates in various English dialects, had some part to play in the history of *quim*.

The American Ass

It is possible that American *ass* is coincidentally an animal name as well as a body-part term, just as the homonymy of *tits* "birds" with *tits* "breasts" is apparently fortuitous—although it may get some support from the bird as sexual metaphor. Both *ass* and *arse* have, since at least the sixteenth century, been used to describe someone stupid: hence in British English there is the doublet *silly ass* and *silly arse*. The dysphemism arises because (1) the ass/arse is a tabooed body-part; and (2) this part of the anatomy is blunt and at the opposite end from the head, which renders it appropriate for the seat of stupidity. The British doublet, and the collapsing of *ass* and *arse* in American, are presumably a result of their phonological similarity: *ass– arse* is one of the set of doublets listed earlier.

The most likely explanation for the use of *ass* to mean "vagina" has nothing to do with animals, but is part of a regular series of terms that are used, though not necessarily in the same dialect of English, for both the "vagina" and "backside."

Fish and Bivalves

From time to time we have mentioned 'fishy' associations for the vagina. This is because a vagina may sometimes have a 'fishy' odor; a fact that has given rise to expressions such as *fishfinger* "digital stimulation of a woman", *fishing* or *angling* which mean either that or "copulation," and *fishbreath* "oral sex" (see earlier and Allen 1987). There is also the image of the (?horny) vagina as a *pond*, and the old terms like the fish name *ling* for "vagina" and *hook* for "penis." The visual metaphors of the vagina as *clam* and *oyster* are fairly commonplace, and link up with the food metaphor. There is also a link, probably fortuitous and irrelevant, with *hen*, which is the common name for at least one species of bivalve, *venus mercenaria*. These bivalves also fall in with a metaphor for "vagina," which suggests a mouth ready to snap up a penis (e.g., *bite*, *snatch*, and *snapper*—the last of these being, perhaps coincidentally, the name for a species of fish).

Although the connections here are not so extensive as they are for cunnies, bunnies, hares, and pussies, they nonetheless reveal a complex associative network of interrelationships within a semantic field, a network that is sufficient to reinforce the perceived similarities between the vagina and fish or bivalves.

Concluding Remarks on Animal Names and Body-Parts

In the foregoing pages we have sought to provide some explanatory detail of the animal terms that English speakers use to talk about sexual organs. Much of the detail comes from conceptual and perceptual associations between denotata: on the one hand genitals; on the other, the creatures whose names are used to nonliterally denote genitalia. We began by examining the history and motivation for the wide-spread use of *cock* for "penis," finding it to be one instance of the far-reaching 'bird as a sexual metaphor.' We found evidence of a number of influences from several different sources coalescing to establish the relevant figurative meaning of *cock*. Much less common than *cock* are the dysphemistic terms *rat* and *ferret* for "penis"; and there is a corresponding lack of networked associations to support widespread conventionalization of these particular terms. We then moved on to the semantic field encompassing the use of names for furry animals to denote the human female pudend, which led to an examination of the history and motivation for such words as *pussy, beaver, cunny* and *bunny*; and then various 'fishy' terms for the same body-part. Where there is a network of influences coming together, as in the case of *pussy*, the figure has much greater chance of surviving and being widely used—thus rendering it (paradoxically) a dead metaphor. Conversely, where there is a less supportive semantic network, as in the case of vagina as *beaver*, the terms have much narrower currency. It appears that the *cunny, bunny* set we discussed, is an exception to this hypothesis; but it would seem that when *cunny, coney* "rabbit" dropped out of use, *cunny* "vagina" gave way to the co-existing *cunt*, whose etymology we also discussed. It was argued that *ass* "backside; vagina" is coincidentally an animal name; but the correlation of *ass* with *arse* is part of a regular series of doublets in English, and the naming of the vagina with the same term as the backside is not uncommon.

Summary of Chapter 4

In this chapter we examined the English lexicon for circumventing or defying taboos on mentioning certain body-parts and their functions. This began with the vocabulary of personal freshness and personal hygiene, moved on to the lexicon of defecation and urination, before embarking on an extensive analysis of the lexicon, past and present, of menstruation. We briefly looked at other aspects of gynecological euphemism and dysphemism before turning to the vocabulary for sex, tabooed body-parts, and body-part functions of both sexes. The survey surely justifies our opening remark that there is more synonymy in this area of the English lexicon than in other parts of it, and that the terms used range from the keenly inventive to the crass.

We have tried to impose some order on the wealth of terms by grouping them into nondiscrete classes. We sometimes included within our classification, group-ings into 'euphemisms,' 'neutral terms,' and 'dysphemisms.' Recall that these attri-butions are given according to the middle class politeness criterion (see Ch.1). This would be the criterion operating for the 'shy ladies' written of in this passage (TOTO is Japan's largest maker of toilets):

> For shy ladies who do not want to waste water but still maintain decorum—
> according to TOTO's investigations, [Japanese] women flush an average of 2.5

times a visit to drown out potentially embarrassing or offensive noises—there is the *Oto Hime* (Sound Princess), which plays a recording of flushing water.

('King for a Day In a Small Room With a View: *High-tech toilets turn Tokyo into a "Bottom Heaven"* ', *Time*, November 28, 1988:68)

In the course of this chapter we have sought to provide some explanatory detail of the terms English speakers use to talk about bodily effluvia and body odors, sex acts, and sexual organs. Much of this detail comes from conceptual and perceptual associations between denotata, from semantic associations between items in the same semantic field, some from phonetic or morphological similarities, and some was simply historical. We hypothesize that widespread conventionalization of this vocabulary derives from the coalescence of a demonstrable network of associated influences. If this hypothesis is mistaken, there are some amazing coincidences to be explained away. If, on the other hand, it is correct, then historical semantics should look out for a number of sources coming together to establish the meaning of a term.

We hope that this chapter has thrown some light into an area of vocabulary which is principally part of oral culture, which most people find fascinating, and yet which has received comparatively little scholarly attention.

CHAPTER
—5—

Ways of Being Abusive:
Taboo Terms as Insults, Epithets,
and Expletives

The ordinary reaction to a display of filth and vulgarity should be a neutral one
or else disgust; but the reaction to certain words connected with excrement and
sex is neither of these, but a titillating thrill of scandalized perturbation.

(Read 1977:9)

Why Taboo Terms Are Good for Insult, Dysphemistic
Epithets, and Expletives

In this chapter we examine the dysphemistic use of taboo terms in insults, epithets,
and expletives; and we begin by seeking the reasons why taboo terms should prevail
in these functions.

Insult affronts the target's face and thus destroys social harmony; therefore, terms
of insult are not only dysphemistic, but are also socially tabooed. If insult is taboo,
then terms tabooed for other reasons are ready-made components for insults. Consider imprecatives like *Eat shit motherfucker!* or *Why don't you fuck off!*, which are,
by all the usual criteria, kinds of directives (cf. Searle 1979; Bach and Harnish 1979;
Allan 1986a): they purportedly seek to cause the addressee to do something, even
though they are not expected—not even meant—to be taken literally; it is their
illocutionary point that is of primary relevance, which in the case of the first
example is simply a recognition by Hearer of Speaker's dysphemistic attitude to
Hearer. The same presumably holds for *God damn you!*

The very use of taboo terms—whether as insults, epithets, expletives, or even
descriptives—will often insult Hearer, particularly where there is a difference in
status and/or wide social distance between them and the user. For example, at least
one reader of a draft version of this book has been upset by our occasionally USING
terms like *cunt* as well as just CITING them. To be upset in this way is to feel insulted:

there has, of course, been no affront directed to a reader personally; in the view of such a reader, the affront lies in us not having observed appropriate politeness conventions for this kind of expository text.

Taboo terms make good dysphemistic epithets for the same reason that they make good insults. At least one occasional reason for using taboo terms—whether for insult, as epithets, or expletives—is to savor Hearer's adverse reaction. A related reason is for Speaker to flaunt his or her disrespect for social convention (this is presumably one motivation for writers of graffiti); though in verbal duels such as 'the (dirty) dozens' (see later discussion), this inverts to a respect for the social convention of the game.

There is presumably no need to try to account for why people (deliberately) use insults like *You are a stupid little shit!* or dysphemistic epithets like *It's a pain in the arse*: it is because they do not like who they are addressing, or who or what they are talking about. It is not so obvious why people use expletives, and in particular why so many taboo terms have expletive function. Expletives are kinds of exclamatory interjection; and, like other interjections, they have an expressive function (cf. *Wow!*, *Ouch!*, *Oh dear!*, *Gosh!*, *Shit!*). Unlike typical expressives such as greetings or apologies, interjections (including expletives) are not normally ADDRESSED to Hearer; at best Hearers are treated like ratified participants, and at worst as over-hearers (bystanders) and not Hearers at all. Instances of expletives, and other interjections uttered without an audience, are expressions of autocatharsis. Where they are used with an audience of ratified participants or bystanders they are displays of autocatharis (i.e., the illocutionary point is to display a particular attitude or degree of feeling). Even private autocatharsis can be regarded as a display to oneself; so that a private expletive has the same illocutionary point as a public one.

Because taboo terms make good dysphemisms they make good expletives. Furthermore, the very fact that a term is taboo may improve its value as autocathartic: the breaking of the taboo is an emotional release (cf. Bailey and Timm 1976:444; Brain 1979:89f; Eckler 1986/7; MacWhinney et al. 1982). It is what Read (quoted at the beginning of this chapter) so aptly described as the 'titillating thrill of scandalized perturbation' that provides the autocatharsis Speaker wants in order to cope with the situation that provoked the expletive. This very strong motivation no doubt accounts for the consistent historical failure of legislation and penalties against swearing; such measures 'have only had the effect of driving it under the cloaca of those more noisome regions, where it has flourished and luxuriated with the ruddiness of the poppy's petals and blackness of the poppy's heart. It has never been successfully repressed.' (Montagu 1967:25, getting a little carried away.)

It should be said that autocatharsis through swearing (the use of taboo epithets) is regarded as a conventional way of violating a taboo: a convention that is not socially approved of, but one that is grudgingly excused by society. In both public and private, an individual's self-control will determine the choice of vocabulary used. Where the situation provoking the autocatharsis is pleasing and there is no call for dysphemism, it is less likely that a taboo term will be used. However, there are situations under which euphemistic uses of taboo terms normally reserved for abuse are appropriate, and we discuss these later. Where a situation provokes dysphemism,

Speaker can choose between using a euphemistic dysphemism like *Oh dear!* or a straight dysphemism like *Fuck!*

There are people who use expletives and taboo epithets so frequently that one cannot persuade oneself they are autocathartic. Typically, such a Speaker will be male (cf. Coates 1986:20–22,108f), and numerous surveys leave no doubt that in nearly all societies, if not all, males 'swear' more and use more 'obscene language' than females (e.g., Brain 1979; Ide 1982). Against this backdrop, it is apparent that the use of taboo terms in epithets and expletives is sometimes a display of macho—a sign of masculinity. Brophy and Partridge (1931), writing of the use of *fuck* by British soldiers in World War I, said:

> so common indeed [was *fuck*] in its adjectival form that after a short time the ear refused to acknowledge it and took in only the noun to which it was attached. . . . Far from being an intensive to express strong emotion it became a merely conventional excrescence. By adding -ing and -ingwell, an adjective and adverb were formed and thrown into every sentence. It became so common that an effective way for the soldier to express emotion was to omit this word. Thus if a sergeant said, 'Get your -ing rifles!' it was understood as a matter of routine. But if he said, 'Get your rifles!' there was an immediate implication of emergency and danger.
>
> (Brophy and Partridge 1931:16f. Quoted in Read 1977:14)

Edward Allen has referred to such speech behavior as 'Fuckinese':

> From the doors of meat companies, the cigarette-wounded larynxes of overweight men shouted utterances such as, "We gotta hurry the fuck up and get these fuckin' boxes on that fuckin' truck or we're all gonna be fucked!" The language of 14th Street is so dependent on the all-purpose word *fucking* that it can't really be called English; rather a separate dialect best referred to as "Fuckinese."
>
> (Allen *Straight Through the Night* 1989:34)

Where a taboo term such as *fuck* is bleached of its taboo quality, it becomes a mere particle and loses all its standard force. One must presume that under such circumstances the autocathartic value of the corresponding epithet is also reduced, and that either alternative expressions will be invented, or some other form of catharsis will be sought. We are put in mind of Shakespeare's aphorism:

> If all the year were playing holidays,
> To sport would be as tedious as to work.
> (*Henry IV Pt.1* I.ii.192)

One widely held view of people who use Fuckinese (i.e., swear a lot) is nicely brought out in the following passage from Edward Abbey's novel *The Monkey Wrench Gang*:

> "Why," [Bonnie] said, impaling Hayduke the oaf on the beam of her laser glare, her casual scorn, "why is it"—blowing smoke rings into his hairy face—"that you can never speak a single complete English sentence without swearing?"
>
> Smith laughed.
>
> Hayduke, under the hair and sunburned hide, appeared to be blushing. His grin was awkward. "Well, shit," he said, "Fuck, I don't know, I guess . . . well, shit, if I can't swear I can't talk." A pause. "Can't hardly *think* if I can't swear."

"That's exactly what I thought," said Bonnie. "You're a verbal cripple. You use obscenities as a crutch. Obscenity is a crutch for crippled minds."

(Abbey 1975:153f)

There are people, and Hayduke may be among them, who use obscenities as 'filled pauses'—where other people might use *ums*, *ers*, *you knows*, and the like. Judging from other parts of *The Monkey Wrench Gang*, Bonnie is probably not especially distraught by the obscenities themselves (though some people would be); what seems to motivate her here is Hayduke's violation of the cooperative maxim of manner (and perhaps also quantity, though we think not). When she accuses Hayduke of being "a verbal cripple who uses obscenities for a crutch," she is reacting to the fact that his frequent use of obscenities distracts her from readily and easily understanding the rest of what he is saying. Hers is a fairly common reaction to the impositive face affront she perceives to be inflicted on her.

Dysphemistic Forms of Naming and Addressing

In Ch. 2 we discussed euphemism in naming and addressing; here, we look at some aspects of dysphemistic behaviors in these areas. In other sections of this chapter we shall look at the dysphemistic use of tabooed bodily organs and effluvia in naming and addressing, and dysphemistic ascriptions of mental and physical inadequacy in naming and addressing. We mentioned in Ch. 1 that Hearer-or-Named may be named or addressed dysphemistically using animal names, most of which have their own peculiar denotation. There are constraints on the application of these metaphors. Names of female animals can normally be used only in naming or addressing women and male homosexuals: for instance, a *cat* is typically a "vicious and/or scratchy woman"; a *bitch* is a "(usually nasty) woman held in contempt"; a *vixen* is a "cunning, perhaps sneaky, woman"; *cow* and *sow* don't differ much, they generally denote a "woman disliked, who is typically doltish"—and there are connotations of being fat, too (cf. the commonly used *fat cow/sow*). (*Silly*) *old bat* would normally be used of a woman past middle age; *bat* in this sense does not occur unmodified. (The predicative *bats*—e.g., *You're bats*—is used of either sex, and probably derives from the figure *have bats in the belfry*; i.e., "be mad, nuts".) Some animal names are typically used of men: *mongrel*, *cur*, or *swine* denotes a "vicious, nasty fellow, held in contempt" (cp. *cat* and *bitch* of women); a *fox* denotes a "cunning man" (cp. *vixen*); a *bull* is for a "big, often rather clumsy, man"; a *goat* or *ram* can be used of a "horny/randy man." Most animal names can be applied to either men or women. According to *Webster's New Collegiate Dictionary* one sense of *dog* is "worthless person"; we have attested *dog* and *bow-wow* used of a woman otherwise described as "bone ugly." A *louse* is "someone unpleasant, irritating, someone Speaker wants to be rid of." Used of men, *louse* and *rat* usually denote "unfaithful cad." A *mouse* is more often applied to women then men, partly because it is not normally applied to big people; it denotes "someone insignificant and timid." *Coot*, *turkey*, and *galah* are used of someone stupid, so too are *goat* and *ass* (or *donkey*); *donkey* and *mule* denote "someone stubborn;" a *chicken* is a "scaredy cat, coward," mostly used of men. A *pig* is "someone rude, uncouth, slovenly," and for two centuries or more has been used to denote "policeman, constable" (cf. Grose 1811). In some U.S. dialects *pig* and *oinker* are used of "an overdressed (or perhaps underdressed) woman with too much make-up, who looks like a hooker." An *ape* is "someone uncouth"; while a *monkey* is "someone mischievous," usually a child—it is a tease rather than a dysphemism. A *snake* is "untrustworthy, sleazy, someone who will spread poison about other people." Worms and toads have always been despised, even loathed, perhaps because of their association with dirt and decay, perhaps because they are unpleasant to touch. Applied to humans the words *worm* and *toad* imply "someone who is loathsome, who crawls, is sycophantic"; hence *toadyism*. Similar meaning attaches to *creep* and *crawler*, terms that derive from animal behavior. An *insect* is "someone insignificant, beneath contempt"; a *parasite* "someone who lives on others"; and *vermin* "someone loathsome and contemptible." Neither *dove* "peace-worker, someone who is antiwar" nor its contrary *hawk* "someone who favors the military" is intrinsically dysphemistic, but they are used dysphemistically by those of opposed ideology.

It is clear that these metaphorical uses of animal names take some salient characteristic from the folk concepts about the appearance and or behavior of the animal (cp. the fables of Aesop and La Fontaine), which is then attributed to the human named or addressed.

We continue this section with a category of dysphemisms that we will call '-IST' dysphemisms (i.e., racist, sexist, ageist, etc.) putdowns. All of these have the same dysphemistic pivot: they fail to demonstrate respect for some personal characteristic that is important to Hearer-or-Named's self-image. For instance, racist dysphemisms to Hearer-or-Named arise when Speaker refers to or implicates Hearer-or-Named's race, ethnicity, or nationality in such terms as to cause a face affront to Hearer-or-Named, or perhaps to Hearer on behalf of (the person) Named. Sexist or ageist dysphemisms to Hearer-or-Named arise when Speaker refers to or implicates Hearer-or-Named's gender or age in such terms as to cause a face affront to Hearer-or-Named, or to Hearer on behalf of Named. And so forth. In general, then, -IST dysphemisms cause affronts to Hearer-or-Named's positive face by picking on a personal characteristic and, in Hearer-or-Named's perception, downgrading it. Later on we turn our attention to status putdowns, and other markers of Speaker's dysphemistic attitude toward Hear-or-Named that are revealed in the process of naming or addressing.

We are going to use racist dysphemisms to exemplify -IST dysphemisms in general. So far as we know, all human groups have a derogatory term available in their language for at least one other group with which they have contact. We will classify this as racist vocabulary, and not subclassify it into nationalist versus ethnicist, or the like. Among the racist dysphemisms of English, are: *mick* for Irish person (or in Australia, a Roman Catholic), *frog* (Cockney *jiggle and jog*) for a French person, *kraut* and *hun* for a German, *chink* (Cockney *widow's wink*) for a Chinese, *jap* or *nip* (Cockney *orange pip*) for a Japanese, *paki* for a Pakistani, *polaks* for a Pole, *wop* (Cockney *grocer's shop*) and *eyetie* for an Italian, *ayrab*, *dune coons*, and *camel jockeys* for an Arab, *kike*, *yid* (Cockney *dustbin lid* and *four-by-two*) for a Jew, *chief*, *Hiawatha*, and *Geronimo* for male North American Indians and *squaw* for their womenfolk, and so forth. English whites may use *black*, *nigger*, *nignog*, *wog* (Cockney *spotty dog*), *coon* (Cockney *silvery moon*), and so on for people of African ethnicity and for other people with similar skin color to Africans, such as Australian Aborigines and south Indians. In Australia we hear *boong* and *abo* for Aborigines; *gin* for an Aboriginal woman; and for people from east and southeast Asia, *slants/slanties*, *slopes*, *gooks*, *RGBs* (Rice Gobbling Bastards), *UFOs* (Ugly Fucking Orientals), *kanardles* (spelling uncertain) from *can hardly see*. And so forth.

Some of these racist terms are not intrinsically dysphemistic, and can be used without prejudice: for instance, *blacks* is not necessarily any more dysphemistic than *whites*; and in Australia, *boong* and *gin* are not invariably dysphemistic, no more so in fact than are *lebo* "Lebanese," *wog* "Caucasian Australian who is not Anglo-Celtic" and *skip(py)* "(young?) Anglo-Celtic Australian" (perhaps from the television series about 'Skippy the bush kangaroo'). Practically all these "racist" terms can be used without irony in neutral or euphemistic illocutionary acts. For example, Folb (1980:248) glosses *nigger* as follows:

> **nigger** Form of address and identification among blacks (can connote affection, playful derision, genuine anger, or mere identification of another black person; often used emphatically in conversation).   (Folb 1980:248)

Something comparable holds true for almost all racist terms—indeed, all -IST terms—in English. Nonetheless, the lexicon entries for the racist terms exemplified earlier will need to mark the degree to which they are dysphemistic: for instance, *black* should probably NOT be marked as dysphemistic, but rather as neutral; on the other hand, *nigger* should be marked as typically dysphemistic. A nondysphemistic use of *nigger* will be a function of the illocution; that is, it will at best be a euphemistic dysphemism. It could be that, used between African-Americans, as distinct from its use by speakers of other dialects, *nigger* is not dysphemistic, but neutral. We know that Australian Aborigines find terms like *black*, *blackfella*, *darkie*, *Abo*, *nigger*, and *boong* offensive when used by white Australians; however, they reportedly show no strong aversion to using these terms among themselves (cf. Eagleson 1982:157).

Hearer-or-Named's self-image depends in part on the perception of his or her own status relative to Speaker and of the social distance between them at the time of utterance. If Hearer-or-Named perceives that either the social distance or the relative status is significantly distorted by Speaker, there will be an effect on their positive face. The kind of effect we are concerned with here is a detrimental one, in which Hearer-or-Named's positive face is affronted. The following, taken from a letter to the editor of a German-Swiss journal *Der Schweizerische Beobachter*, July 15, 1976, reports just such an affront:

> In fact, we have a little problem. Recently my wife and I offered the fiancé of our daughter to say "Du" to him. He then asked how he should address us. We explained he could address us just as he liked; we thought he would say either "Mutter" and "Vater" or "Mama" and "Papa". However, we are greatly surprised to see that the young man calls us now by our first names. Whilst our daughter thinks this is all right, modern and colloquial, we ourselves find this world hostile and confounded. Being both about fifty years old, are we already hopelessly antiquated?   (Adler 1978:202)

In English, Hearer-or-Named will more often than not regard it as dysphemistic if a consanguineal kinsperson from a descending generation uses their given-name instead of a kin title like *Dad*, or kin title + name such as *Aunt Jemima* (whatever is appropriate, see Ch. 2). This reaction is comparable to Hearer-or-Named's reaction when Speaker names or addresses them using forms from a contextually inappropriate style; in fact it is motivated by precisely the same conditions, which we discuss immediately below. Before doing so we would remind readers that we are using a five point scale for style: frozen > formal > consultative > casual > intimate. Also that one very important means of maintaining face all round is to use an appropriate style, see Ch. 2.

Where Speaker should conventionally defer to Hearer-or-Named because there is considerable social distance between them, or because of the latter's superior status, it is dysphemistic for Speaker to employ the naming or addressing forms from a lower style than convention warrants: such behavior will affront Hearer-or-Named's positive face (positive face is the want of a person that they themselves, and things dear to them, be valued by others). And it is not only the person of superior status who can be affronted in this way: if the leering middle-aged man calls his young female secretary 'sweetheart' she might well object that the social distance between them requires at least the 'casual' level of formality, and consequently she finds this 'intimate' style offensive. This particular affront would often today be classed as 'sexist', which, strictly speaking, is inaccurate: it is the inappropriate 'intimate' style that is the basic source of the problem here. (We cannot guarantee that this nicety would sway the court if the secretary brought a sexual harassment charge.)

The last example is interesting: one might imagine that if Speaker is superior in status to Hearer-or-Named and chooses nevertheless to adopt a 'casual' or 'intimate' style, then Hearer-or-Named should be flattered (i.e., find this a positive face enhancement); but in practice this is only possible where Speaker's behavior is welcome. Face is grounded in a person's WANTS; and, despite there being conventional beliefs about what a person might be expected to want, ultimately a want is a personal idiosyncrasy. There are people (of no matter what status and social distance from oneself) whom one is pleased to have the opportunity to come close to (or even appear to come close to), and concomitantly to have the (apparent) closeness marked by the use of 'casual' or 'intimate' style; and there are other people whom one wishes to maintain a social distance from, a distance that is demonstrable through linguistic marking.

Anger with Hearer-or-Named will often lead Speaker to use dysphemistic means of naming and addressing. In addition to all the terms of abuse we consider in this chapter, Speaker can change his or her normal mode of addressing or naming in order to display antipathy to Hearer-or-Named. The default mode of naming and addressing (i.e., Speaker's normal mode) is euphemistic: Speaker is polite to Hearer and about Named. However, it needs to be said that while this convention holds true for most Speakers to most Hearers and about most Nameds on most occasions, it is not universally true. Suppose that Speaker is a person of firm sectarian ("religiousist") beliefs: for instance, Speaker is a Northern Ireland Protestant who is normally dysphemistic about Roman Catholics; dysphemism will be Speaker's normal mode for naming or addressing that group of people, but not other groups.

Consider occasions where Speaker wishes to display anger by using a linguistic form that distances him or her from Hearer-or-Named in an abnormal way. For instance, a Speaker who normally uses an 'intimate' or 'casual' style with Hearer-or-Named will typically shift to a more formal style, such as switching from given-name only (*Fred*) to given-name + surname, title + surname, or surname alone (*Fred Whimple* or *Mr. Whimple* or *Whimple*). In some other languages, such as French or Japanese, Speaker can display anger by substituting out-group forms for normal in-group forms (e.g., by using *vous* in place of *tu* in French; *anata* instead of *kimi* in Japanese, and so forth). Social distance marking in forms used for naming and especially addressing can be achieved in many languages through sarcastic use of 'intimate' terms. In English the angry Speaker who is inferior in status to Hearer-or-Named may use title alone (*Mr.*) or an inappropriately familiar term (*bud*, etc.). In French or Japanese, Speaker can insult a socially distant (out-group) Hearer by using an in-group address form (such as French *tu*); in Japanese a socially distant third person can be insulted in a similar way by Speaker using an in-group pronoun or verb form. Undisguised sarcasm in naming and addressing is of course dysphemistic, whatever style Speaker adopts: it is just as offensive to address a superior Hearer with a sarcastic 'Sir,' as it is to address someone of similar status of the oppositive gender with a sarcastic 'Dear.' All languages have conventional strategies of one kind or another that allow Speaker to display antipathy by varying the normal mode of address or naming.

Profanity/Blasphemy

The ancient Greek roots of *blasphemy* give *blas-* "evil, profane"—*phēmos* "speaking" and the Latin roots of *profane, pro-* "before, outside of"—*fanus* "temple" are carried through into the present day English understanding of *profanity/blasphemy* as "bad, unseemly, irreverent, impious" (see Ch. 2 for discussion of the precise meanings of each of these words). In this section, however, we will extend the meaning of *profanity* to comprehend not only "taking the Lord's name in vain", but also invocations of hell and the Devil. Expletives run the gamut of reference to God, Christ, heaven, hell, and the Devil; curiously, though, dysphemistic epithets are restricted to invoking either the Devil or hell.

Profanity is dysphemistic because it is confused with blasphemy, which violates a religious taboo whose details were discussed in Ch. 2. *God!, Christ!, Jesus!, Jesus Christ!, Christ almighty!* are the classic profane expletives. A little more exotic is *Jesus H. Christ on a (rubber) crutch*; the middle initial 'H.' probably comes from the 'H' in *IHS* discussed in Ch. 2, and 'crutch' is, of course, a remodelling of *cross*. This phrase has traits of a euphemistic dysphemism. One euphemistic dysphemism not discussed in Ch. 2 was (*Oh*) *Boy!*: it is possible that this is based on the notion "son of God", and gains impetus from the *bl-* phonestheme we shall discuss shortly, even though it does not have the full *bl-* onset.

In addition to all these expletives there are such phrasal ones as in *For Christ's sake, what are you doing?* and *What in God's name are you doing?* and variants on these. *For chrissake* is an orthographic remodelling from *Christ's sake*; lexical variants are *For God's sake*, *For Jesus' sake* and the euphemistic *For heaven's sake*. *What in God's name* . . . has variants *What in Christ's name* . . . and the euphemistic *What in heaven's name*. . . . If the expressions like *Holy Mary!*, discussed in Ch.2, are euphemistic, *Holy Jesus!* is outright profane—and some people would say blasphemous.

Turning to epithets: only euphemistic epithets invoke God, Christ, or heaven (cf. the adjectives *godlike, christlike, heavenly*). Dysphemism can be achieved by negating such invocations as in *ungodly* or *unchristian*. Otherwise, profane epithets invoke the Devil or hell in one way or another (cf. *devilish, hellish, He's a devil*). The latter is of course ambiguous between the dysphemism "He's wicked," and "He's a dare-devil," which is a term of approbation if not outright praise. Note that *He's a little devil/demon* is an affectionate description for a naughty boy, for example, with only very slight disapprobation; it is comparable with *dare-devil*, and perhaps *devil-may-care*—which are not at all dysphemistic. *It's a devil/hell of a nuisance* uses 'a devil / hell of' as a dysphemistic intensifier (cf. *I had a devil/hell of a job reaching him*). In these environments *heck* is the standard euphemistic dysphemism for *hell*, as it is too in expletives like *What the devil/hell/heck [are you doing]?* (cp. *What the fuck* . . . later). Semantically associated with hell is the euphemistic epithet *flaming* (cf. *flaming hell/heck*), and what is probably a remodelling of it: *flipping*. The initial *f-* suggests that these may link up with *fucking* (cf. *fucking hell*), whose illocutionary point they share.

Hell!, *Hell's bells!*, and *Heck!* will stand alone as expletives, where they have about the same force and meaning as the expletive *Damn!* and its cohort *Damnation!* Although *dash* could stand for any expletive, it probably gets some motivation from *damn* (even *Oh dear*, though it presumably derives from *(Oh) dear God*—possibly with help from French *dieu*, might get some push from *damn*— compare *Dear me!* with *Damn me!*). Except for *Oh dear*, all of them condemn a provocative situation to hell. *Damn!* is based on a verb meaning "condemn" (*condemn* is essentially *con-* + *damn*), which has become a zero-derived expletive along the lines of *Shit!* or *Fuck!* (see the following); this is more likely than that it is an end-clipping from *damnation*. All of these are modelled on a clause in which God is the agent: *(May/Let) God damn NOUN PHRASE* (giving rise to the euphemistic dysphemism *Dagblag it!*). According to Montagu (1967:124, 282f), between the fifteenth and nineteenth centuries, the English were known in France as *Goddams* because of their ubiquitous use of this oath; and it is still going strong. In the same vein is *Strike!* (euphemized to *Stripe!*) from *God strike me dead [if I'm not telling the truth]*.

As an expletive, *Damn!* may have been partly motivated by *dam* meaning "mother-of", which is reminiscent of the archaic *By our lady* referring to Mary, mother of God; her name was the source for the remodelled expletive *Marry!* found in the writings of Shakespeare and his contemporaries. Consider, however, the expression *It ain't worth a damn*. We are inclined to think this straightforwardly means "It's not even worth (such a valueless thing as) a cuss" (cf. *It ain't worth a tinker's cuss*); Montagu (1967:92), on the other hand, claims that it 'is purely Indian. In India the coin of least value is known as a dam.' We are not convinced that Montagu is right, though it is quite possible that there has been a network of disparate but mutually reinforcing motivations for its current meaning and use, a network of a similar kind to the one that gives rise to *pussy*.

In place of the malevolent curses *Damnation!* and *Damn!* there are the corresponding euphemistic dysphemisms *Tarnation!*; *What in tarnation?!*; *Consarn it!* and *Darn!*, *Dang!*, *Drat!*—all of which are remodellings. There are also *Blast!* and *What the blazes!* (see the upcoming section on the *bl-* onset), both of which clearly invoke hell. *Blast it!*, and the even more euphemistic *Bother it!*, presumably mean "(May) God blast/bother it!" and so euphemistically omit God's name. The (very) euphemistic dysphemism *Blessed thing!* (A favorite epithet of an aged middle-class aunt) does the same. The expletive *Bother!* has a cohort *botheration* (cf. *damnation*, *tarnation*); it also has a variant, *Brother!* whose profane credentials may be affected by *brother in Christ*.

Profanity/blasphemy is still quite a strong force within the community, despite our increasing secularization during the course of the twentieth century. Many people regularly use euphemistic dysphemisms in place of terms invoking God or Christ. Predictably, *heaven* is euphemistic in expletives and epithets whereas *hell* is intrinsically dysphemistic; furthermore, it almost always appears in dysphemistic illocutions. There seems to be rather less community concern over invocations of the Devil than invocations of God or Christ—unless the latter are heavily disguised as in, for example, *Oh dear!*; this could explain why the epithets *dare-devil* and *devil-may-care* are not dysphemistic at all.

Perhaps we should end this section with Barry Blake's story of how, as a kid brought up not to blaspheme or be profane, he refused to say *Amsterdam* and instead used to talk about *Amsterdash*. We live and learn.

Bodily Effluvia in Maledictions

We turn next to the dysphemistic use of taboo terms, starting with RRR-rated bodily effluvia (see ch. 3). The principle waste products of a body are shit and piss, and the former is widely used in maledictions. Vomit is only infrequently referred to, but there is *You make me sick/throw up/wanna spew/wanna puke* and comparatives such as *It looks like dog vomit*. Menstrual blood is not explicitly used in figurative maledictions; indeed, it may not be invoked in any way (see later discussion). The rare *codswallop* (not in Farmer and Henley 1890–1904, nor Partridge 1961, 1970) is presumably based on "spunk, ejaculate"; it is apparently compounded from *cod's* + *wallop*: *cod* means "scrotum, balls," *wallop* "bubbling liquid"; it is used in much the same way as *bullshit* to mean "rubbish" (see the next section) or perhaps more derivatively, "verbal masturbation" (cp. the gerund *rodwalloping* "masturbation," compounded on the verb *wallop* "beat, strike"). Lower r-rated taboo terms do not generally give rise to taboo epithets. For instance, RR-rated *snot* gives rise to dysphemistic, though nontaboo, epithets. *Snotty-nosed* suggests "snotty-nosed like a kid" (i.e., that the target is immature and inexperienced); the adjectives *snotty* and cognate *snooty* denote "someone who turns their nose up, i.e. acts superior." There are, however, the putdowns *He's got a fart in a windstorm's chance (of succeeding)!*, *He's about as effective as a fart in a windstorm!*; and also the descriptive use of *fart* as in *Cedric's a silly old fart*. The phrase *silly old N* is dysphemistic, and the insult is increased if 'N' is a taboo term. In *silly old fart* 'fart' is associated with "hot air" and *Cedric's a silly old fart* implies that "Cedric talks a lot of hot air" (i.e., the content of what he says is insubstantial and inconsequential).

These preliminary remarks aside, we turn to a closer examination of RRR-rated bodily effluvia in maledictions.

Maledictory References to Feces

> A horson filthie slaue, a dunge-worme, an excrement!
> (Ben Jonson *Everyman in His Humour* III.v.127)

The expletive *Shit!*, in some dialects (e.g., northern British English) *Shite!*, along with the remodelled euphemistic dysphemisms such as *Sugar, Shoot, Shucks*, or *Shivers*, typically express Speaker's anger, frustration, or anguish. However, with appropriate fall-highrise intonation and lengthening during the rise it can express wonderment (cf. /ššìíᵗ/). In line with the mood of the typical expletive, are epithets like X *gives* Y *the shits!* "X makes Y angry, Y doesn't like X" and its near-antonym X *doesn't give a shit* "X doesn't give a damn, X doesn't care." Also, *be shat/crapped off with; be in a shit, be in a shitty mood, be shitty,* all meaning "be in a bad mood, feel bad." A euphemism for the latter is *be shirty*—a euphemistic dysphemism created by remodelling. Less obviously euphemistic is the dysphemism *be browned off with; brown* is a substitute term for *shit* in expressions like *brown-nosing*, too.

Tough/stiff shit means something like "however shitty you feel about it, you've got to put up with it, so there"; and it may well be based on the discomfort of constipation. One should not overlook the contribution to this of the unsympathetic monosyllabic comment *Tough!*—which might be a euphemistic abbreviation of *Tough shit!*; and the associations with *That's tough* and *Tough luck*. The last of these, however, is the only one that typically expresses sympathy for Hearer's predicament. Similar in meaning to *tough shit* is the alliterative *tough titty*. There seem to be a number of euphemistic versions of *Stiff shit* clustered about *stiff/hard cheese*; they include *stiff cheddar* and at further remove (Australian) *stiff biscuits/ bikkies*—presumably through a cheese and biscuits (i.e., crackers) scenario. Why is *cheese* substituted for *shit?* Because (1) there is a degree of phonetic similarity in the palato-alveolar onset followed by a high front unrounded vowel; (2) a degree of physical similarity in the consistency of stiff shit and hard cheese; and (3) both have a reputation for pungent odor; (both are also animal products, but their similarity in this respect is inconsequential).

Expressions like those above are prompted by the very reasonable notion that feces is something to be avoided. This presumably explains why *Shit on you!* works as an imprecation: the analogy with *Damn you!* and the archaic A *pox on you!* (see later) suggests it should be interpreted along the lines of "May God [heap] shit on you," condemning Hearer to being *in the shit/pooh. In the pooh* and (Vice-President George Bush's) *in deep doo-doo* are euphemistic dysphemisms; this may be because they are based on nursery terms, suitable for the ears of mothers, nannies, and children. Other maledictions with literal fecal associations include *up shit creek without a paddle (in a chickenwire boat/barbed-wire canoe)*, and *when the shit hits the fan*. By invoking the revolting image of being covered in shit, they all derive the meaning "(be) in bad trouble." Euphemisms for these include *What a mess!* (in former times possibly influenced by anti-catholic feeling, *mess(e)* = *mass*); also the now-archaic *Here's a how-do-you-do!* (as in Gilbert and Sullivan's 'Mikado'). In the imprecative *Eat shit!* the noun phrase is intended to be UNDERSTOOD literally even though a literal perlocution is not expected to result.

The fact that fear may cause the sphincter muscle to relax is alluded to in *He shat himself/his pants*; also suggested by the Cockney *He lost his bottle*, from the rhyming slang *bottle and glass* "arse" (*lose one's bottle* also has the meaning "get angry, get into a shit,*" see earlier). Perhaps the same thought lies behind *beat/kick the shit out of*—an additional motivation for which is no doubt that whoever has the shit beaten out of them was *full of shit* in the first place (see the following).

The pejorative epithets *It's a load/pile/heap/crock of shit*; *It's shithouse*; *You're a shit* (like many dysphemisms, this one CAN be used affectionately as in *He's a cute little shit*; see later); *That's a shitty thing to do*, where 'shit,' 'shitty thing,' and the like mean variously "awful, bad, rotten, rubbish, etc." Euphemisms for these include *It's (a load of) muck* and *It's a mess*. Related to them is the dysphemistic euphemism *It stinks!*, which trades on the noxious odor of feces. The epithet *It's cacky* (ultimately from the Indo-European verb **kak(k)-* "shit") is close to being understood literally, and means "It's messy, revolting, (?shitlike)". Partridge (1961) glossed the epithet *cackhanded* as "lefthanded," which has given rise to the prevalent current meaning "clumsy" (compare English *gauche* from the French word meaning "left"); the earlier meaning of *cackhanded* may come from the muslim practice of wiping oneself with the left hand after defecating.

Then there are the intrinsically dysphemistic epithets *Crap!*, *Bullshit!*, *Horseshit! Ratshit!*, and *Chickenshit!* meaning "Rubbish!". For *bullshit*, but for none of the others, there is apparently an end-clipped euphemism *bull*; the reason for this may well be, as McDonald (1988) suggests, that *bull* in this sense in fact derives from *papal bull*, which, after the English Protestant reformation, would have been regarded as worthless rubbish by the majority of the population. Against this however, one hears *Are you shitting me?* a macho version of "are you kidding?" (cf. Fox 1987:42), which is a foreclipping of the transitive verb *bullshit*. *Ratshit* and *chickenshit* indicate something trivial and beneath contempt (cp. the uses of 'rat' and 'chicken' in *You rat!* and *You're chicken!*). The judgment *It ain't worth shit* means "It ain't worth nothing, it's worthless". *No shit?* is "Are you telling the truth (i.e., This is no bullshit?)"; to which one answer is the avowal, *No shit*. Imprecations like *You're full of shit (that's why your eyes are brown)* imply "You're full of rubbish, you talk a lot of bull"; as do the corresponding epithets. An expression like *He thinks he's such hot shit* "He thinks he's really good because he's got new ideas"—implies the scatological simile 'new like fresh shit.' The latter comparison identifies Speaker's judgment of the "new ideas," and the motivation for using such a dysphemism. The expression *hot shit* is formally associated with *hot stuff*—which is not a euphemism for it, but derives by extension from *hot* "sexual ardor" to mean "good." Finally, there is the sarcastic *He thinks his shit don't stink*—to describe someone who has, in

Speaker's judgment, much too high an opinion of himself. (This is reminiscent of the German aphorism *Eigener Dreck stinkt nicht.*)

To sum up: *shit* is used nonliterally to express Speaker's anger, frustration, or anguish. It is used in metaphors that exploit perceptions of the effluvia denoted by literal uses of the term to denote something unwanted, revolting, messy, and stinking, and in addition it is associated with fear and with trouble. The expression *take a squat* means "defecate": is there any connection between this use of *squat* and its use in idioms like *I didn't do squat* "I didn't have to do anything" and *He doesn't know (diddly) squat?* These uses of *squat* seem to correspond to the "worthless, nothing, trash" sense of *shit* (cf. *It ain't worth shit; He don't know shit;* and, of course, *It's shit* meaning "It's trash/worthless").

Interpreted literally as denoting effluvia, *shit* and *crap* are used in uncountable noun phrases, hence the use of a classifier (Cf. Allan 1977; Lehrer 1986) such as 'a load of' in *a load of shit*. In periphrastic verbs like *to have a shit* (= to shit), *to have a crap* (= to crap), however, these terms appear in countable noun phrases. *Turd*, however, normally occurs in countable noun phrases, and there is no periphrastic verb *to have a turd* nor a verb *to turd*. Maledictions about people use countable noun phrases (cf. *You are a shit; It's a shit of a day,* etc.). A person may be described as *a turd* but other entities may not, and nothing is described as *a* (nonliteral) *crap*.

Maledictory References to Urine

Corresponding to the insult *Shit on you!*, there are *Piss on you!* ("May God [heap] piss on you") and the imprecatives *Piss off* and *Go piss up a rope;* but otherwise *piss* is not generally used as an insult or expletive. However, *pissing* is used as a dysphemistic intensifier in *I'm not pissing going,* where it is an alternative to, say, *bloody,* or *fucking* (see later). *To be pissed (at)* and *to be pissed off with* both mean "[caused] to be annoyed with" ('Mad as a fly pissed off a toilet seat'); the latter is comparable with *shat/crapped off with*. In the dysphemistic epithets such as *piss poor* or *piss weak,* the dysphemism of the attributes 'poor' and 'weak' is significantly intensified by the dysphemistic taboo term 'piss'; *piss poor* gains additional force through the alliteration. Where a contrast can be made between nonliteral *piss* and nonliteral *shit,* the former expresses Speaker's annoyance; whereas the latter expresses Speaker's anger, frustration, or anguish. This lesser force correlates with the lower revoltingness-rating that we ascribed to piss in Ch. 3.

There are references to urine in dysphemistic epithets for drinks (e.g., comparing beer to *warm piss* or the strength of beer or tea, etc. to *gnat's piss*). In British English there is an 'effect for cause' metaphor, such that *to be on the piss* means "to be drinking [alcohol]" and in all dialects *to be pissed* means "to be drunk." The latter gives rise to euphemistic expressions like *He was Brahms last night,* where *Brahms* (or *Mozart*) is end-clipped rhyming slang from *Brahms/Mozart and Liszt* (= pissed). The connection between piss and drunkenness is obvious. Notice that to accuse someone of *being pissed* is dysphemistic, whereas to claim that someone is *merry* or even *tipsy* is euphemistic—despite the latter being associable with *tip over, tipple, topple,* and *topsy-turvy*.

The *BL*- Phonestheme and Some of Its Side-Effects

A sunburnt . . . stockman stood,
And in a dismal . . . mood,
Apostrophized his . . . duddy:
"This . . . nag's no . . . good,
He couldn't earn his . . . food—
A regular . . . brumby,

. . . !"

(From '. . . (The Great Australian Adjective)' by W.T. Goodge
Sydney Bulletin 1899, quoted in Montagu 1967:355)

Menstrual blood is never named as such in epithets, but it is plausible that the dysphemistic adjectives *bloody, bleeding,* and the milder *blooming* and *ruddy* are associated with the Judeo-Christian taboo on menstruating women (see Chs. 3 and 4). *Blooming* is part remodelling of *bloody* and could be part derived from *bloom* "flower, menstruate." There are three relevant kinds of *bloody,* the first two being closely related: (1) the intensifier as in *It was a bloody good party*; (2) the strongly pejorative adjective as in *You've got a bloody cheek!*; and finally (3) a figure from the substance blood, used, for example, in responding to the question *How did it go?,* answer *It was bloody,* which is understood to mean "It was bad". (1) and (2) have the same source, and for the rest of the discussion we can amalgamate them as intensifiers. As one would expect, *bleeding* and *blooming* can be used in place of *bloody* in both (1) and (2). This is not the case in (3), which may feed on a figure from menstruation, from battle, or simply be an abbreviation of *bloody awful.* It would appear that intensifier *bloody* has nothing to do with menstruation, and Montagu (1967) suggests that it originated among soldiers in the low countries during the sixteenth century, who modelled it on German *blütig,* Dutch *bloedige.* He also argues that it was not an impolite term:

> Swift, writing to Stella on May 28, 1714, informs her that "it was bloody hot walking today." In 1727 he describes himself as "bloody sick," and later, "It grows bloody cold." All these *bloodys* in letters to a gentlewoman.
>
> (Montagu 1967:245)

One has to agree that Swift seems to be using 'bloody' with the same freedom that gentlemen and ladies of yesteryear would have used something like *awfully, frightfully, terribly,* or *dashed.* This fits in very well with much of what Farmer and Henley have to say about early uses of the word.

> The origin is not quite certain; but there is good reason to think that it was at first a reference to the habits of the bloods or aristocratic rowdies of the end of the 17th and beginning of the 18th cent. The phrase bloody drunk was apparently as drunk as a blood (cf. be drunk as a lord); thence it was extended to kindred expressions, and at length to others; probably in later times its associations with bloodshed and murder (cf. a bloody battle, a bloody butcher) have recommended it to the rough classes as a word that appeals to their imagination. Compare the prevalent craving

for impressive or graphic intensives as seen in the use of *jolly, awfully, terribly, devilish, deuced, damned, ripping, rattling, thumping, stunning, thundering,* etc.

(Farmer and Henley 1890–1904)

We would suggest that Farmer and Henley have identified the way that *bloody* declined in respectability. According to Montagu it came to be perceived as a lower-class (i.e., vulgar) word; consequently, it only gets recorded in the language of footpads, flowergirls, and their ilk (cf. Byron's *Don Juan* 1824, canto 11 and Shaw's *Pygmalion* 1914 quoted in Ch. 1). We would endorse Farmer and Henley's etymology, appending the associations with menstrual blood to those of 'bloodshed and murder'; hence the sense of anger or disapproval, frustration or anguish that this epithet carries. No credence whatsoever can be given to the claim that *bloody* is a euphemistic remodelling of *by our lady* nor of [*God*]'s *blood*, since these two last are expletives and *bloody* is an epithet: this particular claim is merely a euphemistic explanation for the etymology of *bloody!*

Bloody and *bleeding* are members of a set of fairly mild maledictions that are marked by the onset phonestheme *bl-*. A phonestheme is a cluster of sounds, located at either the onset of a word or its rhyme, which symbolize a certain meaning; they are typically incomplete syllables, and their meaning is much more difficult to pin down than that of morphemes; furthermore, exactly the same cluster of sounds that constitute the phonestheme will appear in many vocabulary items that have no trace of the meaning associated with the phonestheme! Consequently, phonesthemes are often said to manifest a SUBREGULARITY in the language (see the Glossary for more). In addition to the blood-linked members of the maledictions marked by the onset phonestheme *bl* , there are those with profane/blasphemous implications invoking the fires of hell or the wrath of the old testament God (cf. the expletives *Blimey!, Blast!, Blow!* and *What the blazes!?*; and the epithets *blessed, blamed, blinking, blinding, blasted, blighter,* etc.). There is an innocuous euphemism created in this pattern: *blankety(-blank)*. The verb *bleep*—as in bleeping out indelicate utterances from a broadcast tape—denotes a sound known in computer systems as *beep*, and it would seem that the *bl-* onset to *bleep* is deliberately phonesthetic. The noun *blooper* seems to have originated with people inadvertently swearing on soundtrack when something goes wrong during filming—also known as a *blue* (cp. *blue film* "a pornographic movie"; this sense of *blue* has passed its sesquicentennial). *Blooper* looks like some blend of *blue* and *bleep*, the suffix *-er* being agentive "what causes the bleeping out."

Possible side-effects or side-influences of the *bl-* phonestheme are the epithet *bally*, and the expletives *Bother!, Botheration!, Brother!, Boy!* and even *Bugger!*; all of which have an initial *b-*. As shown earlier, the expletive *(Oh) Bother!* and the imprecative *Bother it!* may be euphemistic remodellings of, for example, *blow, blast,* etc. all of which have a semantic relationship with *Damn (it)*. The justification for such speculations as these is circumstantial: in Ch.4 we discovered intricate networks of relations that link items of vocabulary by filigrees of semantic and phonological similarities into semantic fields; it is not unlikely that these mechanisms contribute to the development of the vocabulary associated with the *b(l)-* onset phonestheme.

Maledictory Reference to Sexual Acts and Anal Pastimes

[A *doubtless apocryphal tale, reputedly told by a British soldier charged with
assaulting a man during World War II*] 'I come home after three fucking years in
fucking Africa, and what do I fucking-well find?—my wife in bed, engaging in
illicit cohabitation with a male!'

(Richard Hoggart 'Introduction' to *Lady Chatterley's Lover* 1961:ix)

The expletive *Fuck!* and its remodelled euphemisms *Fudge, Muck,* and *Frig* (which
is, of course, the archaic word for "wank" or "masturbate") are used in similar
circumstances to *Shit!* to express anger, frustration or anguish. So too is *Bugger!* The
epithet *fucking* has for cohorts the euphemistic dysphemisms *mucking, frigging,
freaking, flipping,* and even *flaming;* all are used under similar circumstances to
bloody, bleeding to express anger or disapproval, frustration, or anguish though
fucking is regarded as the most strongly dysphemistic. These epithets complement
the expletives *Fuck!* and *Shit!*, whereas *shitty* means something rather different.

There is a meaning contrast between the verbs that Quang (1971) distinguished
as *fuck₁* and *fuck₂*: *fuck₁* is the literal sense of *fuck* and means "copulate," while *fuck₂*
is nonliteral, and occurs in the imprecative *Fuck you!* and the like. It seems likely
that *Fuck you!* is analogous to *Damn you!* where the understood agent is God (cf.
earlier for the discussion of *Shit on you!* etc.) We have no evidence that **God fuck
you!* is attested; however, if 'fuck' is understood not as "copulate," but in the sense
"spoil, wreck, ruin" then *Fuck₂ you!* means "May you be wrecked, ruined." This is
very much the same sense as once attached to *God damn you!*, which surely
confirms our explanation for this otherwise curious imprecative.

The difference between *fuck₁* and *fuck₂* has legal status in the state of Queensland, Australia. A certain David Pearson was charged with using obscene language
for allegedly saying *Get a fucking hurry on* while on stage in a Brisbane nightclub in
March 1989. In August 1989 the magistrate ruled that *fuck₁*, denoting sexual intercourse, is considered obscene, but the expletive *fuck₂* is not, and thereby cleared
Pearson of the charge. The Brisbane *Courier Mail* began its report of the decision
thus: 'Australia's most common four-letter word is no longer considered to be
obscene.'

There are also meaning differences between the phrasal verbs *fuck₂ around, fuck₂
off, fuck₁ with, (NEG) fuck₂ with,* and *fuck₂ up*, most of which utilize *fuck₂*. (*Fuck₂
with* nearly always occurs in the scope of a negative, cf. *Don't fuck with me, honkey!*
and *Stop fucking with my things;* but also *I hit him because he was fucking with my
things.*) The semantic contrast between these phrasal verbs derives from the particle
(or its absence). Thus, for instance, 'around' implies "going round in circles"; hence
be fucking around means "doing nothing in particular; not doing what ought to be
done," as in the complaint *They're just fucking me around.* Although *fuck₁ around* is
a possibility, the more usual expressions would be either *screw₁ around* or *sleep
around.* The 'off' in *fuck off* has the same ablative sense as it does in *go off* "depart";
hence the usual interpretation of *fuck off* is "go away"—though it is much more
forceful than this gloss suggests. The same meaning is found in other imprecations
containing *off* (cf. *Piss off!, Bugger off!, Sod off!,* etc.). The 'with' of *fuck₂ with* marks
the affected object, as in *What are you doing to do with the money?;* by contrast, the

'with' in *fuck₁ with* is comitative, as in *meet up with*. The 'up' of *fuck up* marks completives (cf. *chop up, finish up, settle up*, etc.); this is what gives *fucked up* the meaning "be spoiled or wrecked" (cp. *It's a fuck up* and *It's a bugger-up/cock up*). As for a number of phrasal verbs with the *up* particle, where the contexts allow for the intended meaning to be readily inferred, the particle can be omitted without changing the sense (cf. *It's fucked*, which has the basic sense of *fuck₂* "spoiled, wrecked, ruined"). Hence complaints like *I'm fucked/shafted/rubber dicked/buggered* mean "sorely mistreated; my hopes are dashed". The same expressions are used hyperbolically for "be tired". Whereas the particle *up* in phrasal verbs like *fuck up* and *finish up* is completive, in *Up yours!* and *Shove it (up your arse)!*, and their like, *up* is a directional preposition. The epithet *He's up himself* is very widely used in Australia to describe someone thought to be opinionated (cf. *He's so far up himself he needs a snorkel to breathe*).

Imprecatives like *Go fuck yourself!, Sod off!, Screw you!* or *Stuff him!* are used much like *Shit on you* (but by more people). *Bugger you!* and *Bugger off!* are used in much the same ways as *Fuck you!* and *Fuck off!*, respectively, but are regarded as less objectionable—despite the implications of unnatural sexual behavior. Self-as-patient expressions like *Bugger me!, Fuck me dead*, and *Fuck me backwards through the eye of a needle* are used to express real or pretended incredulity (cp. the less forceful *Are you having me on?*—on a string like a monkey; this may have looked obscene, but it isn't). Some other expressions with a similar meaning are the quasi-reduplicatives *fuck a duck, frig a dig, root me boot* (the latter being Australian, of course).

For all the *fuck₂* examples, there are corresponding examples with the euphemistic dysphemisms *screw₂* and *stuff₂*. Note that the epithet *fuck*, and in lessening degrees its cohorts *screw* and *stuff*, are strongly dysphemistic: they indicate Speaker's anger, frustration, anguish, contempt, or the uselessness, or destruction of something—meaning extensions that cast our attitude toward copulation in a curious, perhaps paradoxical, perspective.

The negative implications of *fuck₂* are captured precisely in the quantifier *fuck all*, which is a synonym for *nothing* (cf. *He knows fuck all about linguistics; It's fuck all to do with you; There was fuck all left to drink after Bruce had his buddies come for a poker game*). Slightly milder alternatives to *fuck all* are *sod all* and *bugger all*.

The noun *fuck* is almost always *fuck₁* (i.e., meant literally, except in the idioms *For fuck('s) sake* "For God's sake" expressing irritation or anguish, and *What the fuck!?* "What the hell!?" expressing dismay or incredulity. At the end of The Who farewell concert in Dallas, Texas, September 3, 1989, Pete Townshend said, among other things, 'We want to thank all of our magnificent crew [. . .]. They get paid shit. And they're gonna get shitty bonuses, too. But then, you know, we're just mean fucks.' Here the descriptive noun *fuck* has the same sense as *fucker*; these nouns, like *cocksucker*, seem to have much the same range as *cunt, asshole*, and other taboo body part terms (TBTs), discussed shortly; they indicate the user's contempt and/or dislike for someone. Epithets ascribing homosexuality are often used dysphemistically particularly by heterosexuals; vocatives like *You sod!, You bugger!, Faggot!, Puff!, Poofter!* and *Dike!* are normally dysphemistic. For anyone, especially a man, to insult a woman with the term *Lesbian!* is a dysphemistic use of a neutral locution.

To call someone an *arselicker* or to accuse them of *brown-nosing/brown-tonguing* implies that they are crawlers, smarmy, sycophantic. *Cocksucker* and, for example, *She sucks*, sometimes have a similar meaning. Yet the verb *suck* is often used, especially in the impersonal *It sucks*, simply to express disapprobation—which led *Time* Magazine October 17, 1988:87, to describe it as 'the all-purpose jeer.' These are all regarded as euphemistic dysphemisms, and they compare in meaning, though not in effect, with *She's a shit* and *It's shitty*, respectively.

It is peculiar that there is no euphemism for *incest* in English; perhaps because the topic is rarely mentioned—true taboo. There is just one incest epithet, *motherfucker*, and its euphemistic dysphemism *mother*, and they are principally American. Why this should occur, while *fatherfucker* and *daughterfucker* are neologisms, is anybody's guess, but may reflect its greater abnormality. It should also be taken into account that insults denigrating the target's mother are common in many of the world's languages (cf. American *son of a bitch*, euphemized to *S.O.B.*): Romance variations on "Your mother's a whore" (e.g., French *fils de pute*, Spanish *hija de puta*, etc.); and in other languages "Your mother's cunt" (e.g., Arabic *kos omak*, Kiswahili *kuma nyoko*, etc.).

In English *to have a wank* is "to masturbate". In Australia and perhaps nowadays in Britain, too, there has been a nice figurative meaning extension such that *a wank* is "a self-indulgence," and *a wanker* is "a charlatan, a humbug, a prick." In Australia, both nouns are readily quoted in newspapers from the utterings of public figures—which indicates that they are not strongly tabooed. They are, however, dysphemistic.

We have considered dysphemistic uses of all sorts of terms including *bugger*, *cocksucker*, and *wanker*; but our major concern was *fuck₂* and related lexemes. The predominant meaning of the nonliteral *fuck₂* is "spoiled, wrecked, ruined." At first sight this seems to have little connection with the meaning of *fuck₁*, "copulation", but there are associations with adultery that give rise to such concepts as *a fallen woman*— one who would probably be regarded (or would once have been regarded) as "spoiled, wrecked, ruined." Certainly *fuck₂* displays a very negative attitude on Speaker's part, and there are a number of expressions (e.g., *What the fuck?!*), where the function of 'fuck' is to mark strong dysphemism without any additional meaning.

Body-Part Terms and Abuse

> Horson connie-catching raskall
>   (Ben Jonson *Everyman in His Humour* III.i.181)

> Life is like a penis: when it's soft you can't beat it; when it's hard you get
> screwed.   (Shem *The House of God* 1978:9)

We now turn to descriptive dysphemisms using those body-part terms that are highly restricted as to mentionability. The syntax of such English body-parts is arguably their least interesting feature, even when they are used dysphemistically as insult terms. We exclude from consideration here uses of body-part terms that are offensive because they treat the target as a sex object (e.g., *That's a nice piece of ass* or *You've got a nice pair of tits, love:* -IST dysphemisms of this kind can involve any body-part term).

BLS and TBT

There are two categories of abusively used tabooed body-part terms in English; we dub them BLSs and TBTs. They are demonstrated in (1–2). BLS in (1) is restricted to *balls*, *bollocks*, and the euphemistic dysphemisms *nuts*, *rowlocks/rollicks/rollocks*, and *cobblers* (derived from rhyming slang *cobblers' awls*). TBT in (2) = "taboo body-part terms excluding BLSs." These are all normally uttered with falling intonation. This is the case even in dialects like Australian, which are noted for high-rising terminals to declaratives (sometimes miscalled 'the question intonation') (cf. Allan 1984; Horvath 1985 Ch.8, and works cited in them).

(1) (i) BLS! *(To you!)*   E.g. *Bollocks!*
    (ii) *That's (just)(a load of)* BLS.   E.g. *That's a load of balls!*

Nuts, however, is not used in constructions like (1,ii).

(2) (a) (i)   TBT!   E.g. *Asshole!*
        (ii) *You* TBT!   E.g. *You prick!*
        (iii) *The* TBT!   E.g. *The cunt!*
        (iv) *What a* TBT!   E.g. *What a prat!*
    (b) (i)   NOUN PHRASE *be a* TBT.   E.g. *Tom is an asshole!*
        (ii) NOUN PHRASE *be the*
              world's biggest TBT.   [Etc.]

Any TBT or BLS can be preceded by an intensifying phrase such as *bloody fucking*, and in complement NPs by *real*. The function split between TBT and BLS is that TBTs are descriptive of Speaker's attitude to certain people or objects, whereas BLSs are used to indicate disagreement with a statement, idea, or opinion, and so overlap in meaning with such epithets as *bullshit*.

One common insult in American is to describe someone as *a scumbag* or *slimebag*, which is said to literally mean "used condom"; it is a moot point whether this is generally recognized, because *scum-/slimebag* is used to express strong dislike and contempt for the target. It therefore seems equally probable that it was compounded from the traditional epithets *scum* or *slime* and *bag*, as in *old bag*, and was extended from women to men at some point in its history. The TBTs *scumbag*, *dirtbag* and *douchebag* (the latter applied only to women according to some informants) are less dysphemistic than calling the person *a shit*; even so, they have been rather jokily euphemized to *hairbag*. This may not be such a euphemism, though: like *hairburger*, *hairbag* suggests "cunt," via the significance of public hair on a woman, discussed in Ch. 4, and *bag* as a classic figure for vagina. Also, in American, *asshole* is a weaker dysphemism than *cunt*, which is not used so widely as in Britain or Australia. In other respects the two have much in common. Not included above is the expletive *Oh bum!* or just *Bum!*, used in Australia mostly by women; it is a euphemistic dysphemism roughly equivalent to, and perhaps deriving semantically from, *Shit!*; it is not a form of the American adjective *bum* as in *a bum rap*, even though they probably have the same source historically. It is neither a BLS nor a TBT.

A *Curious Difference in Semantic Relations Revealed by a Comparison of Literal as Against Nonliteral Usage*

We have already remarked, in Ch. 4, on the vast number of synonyms for genitalia in the lexicon of English. We take it that where there are true synonyms, the expressions have identical senses. For instance, the semantics of *twat* and *cunt* are identical in literal statements like (1) and (2):

(1) *Your twat needs a wash.*
(2) *Your cunt needs a wash.*

Both have the sense "Your vagina needs a wash" (we won't elaborate on the semantics of cotextual material such as 'your' and 'needs a wash'). Apparently, the senses of *twat*, *cunt*, and *vagina* are identical—at least, they are in this context—and so they presumably have the same semantic prototype. This semantic prototype will differ from that correlated with a semantically distinct term such as *prick* in (3):

(3) *Your prick needs a wash.*

Details of the two semantic prototypes just mentioned are given in Allan (1990).

When we look at the use of terms for genitalia as insulting epithets, a curious difference in semantic relations is discovered. (Like many other such epithets, they can be used upside-down as markers of friendly or affectionate disapprobation.) In at least some dialects of English, (4) and (5) mean the same:

(4) *You twat!*
(5) *You prick!*

They mean roughly "You are stupid" or better "You contemptible idiot!" (cp. Yiddish/Yinglish *putz* and *shmuck* in Rosten 1968). These interpretations, along with all the others on which this paper is based, hold for our, southeast British (London) and West Australian (Perth), dialects, respectively (cf. Partridge 1961, 1970). In some other Australian dialects and for many Americans, nonliteral *twat*, *prick*, and *cunt* are synonymous (cf. *Macquarie Dictionary* and Wentworth and Flexner 1975). We comment on these dialect differences later. Insulting epithets (4–5) exemplify nonliteral uses of *twat* and *prick*, whereas in (1) and (3) these words are used literally. The problem for any theory of meaning that would derive nonliteral meaning from literal meaning is that these terms are, if not antonymous, then incompatible, but their nonliteral meanings seem identical—though there is a difference that we will be discussing later. One cannot bypass the difficulty by claiming that the nonliteral interpretations do not result from sameness of sense, but from sameness in reference, because this judgment is obviously incorrect: for anyone to whom (5) applies, (4) applies: there is a sense relation between *Fred is a prick* and *Fred is a twat* where these are meant nonliterally as derisive epithets. And we have already seen there is quite a different sense relation between *Hers is a twat* and *His is a prick* where these are meant literally as statements labeling genitals.

We should comment on our assumption that the sense "penis" can properly be called a 'literal' sense of *prick* rather than a nonliteral one. Some readers may be swayed by the following notion: The verb *prick* names the effect of a certain kind of event in which a sharp object penetrates a membrane; with imperceptible stretching of the imagination, it thus describes the effect of inserting the penis into the vagina: intromission is an act in which the penis is the instrument of pricking (i.e., a prick) that makes this noun deverbal. If this derivation is correct, it is also prehistoric: according to the *O.E.D.*, the noun and verb have coexisted since the earliest records in English. In addition to its current meaning, the verb *prick* has meant "to spur or urge a horse on" (*O.E.D.* 9–12; cp. the copulation-as-riding metaphor, with the man as rider); "to thrust a stick (or pointed object) *into* something" (*O.E.D.* 25; cp. the metaphor of the man as *tailor stitching* the woman that turns up in many bawdy folk songs, and in Grose 1811, as mentioned in Ch.4); and *prick up* still means "to rise or stand erect with the point directed upward" (*O.E.D.* 28). It is hardly surprising, then, that the noun *prick* was used variously for (1) a thorn, a sting, and figuratively as a vexation or torment (*O.E.D.* 12; this could be partly responsible for the current American and Australian interpretations of nonliteral *prick*); (2) a dagger or pointed sword (*O.E.D.* 15; cp. the penis as weapon metaphor); (3) the upright pole of a tent (1497, *O.E.D.* 16); (4) and it has long been a term for the penis. Although the earliest record for *prick* "penis" in the *O.E.D.* is 1592, there is the record of it being used as a term for a lover in 1540 (*O.E.D.* 17b), which suggests at least a contemporary sense "penis," too. In all probability, *prick* was used much earlier in this sense. Even if *prick* "penis" was originally nonliteral, it has established a separate identity for itself with the passing of time. The original motivation for many other words is nonliteral too, yet they are now taken to be literal: for instance, the noun *crane* "lifting device" was based on its visible likeness to the bird; the *pupil* (of the eye) was, before being adopted from Latin, a metaphorical "child of the eye" (cp. *school pupil*); now both senses are taken to be literal. One difference between *prick* "penis" and the words *crane* and *pupil* is that it seems closer to its nonliteral origin than they to theirs.

There can be little doubt that the semantic prototype for the LITERAL sense of *twat* is not identical with the semantic prototype for the NONLITERAL sense of *twat*; and similarly for the literal and nonliteral senses of *prick*. Because there is good circumstantial evidence that the nonliteral dysphemisms are based on the literal meanings, a paradox is created by the fact that the semantic relations between literal *twat* and *prick* are more or less the opposite of the semantic relations between nonliteral *twat* and *prick*. There are three reasons for believing that the nonliteral meanings of epithets invoking sexual organs, such as *cunt* and *prick*, derive historically from the literal meanings: (1) Similar kinds of genital-based nonliteral dysphemisms exist in many languages; (2) When new slang terms for genitalia come into the language, there is often simultaneous importation of both the literal and nonliteral uses; (3) In general, taboo terms function well as terms of abuse. We will consider these in turn.

Similar kinds of genital-based nonliteral dysphemisms exist in many languages, as a review of past volumes of the journal *Maledicta* will verify. Their wide provenance indicates that the literal meanings of the terms for genitals have some characteristic befitting them to function as terms of abuse. It would be too much of a coincidence if in one language after another, this nonliteral function were based on a chance perception.

When new slang terms for genitalia come into the language they are potential recruits to nonliteral usage (e.g., *flange* "vagina" is a fairly recent addition to the lexicon, and it appears in such dysphemistic epithets as *You dripping flange!*). Although there is an analogical basis for this, the simultaneous importation into the language of both the literal and nonliteral homophones suggests an interrelationship (i.e., that they are not homophones at all, but two senses of a single polysemous item). There is no guarantee, of course, that a slang term for genitalia will be used for insult because, in English, not all taboo terms can function as epithets and terms of abuse.

A Sexist Asymmetry in the Vocabulary of Abuse

Terms of insult invoking the female sex organ have a wider range than those involving the male sex organ. For instance, *prick* and *dick* are mostly applied to males and almost never to females, so that we have never heard (1–2) and they strike us as very peculiar, though some people seem to believe they are acceptable as dysphemisms.

(1) *??She's a prick.*
(2) *??She's a dick.* [≠ "She's a detective"]

On the other hand *twat*, *cunt*, or *prat* are applicable to both males and females. *Prat*, like *fanny* and American *ass*, as well as several other terms, has the literal senses "backside" and "vagina" (among others, currently irrelevant). In American, the primary sense of both *prat* and *fanny* is "backside"; in British, and to a lesser extent Australian, the primary sense is "vagina." The contrary is true for *ass/arse*.

(3) *He's a right twat/cunt.*
(4) *She's a stupid twat/cunt.*

Why should females not be called pricks? Alternatively, why should *cunt* be so untrammeled in applicability? According to Jaworski (1984/5) the distinction is not restricted to English, it applies also in Polish: for instance one can abuse a man by either *Ty huju!* "You prick! or *Ty pizdo!* "You cunt!," but a woman can only be abused with the latter; native speakers of Polish that we have consulted, however, disagree: for them *Ty pizdo* can only be applied to a woman.

By social convention, a man is downgraded by ascribing to him the characteristics of a woman. There is presumably no greater insult than to ascribe to him that supremely outward and visible sign of femaleness: the vagina. It is notable, however, that we do not say, for example, *He* HAS *a cunt*, but that *He* IS *a cunt*; so there is more to this story, and we look into it in later. One hypothesis for a woman not being abused as a *prick*, is that a woman is often not downgraded at all by being ascribed the characteristics of a man: to say of a woman that *She's got balls* or *She's ballsy* is to praise her strength of character (for instance, we have heard it said of British Prime Minister Margaret Thatcher). Why do we say *She's got balls* but not *She's got a prick?* The answer, we believe, is the following. Given that it is laudatory to ascribe a woman with the characteristics of a man, and that manhood is principally symbolized by the male sexual organs, which of those sexual organs is the most appropriate, the penis or the testes? The penis is fountain not only of semen, but also of urine, it is therefore a polluted organ. The testes, on the other hand, are not a source of polluting urine; on the contrary they are the source of the very essence of manhood: semen—which is believed by some (misguided) people to be the very source for human life. It is this which supposedly gives men the strength and courage traditionally ascribed to them—often seriously by men themselves, but tongue-in-cheek by women. It is obvious that in our quest for the principal symbols of manhood, the testes fare better than the penis. Consequently, people say *She* HAS *balls*, which defeats anomaly to be understood as "she has the strength and courage of a male"—supposedly given by those figurative 'balls.'

Why (in Some Dialects) 'a Cunt' Is "Nasty, Malicious, Despicable" but 'a Prick' is "Stupid, Contemptible"

In OUR dialects (note this qualification) there is a semantic difference between the epithet *cunt* and the epithets *prick* and *dick* when these are applied to humans (or animals) (cf. Partridge 1961, 1970). *Cunt* is primarily used to ascribe nastiness or maliciousness; that is, *cunt* means "nasty, malicious, despicable"; *prick* and *dick*, on the other hand, mean "stupid, contemptible" (cf. Partridge 1961). Thus the common British term *pillock* (from *pillicock*), which in the past was one of many synonyms of *prick*, is exclusively used today with much the same dysphemistic force as *Idiot!* And then there are *dickhead*, *dickface*, and *dickbrain*, all meaning "jerk"; compare also *he stood on his dick* which means "he behaved like an idiot, messed things right up." Although the word *dick* was long associated with silliness (cf. Grose 1811; Partridge 1961), these terms may have gotten a boost from German immigrants in Australia and America and been influenced by German *dick* "thick" (long a metaphor for stupidity) as much as by English *dick* "prick": nonetheless, they fall in very nicely with the traditional stupidity attached to *prick*. In fact there is some additional circumstantial evidence for German influence in this area: one of the terms for an erection is *a fat*, and there just could be some correlation between the notion captured by this term, and the meaning of German *dick*, "thick": Farmer and Henley (1890–1904:209) list German *Dickmann* as one of the translation equivalents of "penis." Furthermore, it is interesting and possibly relevant that American *fink* "a contemptible person; an undesirable, unwanted, or unpleasant person" (cp. American *prick*) was originally used of strikebreakers and is said to be a remodelling of *Pink* from *Pinkerton* the detective agency (cf. Wentworth and Flexner 1975:183): Farmer and Henley also list '*Pink* or *Finke* (Low German)' as translation equivalents of "penis." German *Fink* means "finch"—is it possible that this is another instance of the bird metaphor for penis? The reason, then, that (1–2) are fine, despite (3–4) being odd for many people, is that (1–2) are principally associated with stupidity, whereas (3–4) are principally associated with the penis (i.e., with their literal meanings):

(1) *She's a dickhead.*
(2) *She's a pillock.*
(3) *??She's a prick.*
(4) *??She's a dick.*   [≠ "She's a detective"]

Unlike any of *You asshole/arsehole!*; *You cunt!*; *You twat!*; *You prick!* or *He's a real, fucking, cunt!*; *He's a real dick!* and so on, the dysphemisms in (1–2) are more readily open to use with rise terminals, both in Australian high-rise terminal declaratives, and in interrogatives in all dialects.

For the purposes of discussion we are using the meanings of *twat*, *cunt*, and *prick* in our own dialects. Earlier we remarked on the asymmetry in semantic relations between, say, *prick* and *twat* when used literally as against when they are used nonliterally. So far as we know, that asymmetry holds for all dialects of English. Where dialects do differ is in whether nonliteral *prick* and *twat* (which is much rarer, and which we will henceforth leave out of this particular discussion) mean "stupid, contemptible" or "nasty, malicious, despicable." The nonliteral meaning of *prick* and *cunt* is compared by Partridge (1970) under **cunt**: '*Cunt* tends to mean "knave" rather than "fool." *Prick* tends to mean "fool" rather than "knave." ' This exactly accords with our dialects. The *Macquarie Dictionary* (Australian) gives for **prick** (7b) 'an unpleasant or despicable person,' and for **cunt** (3) '(*derog.*) any person,' though for **cunthook** 'N.Z. *Colloq.* (*derog.*) an unpleasant or despicable man' (in British English, *cunt hooks* are "fingers," cf. Partridge 1984). Wentworth and Flexner (1975) have an inadequate entry for nonliteral *cunt*, because they fail to take account of the fact that it is applicable to men; they give the meaning as 'a girl or woman' and they further note that the 'connotation may be of a sexually attractive girl or of a mean old woman.' Their entry for American **prick** (2) is 'A smug, foolish person; a knave, blackguard; a heel, a rat.' The first part of this coincides with our dialects, the latter part with the Australian; it covers both the pairs mental defectiveness + contemptibility and moral defectiveness + maliciousness.

There are no neat selection restrictions we can detect for confirming our intuitions about the different nuances of *cunt* and *prick*, *dick* and so on in our dialects, except those T-shirts sold in places like Venice Beach, California, which sport the following caveat (plus appropriate line drawings) on the back:

SEE DICK DRINK
SEE DICK DRIVE
SEE DICK DIE
DON'T BE A DICK

It is apparently modelled on 'Dick and Jane' grade-school readers. Note the double-meaning of the last line, with its pun on "Don't be an idiot." In our dialects we can make the message work in a plodding pedestrian fashion by replacing 'DICK' with 'the prick': *See the prick drink/See the prick drive/See the prick die/Don't be a prick*. One cannot make the message work by replacing 'DICK' with 'the cunt': in *See the cunt drink/See the cunt drive/See the cunt die/Don't be a cunt*. In this version, the last line doesn't work, and the whole thing is merely a pottage of obscenity. Even using the word *fanny* in the primary British and Australian sense of "vagina," doesn't produce the right effect, although like /dɪk/, fænɪ/ doubles as a proper name, for instance:

SEE FANNY DRINK
SEE FANNY DRIVE

SEE FANNY DIE
DON'T BE A FANNY.

In languages other than English, to call someone a *cunt* may be to call them "stupid" (e.g., Polish *Ty masz pizdę w głowie* "You have a cunt in your head = You're stupid"; French *T'es con* "You're a cunt = You're stupid"). Moreover in English, the archaic *coney* meant not only "rabbit; cunt" but also "dupe, simpleton". The British euphemism *berk* (from the rhyming slang *Berkeley Hunt*) always means "idiot," as in *He's a right berk, a real wally.* Even the term *twat* means "stupid, contemptible" rather than "nasty, malicious, despicable," and *prat* does, too (cf. McDonald 1988); which leaves *cunt* alone with this sense. All this suggests a general rule that insults invoking genitalia mean "stupid, contemptible," but that one term, namely *cunt*, has undergone a meaning shift toward "nasty, malicious, despicable" for some reason. There is plenty of evidence that human beings in many societies treat with contempt animals, animal instincts, and people believed no better than animals; the essence of being human is to be rational and to control or resist the animal instincts that lurk within us. Our sexual drive, and the sex organs that respond to it, are often an unruly manifestation of our animal instincts—which is why we hide them away. It is a constant struggle to keep them governed and controlled by reason; it is presumably because the prick has a mind of its own that the Latins called it *mentula* "little mind." The sex organs are unreasonable, they are stupid, and even contemptible; if they are someone else's doing something (or suspected of doing something) which arouses one's disapprobation, they may even be despicable. Legend has it that Casanova was a man governed by his lust; he would be an appropriate target for the Polish expression *Ty myslisz hujem* "You think with your prick": and Casanova is regarded as a bit of a joke and a bit of a jerk; the man was *a prick, a twat, a prat.* To ascribe the characteristics of a sex organ to someone is to dismiss them as no more reasonable in their behavior than a sex organ, and hence as stupid and contemptible.

The epithet *cunt* partakes of this meaning: one can say affectionately to someone who has performed a stupid action *You stupid cunt, what did you do that for?*, but it is notable that a modifier like *stupid* is normally required, whereas it is not necessary (though it often occurs) with the other genital epithets, and it is almost tautologous with *dickhead* and *pillock*. None of the genital epithets is normally ascribed to abstract inanimates like ideas and events, thus we have not heard:

5. ?**That's a cunt/twat/prick* [etc.] *of an idea/thing to happen.*

However, it is conceivable they might be used. All the genital obscenities we have discussed can be heard applied to concrete inanimates:

6. *It's a cunt/twat/prat/prick/dick of a fucking bolt, this is. It's fucking stripped!*

Inanimates are sometimes berated as *horrible* or *stupid* when they won't function the way we want them to, presumably under similar conditions to those applicable in (6), and perhaps the simplest way of dealing with such data is to treat the inanimate object as personified. However, if we look behind the particular terms and consider the general reason for using terms of abuse—whether to animates or inanimates—it is that Speaker is upset with (contemptuous of, irritated, annoyed, outraged at, etc.) the object of abuse. Assuming that Speaker chooses to express the upset using one of the terms we are discussing, then the manner and degree of upset will determine Speaker's choice among these terms. When the object of abuse is (concrete) inanimate, it doesn't matter which is chosen, though a degree of personification is possible.

We have been looking at similarities in meaning that *cunt* has with other terms of abuse invoking sex organs. We now want to consider what grounds there might be for its peculiar meaning (in our dialects) of "nasty, malicious, despicable person." We would argue that this results from the pollution of the vagina, which has given rise to a widespread view characterized by Grose's description of it as 'a nasty thing' (see MONOSYLLABLE, Grose 1811). The reasons were given in Ch. 3 and we just summarize them here. Because women and not men bear children, and consequently menstruate and lactate, women are perceived to be more closely bound by and to their bodies and body functions than are men, which renders women more like (other) animals and therefore closer to nature than are men. Men have had the opportunity to expend time and energy on things of the mind rather than of the body; that is, to control the domain that supposedly distinguishes humans from animals. The association of women with the animal side of humans, and men with culture and the ideas that distinguish humans from animals, quite naturally produces a cultural, and hence social, appraisal in which men are superior to women (we remind the reader that we are reporting this, not endorsing it). The effects of this that are relevant to our particular theme are two kinds of pollution taboos: (1) the pollution of women's unique physiological processes, in particular the vagina and its effluvia; and (2) the downgrading of a man by ascribing to him the characteristics of a woman, in contrast with the converse: a woman is not generally downgraded to a comparable degree when ascribed the characteristics of a man.

The notion that the pollution taboo is responsible for the primary sense of *cunt* as "nasty, malicious, despicable", receives some justification from a comparison of sentences like (7–8).

(7) *He's a cunt, a real shit of a guy.*
(8) *She's a shit, a real cunt!*

These sentences show that if someone is a nonliteral *cunt* it is often appropriate to call them a nonliteral *shit*. The survey of revoltingness of various body-parts and bodily effluvia, reported in Ch. 3, revealed that shit is (together with vomit) the most revolting bodily effluvium; it is reasonable to assume that shit is subject to a very strong pollution taboo. If it were the case that both *cunt* and *shit* invoke pollution of similar strength (shall we say), it would explain why they can be felicitously yoked together in epithets like (7–8). It is relevant to recall at this point that several vocabulary items are ambiguous between "backside" and "vagina": *hole, front bum,* children's usage of *bottom* or *botty,* American *ass* and *tush,* and British *prat* and *fanny.* (And, interestingly, in Middle Dutch, *kont* meant "cunt"; whereas in modern Dutch it means "arse".) Both body-parts are at the lower abdomen, and although none of the terms for "backside" actually mean "rectum," there is a pair of salient and polluting passages. If we add into the argument, first, that the backside is an erogenous zone, and second, the practice of buggery, then there can surely be no doubt that the association is made between backside and vagina. The association between *cunt* and *shit* we are hinting at here is tenuous and indirect—vagina/backside to vagina/rectum to vagina/feces; but the ambiguous terms leave little doubt that it is made. Opinions seem to differ on the meaning of the American epithet *He's an asshole!*: one meaning given to us for *asshole* is "an inconsiderate or abrasive or arrogant or unkind person; stupidity is not particularly involved"; it therefore means much the same as *He's a shit/cunt!* but is less strongly tabooed than either. Since American *ass* = "cunt," this may influence the meaning of *asshole*. However, we have also been told that 'if he's an asshole, then he is a dummy; but if he's prick, he's just malicious' and *assholery* is "behavior of conspicuous, and annoying or inconvenient, stupidity." We will comment on the upshot of this in due course.

We do not claim that the two epithets in (7–8), *shit* and *cunt*, mean the same thing: (7) can be glossed "He's a mean, despicable character, a really obnoxious guy"; and (8) "She's obnoxious, really despicable". However, each implicates the other—not necessarily, but with a high degree of probability; such that: *if X finds Y despicable, then X probably finds Y obnoxious* and conversely *if X finds person Y obnoxious, then X probably finds Y despicable*. Because the epithet *shit* has wider currency than the epithet *cunt*, the following would hold for a smaller population: *if X judges that Y deserves the epithet 'cunt,' then X would probably agree that Y deserves the epithet 'shit'* and conversely *if X judges that Y deserves the epithet 'shit', then X would probably agree that Y deserves the epithet 'cunt.'*

(7) is semantically distinct from (9).

(9) *He's a prick, a real shit of a guy.*

This can be glossed "He's stupid and contemptible, a really obnoxious person"; and Speaker is offering two rather different (though by no means mutually exclusive) judgments, because there is no high probability that a contemptible person will be obnoxious, nor conversely that an obnoxious person will be contemptible. The use of *dick* or *twat* within this framework is more similar semantically to (9) than to (7–8):

(10) *He's a dick/twat, a real shit of a guy.*

Because it is more strongly dysphemistic to call someone a *shit* than to call them a *prick* or *twat* the sequence in (11) is preferred to that in (12):

(11) *He's real prick. In fact he's a complete shit!*
(12) *??He's a complete shit. In fact he's a real prick!*

(12) is odd by comparison with (11) for the same reason that (14) is odd by comparison with (13):

(13) *She's very pretty. In fact she's utterly beautiful.*
(14) *??She's utterly beautiful. In fact she's very pretty.*

The reason for the difference is the normal presumption arising from the maxim of quantity that Speaker will make the strongest claim possible consistent with his or her perception of the facts, and in (11) and (13), the second statement is stronger than the first, follows it, and is therefore taken to offer a correction of it. What we are dealing with here are something akin to scalar quantities. 'If a speaker asserts that a lower or weaker point [. . .] on the scale obtains, then he implicates that a higher or stronger point [. . .] does *not* obtain' Levinson (1983:133). Acceptable alternatives to (12) and (14) recognize this convention by using some locution such as (15–16):

(15) *He's a complete shit. Well, not exactly; but he's a real prick!*
(16) *She's utterly beautiful. Well, that's an exaggeration; but she's very pretty.*

Now consider

(17) *He's a cunt. In fact he's a shit of a guy.*
(18) *He's a shit. In fact he's a real cunt.*
(19) *??He's a cunt. Well, not exactly; but he's a shit of a guy.*
(20) *??He's a shit. Well, that's an exaggeration; in fact he's a real cunt.*

What (17–20) reveal is that there is no difference in dysphemistic strength between nonliteral *shit* and nonliteral *cunt*. We conclude that the primary meaning of *cunt* as "nasty, malicious, despicable" has been determined by the pollution taboo on the vagina, a pollution taboo that is comparable to the one on shit. It presumably need not have been *cunt* that did this, but it would have to have been one of the terms whose literal meaning is "vagina." It may have been the relationship between *cunt* and *queynte* or *quaint* that tipped the balance; the relevant, but now obsolete sense of *quaint* being "cunning, crafty, given to scheming or plotting", (*O.E.D.* 1b).

We said earlier that insults and epithets indicate that the target is socially unacceptable because s/he is subhuman or a physically, mentally, or morally defective human, and/or because the target's behavior is contemptible or despicable. The genital insults and epithets we have been discussing select either mental defectiveness—stupidity— along with contemptibility, or moral defectiveness along with despicability—the latter being the more offensive. It would seem that for some Australian speakers (perhaps a majority), the prototypical genital insult selects only the latter (cf. *prick* and *twat* in *The Macquarie Dictionary*); the situation is more complicated in American if we believe Wentworth and Flexner's *Dictionary of American Slang*. In English, these genital terms are used for the most offensive kinds of insult and epithet; they are more insulting than, for instance, calling someone *a pig*, or *spastic*. If insults and epithets must work within the confines of social unacceptability identified earlier, why not assume for any genital based insult the interpretative strategy envisaged for the inanimates in (6), such that *You prick!* would be interpreted simply as "Speaker is upset with [you]"? The reason is just that it would render all terms of abuse synonymous (and therefore obscure the difference in meaning between *You prick!* and *You cunt!*); and it would deny that any explanation is conceivable for the nonapplication of *You prick!* to girls and women. In order to be insightful, it is necessary for us to consider what lies within the semantics of these genital terms that befits them for use as insults and epithets: and that is exactly what we have been doing!

In Summary of This Discussion of TBTS Used as Terms of Abuse

Even though the nonliteral sense of words like *twat, cunt, prat, prick*, and *dick* is based upon the literal sense, each member of the paired (literal and nonliteral)

senses has its peculiar semantic relations. Literal *twat*, *cunt*, and *prat* are synonymous, and—within the field of human genitals—antonymous with *prick* and *dick*.
Nonliteral *twat* and *prat* are synonymous with each other and with *prick* and *dick*,
except that the latter may not be used of females; nonliteral *cunt* is semantically
different from all of these in our dialects. We have examined the belief system that
underlies the differing usage of terms of abuse invoking male and female sex organs
and sought an explanation for the semantic differences in these beliefs. The constraint against using *prick* and *dick* of females and the concomitant lack of constraint
in using *twat*, *prat*, and *cunt* of males, presumably reflects the convention—not yet
revoked within our community—that it is no abuse to ascribe the characteristics of a
man to a woman, but it belittles a man to ascribe to him the characteristics of a
woman. Consistent with the traditional community view of the vagina as a greater
source of pollution than the penis, is the fact that in our dialects *cunt*, a term that
invokes female genitalia, means "nasty, malicious, despicable" rather than "stupid,
contemptible." It is striking, therefore, that in the same dialects the epithet *cunt* and
the epithet *shit* seem to be mutually implicative.

So far as we know, the asymmetry between the literal and nonliteral senses of the
TBTs we have been discussing holds for all dialects of English. Where dialects do
differ is in whether nonliteral *prick* denotes "a stupid, contemptible fool" or "a nasty,
malicious, despicable knave"—which is the denotation of nonliteral *cunt* in our
dialects. Wentworth and Flexner's *Dictionary of American Slang* has an unsatisfactory entry for nonliteral *cunt*, the relevant part of which they incorrectly restrict to 'a
mean old woman'; but they give for American *prick* (2) 'A smug, foolish person; a
knave, blackguard; a heel, a rat.' The first part of this coincides with our dialects, the
latter part with that of many other Australians; altogether it covers both mental
defectiveness + contemptibility and moral defectiveness + maliciousness. The
Australian interpretation of *prick* may result from the spreading of the pollution
taboo on genitalia from female to male sex organs; or both it and the American
interpretation may have been influenced by the now obsolete figurative use of *prick*
as a goad, thorn in one's side, a pain in the proverbial. Spreading contamination
may also explain why, for many Australians, the sense of nonliteral *twat* is like that
of *cunt*; and also for our suspicion that nonliteral *prat* shares the meanings of *twat*
and *cunt* in our dialects.

Further Constraints on the Use of Taboo Terms in Verbal Abuse

It is interesting that, in English, only certain terms can function as epithets, expletives, and terms of abuse: for instance, learned words for taboo topics generally do
not; consider

(1) *Defecate on you! *Urine off!
(2) *Copulate off! *What a fornicate up!
(3) *You anus! *You vagina! *You clitoris! *You penis! *You foreskin!
(4) *Oh foreskin! *Oh prepuce! *Oh feces!
(5) ?*He's a real fucking vagina!

The only way that vocatives like (3) or (4) could be acceptable is in rapturous address to the organ or effluvium itself (i.e., they would have to be taken literally and not as intended insults). If (5) is ever used, it is exceptionally precious. This is true for present day English, but not necessarily for earlier varieties of the language (cf. Jonson's 'A horson filthie slaue, a dungeworme, an excrement!', quoted earlier) or for other languages (see Crowley [forthcoming] for a Paamese translation equivalent of "You penis!"). In present day English, BLSs are subject to the same verbal abuse constraint as we saw in (1–5).

(6) *?*That's a load of testicles!*
(7) *?Testicles!*

(6) is very odd as an expression of disagreement—the tendency is to take it literally; but (7) can get by as joky—we know, one of us has used it—but, like (5), it is precious. In general, however, learned terms are not stylistically appropriate to the task of insult (i.e., they are not intrinsically dysphemistic enough). The same would appear to be true of the colloquial taboo body-part terms in (8):

(8) **You poontang! *You pussy! *You quim! *You willie!*

Once again we get the effect noticed in (3–4). The terms *poontang*, *pussy*, *quim*, and *willie* are also inappropriate as epithets or terms of abuse; rather, they seem to be terms of endearment.

We are forced to conclude that epithets, expletives, and terms of abuse are idiomatic, and that some terms are intrinsically more dysphemistic than others—a fact that will need to be noted in the lexicon.

Maledictions Naming Disease and Mental or Physical Abnormality

Many languages invoke disfiguring and, more especially, deadly diseases in maledictions. Current English no longer does so, though A *pox on/of you!* (principally smallpox) was used in early modern English. For instance, Falstaff's

> A pox of this gout! or, a gout of this pox! for the one or the other plays the rogue
> with my great toe.   (Shakespeare *Henry IV Pt.2*, I.ii.246)

With his usual aplomb, Shakespeare puns: the first 'pox' is "smallpox," the latter "venereal disease." There is also his *a plague o' both your houses* (*Romeo and Juliet* III.i.92)—invoking bubonic plague with its blotchy red sores, pneumonic problems, and death. We still call someone *a poxy liar* in current English; and also *pest*, a word derived from the French word for "plague" (cf. English *pestilence*). The French formerly used the expletive *Peste!* For example, the Dutch still use *Pestvent!* "Pesky guy," or *Hij is een pest* "He is a plague": these are fairly mild; not much stronger than the English *pest* (adjective *pesky*) with which they are cognate. Also in current English we call someone *a* (figurative) *leper* meaning "a person who is shunned." The same disease is invoked in the Korean insult *muntungisɛkki* "young of a leper," roughly comparable with American *son-of-a-bitch*. In other languages we find cholera invoked (e.g., in the Polish expletive *Cholera!*, which is roughly comparable in function to English *Shit!*). It is also invoked in the Thai imprecations *Tai hàa!* "Die of cholera," *Ai hàa* "Cholera on you" (to a male, *Ii hàa* to a female). These are matched by Yiddish *A kholerye afloyf im!* "May he get cholera!"

It is notable that all the diseases invoked in these maledictions—smallpox, bubonic plague, leprosy, cholera—are deadly (cp. the exhortation to *drop dead*); the first three are also disfiguring. It is hardly surprising that they are utilized in curses of various kinds.

Charged with notions of pollution, corruption—sometimes also evil—disease (see Ch.7) can itself become a very evocative metaphor for talking dysphemistically about the things, even the people, one wishes to denigrate. Lifton and Olson (1976:99), for example, in discussing the effects of the Hiroshima explosion, describe nuclearism as 'a peculiar twentieth-century disease of power.' In 1946 George Orwell described politics as 'a kind of mental disease.' If something is diseaselike, it is at least unsound, undesirable, and distressing, but often also hateful, offensive, disgusting, or contaminated, depending on the nature of the disease invoked. Hermann Ahlwardt, in an 1895 Reichstag address aimed at halting Jewish immigration into Germany referred to the Jews as 'cholera bacilli' (see Dawidowicz 1975:54). The same imagery of filth and disease was to be taken up time and again by Hitler in his descriptions of both Jews and *Mischlinge* (people of mixed racial origin). In this case, though, the language is perhaps deceptively metaphorical, since—given Hitler's twisted sense of reality—it is conceivable that he intended the literal meaning of the words. Sontag (1979:57–85) discusses at length the use of the disease metaphor, particularly as a staple of political polemics. The examples she cites both from literature and life are striking. Trotsky, for example, likened Stalinism to a cholera, a syphilis and a cancer 'that must be burned out of the labour movement with a hot iron.' In 1973, John Dean, special adviser to the U.S. president, in speaking of Watergate problems to President Nixon warned 'we have a cancer within—close to the presidency—that's growing'. Even Sontag herself wrote in reference to America's involvement in Vietnam that 'the white race is the cancer of human history.' On November 15, 1989, the Federal Commissioner of Police in Australia expressed concern over what he described as 'the cancer of corruption' in the federal police force.

Cancer with all its images of decay and corruption is a particularly effective and powerful metaphor. Of course, the image is all the more horrifying because the disease does not necessarily have any visual symptoms nor is its etiology fully understood. And perhaps lurking here still is something of the medieval concept of the disease. The name *cancer* is founded on the early belief that the disease was produced by a mysterious and deadly crablike demon. At the time, frightful treatments were felt necessary to eradicate the fiend; and arguably, cancer treatments today involving chemotherapy and radiation are no less frightful (cp. the adage 'the cure is worse than the complaint'). The picture this creates, though, is a useful one in political polemics, since it implies urgent and radical measures if the malignancy is to be destroyed—an effective image, particularly, if the intention is to rally people to a fight. As Sontag points out (pp. 63–64), even the language of cancer treatment, like that for many other diseases, has its own aggressive militaristic hyperbole attached. We speak of the 'fight' or the 'crusade' against cancer. Treatment involving radiotherapy is said to 'bombard' the body with toxic rays to 'kill' the 'invasive' cancer cells that the body's own 'immunodefense system' cannot 'destroy.' The military metaphor is very apparent in the publicity for groups like the American Cancer Society: their stationary, pamphlets, and posters used to depict a fervant crusader, often St George, taking his sword to a dragon labeled "Cancer." Popular articles also carry titles like 'The Common Enemy,' 'The Great Peril,' 'The Greatest Scourge in the World' (cf. Patterson 1987:91–97). All this makes cancer a far more terrible metaphor than other disintegrating or wasting diseases. In its metaphorical use, it represents the embodiment of all evil, and in a society that no longer has any strong religious conviction of evil, we may find ourselves relying more and more on disease metaphors of this kind. With the mounting anxiety surrounding the Acquired Immune Deficiency Syndrome (AIDS), a new disease metaphor is appearing on the scene: Sontag (1989) quotes Palestinian writer Anton Shamas writing about "the AIDS of 'the Jewish State in the Land of Israel' " (p. 67); and also refers to a computer virus that was dubbed *PC AIDS* (p. 70). Although there is undoubtedly less mystery attached to AIDS (we know, for example, that it is caused by a virus and transmitted predictably via specific types of contact with a carrier), it is still an awful disease and one that carries a fatal prognosis—'slow motion death' one magazine describes it (*Oui*, October 1983:84). The imagery surrounding it, though quite different from cancer, is just as fantastic: notions of pollution, promiscuity, immorality, weakness, and vice are already identified with the disease and, charged with this sort of meaning, it is potentially as powerful a metaphor as *cancer*.

Current English retains a number of imprecations and epithets invoking mental or physical abnormality—arising from the stigma attached to such abnormalities. Epithets ascribing mental subnormality include *Retard!, Moron! Idiot! Cretin! Nincompoop! Halfwit! Nitwit! Fuckwit! Fuckhead!* (these last two are doubly dysphemistic in that they not only ascribe mental abnormality, but do so using a dysphemistic locution which unscrambles as "your wits are (/ head is) fucked$_2$"); *It's stupid*; *It's silly*; and so on. All of these ascribe stupidity—or in the least, they purport to do so. Naturally this reflects the stigma attached to mental subnormality, which requires euphemisms for the genuinely subnormal. *Cretin* began as a Swiss–French euphemism *crétin* "christian" (a charitable recognition in a christian country that even the mentally subnormal are blessed, and therefore christians, according to some theologians); however it has sunk to dysphemism. Hock (1986:303) suggests that the attribute *christian* to someone mentally impaired could result from the observation that a 'person who in true Christian fashion turns the other cheek when attacked' must be pretty stupid! However much one might wish to agree, such a notion would have been heretical in eighteenth-century Europe. Our skepticism seems justified by the fact that *silly* once meant "blessed, blissful" (cp. modern German *selig*) and then changed to something closer to "innocent, helpless" and so was used principally of women, children, and animals (e.g., Chaucer's *sely widwe* in the 'Nun's Priest Tale'). It is a short step from "helpless" to the current, only mildly dysphemistic, meaning of *silly*. We shall see in Ch. 7 that most colloquial terms for the mentally disabled start out as euphemisms and degenerate. In fact that euphemistic term we just used, *mentally disabled* may be going the same route to dysphemism as its forerunners, since it is being replaced by *mentally challenged*.

The mental abnormality implicit in *crazy* and *nutter* is a dysphemism that has been usurped as a term of praise among certain macho hooligans: their antisocial behavior is decried as 'crazy' by society in general, but they take such titles as *crazies* and *nutters* as badges of honor in consequence of their revolt against the social norm.

Expressions like *He's a jerk; It's spastic, I was spastic* all suggest the jerky movements of true spastics, who until recently were all presumed to be mentally retarded as well as physically abnormal; hence these expressions all mean "no good, useless, stupid." Terms like *cripple* or *paraplegic* are normally ascribed to someone who has been physically inept in some way or another; similarly with a question like *Are you blind?*, which can be dysphemistic about someone's visual perceptiveness. Dysphemistic epithets like *Fatty!, Baldy!, Four-eyes!*, or *Short-arse!* pick on real physical characteristics that are treated as though they are abnormalities. Most varieties of English have no commonly used terms of insult that refer to the putative abnormality of a person's genitalia; that is, no counterpart to Bislama terms like *Big bol!* "Big balls!", *Bol blong yu i hang wansaed!* "Your bollocks hand to one side!" or *Bol blong yu i saen!* "Shiny balls!" (cf. Crowley 1989:103f). Epithets like all of these merge into racist dysphemisms, and dysphemistic epithets based on behaviors Speaker disapproves of, such as homosexuality or ideology.

Two Quasi-Euphemistic Uses for Normally Dysphemistic Taboo Terms

> Hayduke stumbling around in the dark, blue light glimmering. "All right now, everybody up, everybody up. Drop your cocks and grab your socks. Off your ass and on your feet. . . ."   (Abbey *The Monkey Wrench Gang* 1975:171)

There are two quasi-euphemistic functions for normally dysphemistic taboo terms: one is to stimulate a sexual partner, the other is during friendly banter.

You don't have to delve deeply into pornography to be amazed at what turns people on. This leads us to speculate that items from the full range of abusive terminology, the full range of terms for anal and genital effluvia and other effluvia resulting from any manner of sexual intercourse, plus tabooed body-part terms, will be used in stimulating someone's sexual partner somewhere. Many of the expressions used would, by 'the middle class politeness criterion', be regarded as dysphemistic; but under the conditions just described, they are used in euphemistic illocutions. We therefore classify this activity as a species of dysphemistic euphemism.

Another species of dysphemistic nondysphemism occurs in friendly banter. It is marked by the use of normally abusive address forms or epithets that are uttered without animosity, which can be reciprocated without animus, and which typically indicate a bond of friendship:

(1) FIRST YOUTH   Hullo congenital idiot!
    SECOND YOUTH   Hullo, you priceless old ass!
    DAMSEL   I'd no idea you two knew each other so well!
    (Quoted from a *Punch* cartoon of many years ago by Stern 1965:323)

In (1), and in the quote from *The Monkey Wrench Gang* at the head of this section, the use of a normally dysphemistic term is partly motivated by Speaker's intention to say something striking, and so gain greater effect from his or her words. Here are some recent examples.

(2) [*Two urban working class Australian Aboriginal girls*]
    A Gimme the smoke if you want it lit Eggbert.
    B Here shit-for-brains. [*Passes the cigarette*]
    A Geez you're a fuckin' sook. I swear to God.
    B Shut up fucker . . .

(Allen 1987:63)

    A If I had a pussy like yours I'd take it to the cats' home and have it put down . . .
    B If I had brains like yours I'd ask for a refund . . .
    A Well, if I had tits like yours I'd sell them off for basket balls . . .

(Allen 1987:62)

(3) [*Two working class urban Australian Aboriginal males*]
    A Have you got a match?
    B Yeah, your prick and a jelly bean.

(Allen 1987:66)

These are all examples of what the African Americans call 'cracking on' or 'capping'. Whites and Blacks also call it 'ranking', and *I ranked him out* means "I won." American Blacks also play 'the (dirty) dozens' (i.e., cracking on about the addressee's mother or kin). The following identifies the danger that any tease runs of becoming, or being taken for, insult. (Where necessary, glosses are given in [*italics*]).

> We be sittin' round, get to talkin'. Then, righteous somebody get to sayin' somethin' 'bout some else's momma. "Aw, y'momma!" They be playin'. "You momma's a dog!" Start d'momma's game! I shoot on my pa'tner [*tease one of the guys*], "You momma wash her hair with Babo"—shit like dat. Gets real dirty sometime—"Momma's pussy so big, airplane park in it!" They shoot on you great-granddaddy and you uncle—jus' through the whole generation. Like dat's dirty dozen—shootin' on d'moms, else on kin. Den gets to talkin' 'bout each other. You don' be sayin' somethin' 'bout the family, you talk 'bout him. You see somethin' you wanna talk 'bout, like you see his hair real nappy [*wooly, kinked*], you say his hair's like Brillo—jus' trippin' off [*amusing*] each other.
>
>   We don't play d'momma's dozen too much. That starts confusion. . . . Don' shoot on d'moms less'n you fittin' [*intending*] to fight. You know, when we fittin' to fight, maybe shoot on moms. But mostly we talk 'bout each other. We leave d'moms at home, 'cause she too old to be in the street anyway! Do dat among friends. Way of blowin' off steam, of releasin' yourself. You wanna few laughs. But, it's mostly done by friends. You usually cap with someone who don't take offense. (Folb 1980:93)

As with so many other cultures: you can crack on to your friends about themselves, but it is riskier to crack on about their kin, particularly about their mother.

    Expressions that tease are not properly dysphemistic, nor properly euphemistic; intuition suggests that there is a grading between the two, but we will arbitrarily call everything within this gray area quasi-euphemisms. Dysphemistic euphemisms and quasi-euphemisms of this kind are sanctioned by the following conditions: if A can use a dysphemistic locution to B such that A has no intention (strictly speaking, no reflexive-intention) that *B* interpret it as such a dysphemism, then *A* must be relying on some contextual evidence—such as the tone of voice used, topic to hand, and the nature of the friendship—to implicate that *A*'s apparently dysphemistic locution has (or contributes to) a less dysphemistic, even euphemistic, illocutionary point.

We said at the outset to this section that there are two quasi-euphemistic functions for normally dysphemistic taboo terms: one is to stimulate a sexual partner; the other is during friendly banter. We should extend the latter category to cover expressions of mateship and endearment like *fuckster*, and the epithet 'a cute little shit' in *Have you seen Edna's baby boy? He's a cute little shit isn't he?*. Or 'silly bugger' in *Joe's a silly bugger, he should never have married that woman.*

Summary of Chapter 5

We began by arguing that because taboo terms are by definition dysphemistic; they provide off-the-shelf dysphemisms for use in insults, epithets, and expletives. They are used as expletives because of the cathartic feeling there is in breaking a taboo— even when the act of violating the taboo is recognized as a conventional form of behavior.

Reviewing abusive forms of naming and addressing, we noted that English (like other languages) contains insults and epithets in which some salient unpleasant characteristic from the folk concepts about the appearance and or behavior of an animal is attributed to the targeted human. -IST dysphemisms, on the other hand, pick on some personal characteristic of the individual in order to downgrade it. Then there is the dysphemistic manipulation of style through which Speaker demonstrates disaffection, disdain, or anger with Hearer-or-Named. This is accomplished by distorting the normal relative social status or social distance between Speaker and Hearer-or-Named in such a way as to affront the latter's positive face (i.e., self-image).

Despite the secularization of English speaking communities during the twentieth century, profanity (blasphemy) is still a potent source for dysphemism. Moreover, we postulate that phrases like *God damn* X, which invoke God as an agent of malfeasance, and which are found euphemistically abbreviated to *Damn* X in order to avoid explicit blasphemy (?profanity), provide a model for more potent imprecatives such as *Shit on* X, *Fuck₂* X, or *Bugger* X. Thus we can account for a set of otherwise inexplicable dysphemistic constructions.

Our discussion moved on to terms of abuse naming bodily effluvia, tabooed body-parts, and sexual and anal pastimes. There are several reasons for their use in maledictions. One is that these are taboo terms and therefore off-the-shelf dysphemisms. A second is that the body is the vehicle for many metaphors. Douglas (1966) recognizes that it is a metaphor for society, but it offers many more figures than this: think of nonliteral uses of *head, face, eye, mouth, neck, shoulder, arm, elbow, hand, finger, back, bottom, knee, leg, foot,* and so on (cf. Anderson 1978; Heine, Claudi and Hünnemeyer [forthcoming]; MacLaury 1989). The body's waste products are also potent vehicles for figurative usage, in particular of things that one has no use for, things that one wants to dispose of, get away from, and so on. There is also the body versus mind, the animal versus the intellectual, within a human being: the animal side of man is despised, and those body-parts, body functions, acts and actions regarded as animal rather than intellectual (the response to calls of nature) are the ones that occur in expletives and terms of abuse. We discovered that the most revolting bodily effluvia—feces—and to a lesser extent, urine, fart, vomit, and perhaps sperm, are invoked.

The most severely tabooed body-parts are the anus and genitalia, and they provide the figurative insults, epithets, and expletives based on body-parts. Learned terms cannot be used for such a function though they can be dysphemistic when used in an inappropriate style (see Ch.8). We explained why *cunt* is applicable to males and inanimates as well as women, whereas *prick* is almost exclusive to males. And we guessed at what, in some English dialects, lies behind the differing semantic implications of these two words (and to some extent their abusive synonyms), namely "nasty, malicious, despicable" versus "stupid, contemptible."

We noted that maledictions invoking disfiguring, normally fatal, diseases do not exist in modern English, although they do in other languages. But English does use insults and epithets ascribing physical and mental abnormalities to people, just as it ascribes to them the unflattering characteristics of various animals.

Finally, we looked at exploitations of abusive terms to display friendship markers and to tease friends (and lovers); this was what might be called the upside-down use of abuse, and it gives rise to dysphemistic euphemisms and quasi-euphemisms.

CHAPTER
—6—

Never Say Die:
Death, Dying, and Killing

Ay, but to die, and go we know not where,
To lie in cold obstruction and to rot,
This sensible warm motion to become
A kneaded clod; and the delighted spirit
To bathe in fiery floods, or to reside
In thrilling region of thick-ribbed ice;
To be imprisoned in the viewless winds,
And blown with restless violence round about
The pendent world; or to be worse than worst
Of those that lawless and incertain thought
Imagine howling—'tis too horrible!
The weariest and most loathed worldly life
That age, ache, penury, and imprisonment
Can lay on nature is a paradise
To what we fear of death.
(Shakespeare *Measure for Measure* III.i.118)

Death is a Fear-based Taboo

Death taboos are motivated by the following fears: (1) Fear of the loss of loved ones; (2) Fear of the corruption and disintegration of the body—the body with which one has so long been familiar in life is suddenly to become abhorrent; (3) Death is the end of life, and there is fear of what follows—there can be no first hand experience of death for the living; (4) Fear of malevolent spirits, or of the souls of the dead. To these four fears we will later add a fifth. People's horror of death, and their extreme reluctance to name it directly, has been well documented (e.g., Brain 1979). Ancient Greek and Latin have exact counterparts to the English *if anything should*

153

happen to me, which demonstrates a very persistent euphemism. The reason for its longevity is its subtlety. Taken literally, this idiom would violate the cooperative maxim of quantity, Speaker should make the strongest claim possible consistent with his/her perception of the facts, while giving no more and no less information than is required to make his/her message clear to Hearer: there can be no doubt that SOMETHING will happen to Speaker. Such a blatant seeming-violation implicates that what is denoted will be the ultimate happening in Speaker's life. There is no 'if' about death: in fact, it is the one thing certain in life; the only uncertainty is when it will happen. So this persistent euphemism, *if anything should happen to me*, not only avoids mentioning death, it seemingly pretends that death is uncertain!

Taking out *life insurance* we would seem to be squaring up to the inevitability of personal death; but by utilizing *death*'s antonym *life* in the compound 'life insurance' we deny the reality of what it is we are dealing with. In the documentary 'Breaking the News' (Australian Broadcasting Commission's, *The Health Report*, January 22, 1990) a doctor was recorded advising his patient of terminal illness with the words *I think it's time you got your affairs in order*; and not once during their interview was direct mention ever made of the patient's imminent death. Alan Peterson, a Sydney journalist, reports on the use of this same euphemism: 'There was some talk at the session on retirement about "having your affairs in order." This means being ready to die with all matters arranged to cause the least inconvenience to my friends, my lawyer, the Taxation Office and the undertaker" (Peterson 1986:48). In its failure to come to terms with death, modern society relies strongly on what Fryer (1963:19) called the 'protective magic' of euphemisms like this.

In this chapter we explore the changing attitudes toward death in Western societies, beginning with the Middle Ages. We confine ourselves to the Western experience of death, leaving a study of cross-cultural variation to thanatologists.

Memento Mori—Death and Dying in the Middle Ages

History has surely seen no other period of time so obsessed with death as the middle ages (cf. Huizinga 1924 Ch. 11). We all (well, most of us anyway) have a horror of dying and burial; but two things in particular show the extent of the medieval preoccupation with death. During this time, two books appeared—*Ars Moriendi* "The Art of Dying" and *Quattor Hominum Novissima* "The Last Four Things"—in which were laid out in detail for the reader both the horrors of death and also the most appropriate behavior for the deathbed. They were effectively manuals on 'how to die' and, with the advent of printing in the fifteenth century, became best sellers. A second curious product of the Middle Ages was the 'Dance of the Dead'—a graphic lesson in death that originally took the form of large murals depicting the procession of living figures together with skeletons and cadavers. The Dance of the Dead inspired a number of other modes of artistic works also illustrating human mortality. One was known as *Memento Mori* "remember you will die." Works in this tradition depicted skeletons, usually holding an hourglass on the point of running out, interrupting individuals at their work in order to summon them away to die (cf. woodcuts in Holbein 1538). Such scenes of death can be found throughout Europe, on the pages of manuscripts, in embroideries, on the walls and the stained glass windows of churches—they were once even performed! All levels of society and all ages are represented there and the message is clear: death is everywhere and can strike at any time; no one escapes. And of course this message would have become all the more real as the Black Death swept through Europe bringing death on a scale comparable with a nuclear holocaust. It was made all the worse by a number of subsequent epidemics. In 1430, Europe's population is reputed to have been 50 to 75 percent lower than it had been a century and a half earlier (cf. Gottfried 1983:xvif,134). Such a devastating effect on people is likely the cause of many of the bizarre mass emotional disorders that have been recorded for the time (see Gordon 1959:545–79 on epidemic psychoses during the Middle Ages). It is not surprising that it is the fourteenth century that gives us the word *macabre*—one word that sums up so neatly the whole conception of death at this time. It first occurred in 1376 in Jean le Fèvre's *Respit de la Mort*—the origin of the gruesome *danse Macabre* "dance of Macaber" (the word is believed to be based on a proper name) that became known as the aforementioned 'Dance of the Dead.'

Medieval people were obsessed with the physical aspects of death and dwelled at great length on the decay of beauty and the horrors of the rotting and corrupting corpse. Paintings, woodcuts, and carvings of the period show bodies in various states of decay in frightening detail—skeletons with vestiges of rotting flesh, bellies writhing with worms, snakes, and snails. Chilling inscriptions remind constantly of the thin line between life and death and the ultimate decay of beauty: 'I was that which you are; and what I am, that you will be' (see Boase 1966:240–42 for details of this and similar inscriptions). There is certainly nothing euphemistic about most of the poetry, sculptures, and paintings of this time: death is depicted there with gruesome literalness. Not surprisingly, this literalness is also mirrored in the language, and in the Middle Dutch texts examined the language was usually straightforward. Writers talked explicitly about the various "signs of death" and "of dying" (*teken vander doet/ van steruen*) and were quite blunt about the prospect of "whether the patient would die or live" (*weder een sieke steruen of leuen sol*). Though the contagion of a sick person was cause for euphemism (see Ch.7), there was little attempt to avoid direct mention of death.

> *Alle seide dat hi doot ware ende niet leven en soude toten*
> *avonde . . . dat hij sterven soude*
>
> "Everyone said that he would be dead and wouldn't live till the
> evening . . . that he would die"
>
> *Die mensche starf des sevende daechs*
>
> "The person died on the seventh day"
>
> *Doen seide Meester Colaert die man waer doot*
>
> "Then Master Colaert said the man was dead"
>
> *Sonder twifel die sieke sal sterven*
>
> "Without doubt the patient will die"

There were some occasional attempts at delicacy, like the three following examples, but these are few. On the whole, these Middle Dutch texts are surprisingly free of the evasive circumlocutions so often cited in discussions on euphemism and death.

So sel hij hem verenichen mit gode

"So he will join himself with God"

Vliet van hem ende beveelten gode, dat es miin raet

"Flee from him [the dying man] and let him call on God—that is my advice"

Sonder twifel die sieke en sal niet ghenesen ende daerom vliet van hem ende onderwindes u niet, dat is miin raet

"Without doubt the sick person will not recover and therefore flee from him and don't undertake anything, that is my advice"

One problematic aspect, however, is the curious absence in Middle Dutch of any verb cognate with English *die* and Proto-Germanic **dawjan*; however, there are frequent examples of the deverbal noun and adjective form *doet* "dead; death." In this sense the verb has, in fact, now totally disappeared from Continental West Germanic. The *Middelnederlandsch Woordenboeck* (1885–1952) does cite a Middle Dutch reflex of the verb; namely, *doyen/douwen* which is glossed there "to die or waste away." Very few examples are given, however, and those that are show a figurative application (e.g., to a flame dying); or else they have the meaning of "to languish." It is from this figurative use that Modern Dutch cognate of English *die*, namely *dooien*, derives its meaning "to melt." The verb can no longer be used of humans, except with a negative prefix, when it has the meaning "to come to life"— *ontdooien*. It is also notable that there was no use of the verb "die" in any Old English text, either (cf. Pyles 1971:357). Because of this, some people have assumed that the verb disappeared from English, and was later re-adopted from Scandinavian. With this in mind, how are we to assess the directness of the term *sterven/ steruen*, glossed above as "die"? Was this the blunt term or a euphemism? It is impossible to know for sure, but there is some evidence to suggest that *sterven* at least began as a euphemism with the meaning of "to grow stiff"; hence the use of *starve* in northern English dialects with the hyperbolic meaning "be cold," and similar meaning in the Scandinavian cognates. *Sterven* may have started off as a euphemism for **dawjan*, but there is no way of telling whether or not it remained a euphemism by the Middle Dutch period.

On the other hand, why is it that all the Continental Western Germanic languages retain reflexes of the derived nominal or adjectival forms "death" and "dead"? Perhaps it is because people dreaded the actual process of dying more than they dreaded the state of death. Both were seen as inevitable stages at the end of life's journey, but dying was a particularly perilous one. Certainly people feared the agonies of the deathbed; but more importantly, they believed that it was the state of the soul at this very time which would determine its afterlife in either purgatory or paradise. In the 'how to die' manuals mentioned earlier, numerous deathbed scenes show the devil fighting with an angel for a dying man's soul, thereby seeking to impress on readers the importance of appropriate behavior at this time. Purging one's sins, or carrying out good deeds, even at the last moment, enabled the sinner to escape Hell's cauldron—or at least shorten the sentence. Once dead, however, the outcome had been decided one way or the other.

Comparison with Modern-Day—the Obscenity of Death

Then, suddenly again Christopher Robin, who was still looking at the world with his chin in his hands, called out "Pooh!"

"Yes?" said Pooh.

"When I'm—when—Pooh!"

"Yes, Christopher Robin?"

"I'm not going to do Nothing any more."

"Never again?"

"Well, not so much. They don't let you." Pooh waited for him to go on, but he was silent again.

"Yes, Christopher Robin?" said Pooh helpfully.

"Pooh, when I'm—*you* know—when I'm *not* doing Nothing, will you come up here sometimes?" . . .

Still with his eyes on the world Christopher Robin put out a hand and felt for Pooh's paw.

"Pooh," said Christopher Robin earnestly, "if I—if I'm not quite—" he stopped and tried again—"Pooh, *whatever* happens, you *will* understand, won't you?

"Understand what?"

"Oh nothing." He laughed and jumped to his feet.

"Come on!"

"Where?" said Pooh.

"Anywhere," said Christopher Robin.

(A.A. Milne *The House at Pooh Corner* 1948:177f)

Christopher Robin cannot bring himself to speak explicitly of dying, and this is common. Many people have remarked (e.g., Brain 1979; Gross 1985) that death has now become the great taboo subject—the 'unmentionable' in contemporary polite society. And just as the repression of sex brought with it a thriving industry in pornography during the Victorian times, death has become the pornography of modern times.

> While natural death became more and more smothered in prudery, violent death has played an ever growing part in the fantasies offered to mass audiences— detective stories, thrillers, Westerns, war stories, spy stories, science fiction, and eventually horror comics. (Gorer 1965:173)

Here lies one of the great paradoxes of the twentieth century. While we seek to euphemize death from our consciousness, at the same time we fantasize it as entertainment. In his book on death and dying in the United States, Stephenson (1985:41) cites an incident that pungently demonstrates people's inability to cope with 'natural death.' He describes how the same people who admitted allowing their children to watch violence on television petitioned against the building of a suburban church wall to store the ashes of its congregation. Parents deemed it unacceptable, even harmful, that their children should walk past such a place of death. This contrasts with the Middle Ages, where everyday life was full of symbols to constantly remind of death; and in even earlier times, the Egyptians often placed a miniature mummy on the table while eating. Our society typically shuns such symbols as macabre. Death has become altogether less ritualized. For example, wakes are no longer considered an acceptable practice in most Anglo-communities. There is something rather morbid and shocking about that sort of open acknowledgment of death and we no longer encourage such public displays of grief and mourning. Our society is one where silent stoic endurance is considered the 'proper' way of grieving. Exceptions to this are public ceremonies in memory of the war dead or in memory of assassinated public heroes like U.S. President John F. Kennedy, The Reverend Martin Luther King, Jr., or pop star John Lennon. In all these instances, however, the ceremony has more to do with remembering heroes than with celebrating their deaths.

People in the Middle Ages confronted their own mortality in a way that now seems quite appalling to us. Today, the majority of people are quite removed from much immediate interaction with death. Life expectancy has increased enormously, infant mortality has dropped to very low levels, and there do not exist anywhere near the same number of life threatening diseases. In addition to this, hospitals and other institutions that care for the dying, and the funeral industry that disposes of the dead, ensure that we are well insulated from both these things. Many people (certainly in Australia and mainland Britain) reach adulthood without ever having laid eyes on a corpse, except perhaps on film or video. And—in striking contrast with the medieval fascination for corrupting corpses—we now rely on modern embalming techniques to help create for us 'the Beautiful Memory Picture' (cf. Baird 1976:87). In addition, cremation has now replaced burial as the most usual means for disposing of the dead; and though people still frequently visit grave sites, visits to the urns that house the ashes of the cremated are much less common. For those "left behind" then, cremation is the quickest and most final way of removing the remains—particularly if the ashes are then scattered over a cliff, garden, or some such. We go out of our way to avoid death and anyone tainted by it, and in so doing largely expel it from our consciousness.

Even our cooking has eschewed its once close association with death. The following instructions from early seventeenth- and eighteenth-century cookbooks appear excessively brutal to a modern reader like KB. Recipe for roast duck: 'kill and draw your ducks'; 'To stuff chickens: Make an incision at the back of the neck and blow therein so that the skin may rise from the flesh, draw the chickens, . . . chop off the head'; dry devils 'dissect a brace of woodcocks . . . split the heads, subdivide the wings . . . crush the trail [entrails] and brains'; fried neck of lamb 'cut the ribs asunder, beat them with your knife'; jugged hare 'cutt her in small pieces'; stewed prawns 'pick out their tails. Bruise the bodies'; stewed carps 'stirr ye blood with a little white wine'; boiled pike 'Kill the fish in the head, pierce, take a handful of coarse salt, rub it until the phlegm goes out of the fish, open it at the belly and take out the refete and take out the gall-bladder and strip all the small intestine and . . . save the large gut.' (For additional examples see *English 18th Century Cooking*; Black 1977, Chafin 1979, Hodgett et al. n.d.) Culinary activities are somewhat different today, and so, too, is the vocabulary of cooking: most of us no longer *kill, draw, gut, pluck, clean and bone, hack,* or *mince*; those that still eat any meat at all buy it wrapped in neat cellophane packs, perhaps already crumbed or marinated, so that it bears little resemblance to a creature once living. Death has been pretty well banished from the modern kitchen—except for the occasional fly, wasp, or cockroach that has the temerity to trespass there.

Most importantly, people during the Middle Ages were able to attach some sort of meaning to death. Death may well have been awful but it was at least understandable. Religion provided a reason for living and also for dying; there were the (incompatible) beliefs that death leads to salvation and future bliss, or alternatively, that it results from sins committed. Without such religious conviction, however, death no longer has any metaphysical meaning; and within our increasingly secular society, this meaninglessness has become a great source of anxiety. It provides the central theme, for example, in many contemporary art works and forms of entertainment, and is possibly what lies behind many of the stress-related illnesses we have today. As we saw earlier, death taboos rest on a number of possible fears. To this list should be added one other: (5) the fear of a meaningless death.

Life is very much better nowadays than it used to be, and for most people death no longer comes as a welcome release. It is hard to view human existence solely as preparation for death, and to be consoled by the thought of a better future life. The inevitability of death as a natural event was also something that never used to be questioned. In our technologically dominated society, however, science is exalted to such an extent that many believe our mortality is something which can be overcome: witness the growing interest in cryogenics and the millions of dollars that are regularly pumped into means of prolonging life. All this undoubtedly contributes to the present day fear of, and anxiety about, death; the acceptance of death during the middle ages must surely have provided greater peace of mind. One of the most striking differences between medieval times and today is not death itself, but the process of dying. Today, knowledge about imminent death usually comes from a medical analysis rather than from inner conviction, as would have once been the case. A routine check-up that reveals a malignant tumor, for example, can turn into a death sentence for someone who entered the consulting room feeling quite well. And whereas dying was rarely a long drawn-out process in the Middle Ages, today it can take many years. Having to live with this sort of knowledge (since the process of dying is a part of living) causes a host of problems both for those dying and for the people around them. Medical sociologists have written much on how a dying person's social identity can change as family and friends, unable to cope with the knowledge, begin to drastically modify their behavior toward the dying person (cf. Stephenson 1985, Ch. 4). Withdrawal is a typical reaction. A society that goes out of its way to avoid the topics of death and dying, no longer equips us with the appropriate code of behavior for dealing with them, and we are not trained in how to act or what to say. Expressions too solemnly euphemistic sound unctuous and insincere, but anything less might come across as inappropriately cheerful. We fear confrontation with the dying because they are strong reminders of what it is we most dread: namely, our own finiteness. And of course, mixed in with all this fear, some people feel unworthy to be living when a person they love or admire has died. Small wonder modern-day English appears to have more euphemisms for death than anything else (cf. Pound 1936).

It is perhaps apparent from our earlier discussion of the death motifs of the Middle Ages, that the gradual suppression of direct reference to death is reflected in changes that have occurred to epitaph and gravestone design over the years. A nice example can be found in Deetz (1977: 64–90). In a study of colonial New England gravesites between 1680 and 1820, Deetz notes a succession of basically three designs. The first is that of the winged and grinning death's head, which comes with grim reminders of the transience of life and the inevitability of bodily decay. During the eighteenth century, this gives way to a softer and more optimistic spectacle, that of a winged cherub, whose accompanying message now stresses, not the decay of beauty, but the soul's flight to glory. Toward the end of the eighteenth century, however, religious death is frequently replaced by secular death in the form of an urn and cascading willow pattern with the words "In Memory of . . ." Today, the gravestone has become little more than a commemorative marker—the body is frequently not even present.

Euphemisms as the Instrument of Denial

> Family members look intently at the physician as he speaks. "Scan . . . cytology . . . report . . . primary site . . . malignant tumor . . . adenocarcinoma . . . metastasis . . ." Brows furrow as the family continues to hear: "excision . . . chemotherapy . . . contraindicated . . . radiotherapy . . . palliative . . . any questions?" "Yes," responds the family member, "Can you tell yet whether he has cancer and will he get well?"
> (Cassileth and Hamilton 1979:242)

This sort of barrage of obfuscating terminology unfortunately typifies exchanges between medical professionals and the bewildered public. It reflects the professionals' own fearful reactions to cancer, and the difficulty they experience in communicating the realities of the disease to the lay population, who in many respects seem better able to cope with the information than the physicians themselves. 'Studies have shown that doctors as a group have greater fears of death than others, including seriously and terminally ill patients as well as a healthy control group.' (Stephenson 1985:105; cf. McFate 1979). Doctors are expected to protect us against death and of course the recognition of death means the necessary recognition of failure on the part of the physician. Unfortunately, the exalted position in which our community places the doctor only heightens this sense of failure. For example, one perceptive patient who saw the difficulty her doctor had in talking frankly of her fast approaching death had this to say on the matter: 'He feels he's got to get people well, and I couldn't very well talk to him about not getting better' (quoted in Hinton 1976:310). According to the modern medical model, doctors are skillful technicians; they must be both objective and scientific. But informing a patient s/he has only several months to live is very different from dissecting cadavers in a medical laboratory; and the many years of rigid medical training do not equip those in the profession to cope with patients' feelings. Many doctors hide behind the role of technician and scientific expert, utilizing the associated jargon when confronted by the difficulty of speaking with terminal patients about their condition. In this context, the carefully precise medical terminology could well be described as euphemistic. It represents a strategy to create distance from something that is distressing and painful to both patient and doctor. This message came across very strongly in the documentary

Breaking the News referred to earlier; the topic examined was what doctors tell patients of their condition. The program concluded that doctors' reticence to "break the news" stems largely from their inability to cope with a situation which brings them face to face with their own mortality.

Much has been written on the psychology of denying imminent death (e.g., Weisman 1976:452–68). Euphemisms clearly have an important role to play in this denial process. A terminally ill cancer patient, for example, may prefer to dismiss a *cancer* diagnosis and use instead a term like *tumor* or *growth*. These may all denote the same thing, but whereas *cancer* implies corruption, decay, and inevitable death (recall our earlier discussion of the cancer metaphor), *tumors* and *growths* can be excised to enable a probable recovery. In this way the patient can avoid dealing directly with the threat of oncoming personal death. To take an actual example of this: it was striking that in a television documentary on euthanasia shown on channel 9 in Melbourne during 1989, none of the terminally ill patients who were interviewed referred directly to the act of euthanasia or to their own imminent death. (Incidentally, the word *euthanasia* has taken on negative connotations from *mercy killing*; in literature on the topic the term *humane death* is often preferred.) The following quotations from the interviews are typical of the sort of terminology used: 'If you've reached the end of your tether and you're going to—[*long pause*], doctors do the deed.' 'They should do the other thing.' 'And then the patient sleeps.'

But just as important, the use of euphemistic terminology such as this acts as a soothing strategy for those close to the victim. For a dying patient who fears isolation and the withdrawal of those around—perhaps even more than death itself— euphemism is a means of ensuring that close relationships are maintained. This explains the sometimes contradictory behavior that gravely ill patients display. It is not uncommon, for example, for a fatally ill person to speak quite frankly to a physician or psychiatric consultant about his/her illness and approaching death but make light of the whole situation when dealing with family and close friends (cf. Weisman 1976:457f). As Weisman points out, denial is a way of helping to ensure that significant relationships remain unchanged.

In Memoriam

In a total of 536 'Death' and 'In Memoriam' notices in the Melbourne *Sun* on Saturday May 21, 1988, only one included a use of the verb *die*; all the others used euphemisms of one kind or another. The euphemisms fall into four broad categories: death as loss, worries about the soul, death as a journey, and death as beginning a new life. But there is in addition the occasional use of the ultimate euphemism: not mentioning it at all, for instance,

> SMITH (Ross), Frances Winifred.—On March 8 at Sherbrooke Private Nursing Home. Upper Ferntree Gully, aged 71 years. . . .
> (*Sun*, Melbourne, Thu. March 9, 1989:65)

> THOMAS, Margot—On March 11 in Adelaide. In affectionate memory of our dear Aunt and good friend. . . .
> (*The Age*, Melbourne, Mon. 13 March 1989:21)

It is true that located among death notices, one knows that the subject of the notice has died. Even so, there is usually some mention of the event.

1. DEATH AS LOSS. Both of the euphemisms *losing* X and *missing* X take the point of view of those left alive—or, to use the journey metaphor exemplified in (3), those *left behind*. In expressions like *Our condolences for your tragic loss* (which carefully avoids specifying the loss), *We lost our father last winter* and *Our condolences on the loss of your husband,* the use of *lose/loss* evokes the misfortunate lack caused by an event over which the bereaved has no control: it captures death as malign fate.

The deceased, having been lost, is then *missed* by those left alive to mourn their loss. E.g.

> We lost Mr Bentley last night.
>   (Sunset Hospital nurse to KB, December 1978)
>
> LASSLET.—Jim. Nature's gentleman, sadly missed. . . .
>   (*Sun*, Melbourne, Thu. March 9, 1989:64)

2. Worries About the Soul.

> The sleeping and the dead / Are but as pictures.
>   (Shakespeare *Macbeth* II.ii.52f)

The third and fourth fears listed earlier hinge on there being a metaphysical counterpart to the body: its soul. In Egypt's Valley of the Kings, a relief showing the birth of King Amenophis III depicts two neonates: one is the king's corporeal form, the other is his soul. Belief in this dual character of living things is almost universal. Activity in animals and humans is explained by the presence of a soul in the body; sleep and death indicate its absence—which is why *sleep* is a frequent euphemism for "death." In sleep the soul's absence is only temporary; but on death the soul vacates the body forever; and a soul without a body to reside in must be laid to rest somehow, lest it become distressed and trouble the living. The ancient notion of a displaced soul is invoked in such notices as the following, which may reflect orthodox Roman Catholic belief:

> PARKER.—Prayers for the repose of the soul of Mr Philip Rowe Parker will be
> offered in our Chapel . . .   (*Sun*, Melbourne, Sat. May 21, 1988:70)

As just mentioned, *sleep* is a common euphemism for "die." Our word *cemetery* derives from Ancient Greek *koimētērion* "dormitory." There is an obvious similarity between a sleeping body and a dead one, and sleep has often been regarded as a temporary death, a period when the soul leaves the body to return when it awakens. To describe death as sleep is to pretend that it, too, is temporary. Some recent examples:

> SCOTT.—Wayne. . . . Sleeping peacefully, free from pain . . .
>
> LOGAN.—In loving memory of my darling husband Percival William who fell
> asleep on May 21, 1983. . . .   (*Sun*, Melbourne, Sat. May 21, 1988:68)

3. DEATH AS A JOURNEY. Many peoples have regarded death as the start of the soul's journey into the afterlife and buried their dead with all sorts of paraphernalia, including sacrificial humans, to help them on the way. Accordingly, death is often

represented euphemistically as a journey for (presumably the soul of) the dead person. The stem of our word *obituary* is Latin *obiter* "on the journey." The metaphors that arise from this notion include parting, passing away, passing on, being taken away (either by god or malevolent fate), and arrival at the final resting place—perhaps to be reunited with loved ones already dead.

> One day I went in to see her, and I thought that she'd gone.
> (IC to KA of an old lady who was shortly to die, December 2, 1989)

> . . . Kaye Reeves who departed suddenly on Mar. 5, 1989
> (*Sun*, Melbourne, Thu. March 9, 1989:65)

> MURDOCH.—Jim. We knew the time was coming and soon we'd have to part. . . .

> REID.—Gus passed away May 22, 1985. . . .

> MANSFIELD.—. . . On this day you were taken away . . .

> DAWSON.—Jeffrey. Our darling brother, taken so tragically in New Zealand. . . .

> HARKER.—Maggie, on May 19, aged 94, passed away at DDH. Gran, you slipped away from us peacefully. All your pain has gone, now you're reunited with Grandpa after 53 years. . . .   (*Sun*, Melbourne, Sat. May 21, 1988:68ff)

Somewhat more poetic than these euphemisms are *go the way of all flesh* and *go to the happy hunting ground*. The latter is supposedly borrowed from Amerindian religious mythology and Christian counterparts are *go to meet one's maker, go to a better place* and:

> KARPENY.—Aunty Em. Reunited in God's kingdom with Nan, Pop, . . .

> PIERCE.—David. In ever loving memory of my loving husband, who was suddenly called Home to Glory on May 21, 1987. . . .
> (*Sun*, Melbourne, Sat. May 21, 1988:68f)

The Salvation Army normally uses a phrase like *promoted to glory* as a euphemism for *die*.

Death is also looked upon as the end of life's journey and the notion of death as a state of rest from the pains of life gives rise to euphemisms like X *was laid to rest*; for instance,

> HARDING.—Ronald James. A great man gone to rest. . . .

> KEARNS.—. . . We watched you suffer. God knew you had enough. He took you gently with Him and gave you a place at last. Rest in peace. . . .
> (*Sun*, Melbourne, Sat. May 21, 1988:69)

4. DEATH AS BEGINNING A NEW LIFE. Euphemistic expressions that have to do with notions of an afterlife are used by both the religious and the nonreligious alike when dealing with death. For nonbelievers and agnostics the conventional language is comforting insurance against the possibility that believers have it right, and that the soul might continue to live on after the body is committed to the earth to rot. But what about people of religious conviction, for whom expressions demonstrating belief in a better afterlife are statements of simple fact? Are such expressions

euphemistic for them? According to our definition of euphemism, they are. For all who use them, they are preferred expressions and the image they offer is one of consolation. By glossing over the physical event of death and dying, they also serve to ameliorate the—in this case—abhorrent reality of what it is that causes anxiety. Some examples:

> LIESGANG, Norman Kellner.— Born into eternal life 10 March 1989.
> (*The Age*, Melbourne, Sat. 11 March 1989:35)

> WALSH, Jean Ivy (Kneelock).— Called to a higher life on 9th March 1989, . . .
> (*The Age*, Melbourne, Mon. 13 March 1989:21)

> SARGENT—Aubrey John (Bray), passed away after a short illness on Mar. 10, 1989. Beloved husband of Betty. Alive with Jesus.
> (*Sun*, Melbourne, Mon. March 13, 1989:45)

In the last notice, there is a fiction that the deceased is somehow still alive. In fact the dead are quite often addressed, as in the notices for Murdoch, Mansfield, Harker, and Kearns above, and for Morgan at the end of this section.

In addition to the preceding four categories, and interwoven with them, the death of someone associated with a particular profession or pastime leads to euphemisms like *go to meet the great golfer in the sky, pulled up the stumps for the last time, finally cashed in his chips,* and so forth, which have the recurrent theme of a final act.

> Another sadness, in January, was the departure to the "kennel in the sky" of Flossie, our faithful companion of sixteen years.
> (IS in a letter to MMA, December 1989)

> [*Below Masonic compasses*] GILBERT.—George Richard. Our friend and confidant for thirty years. You lived respected and died regretted. In your promotion to the Grand Lodge above, . . .　　　(*Sun*, Melbourne, Sat. May 21, 1988:68)

> GOODCHILD.—Joan, aged 60. Passed away Mar. 6, 1989. . . . "I hope there's a T.A.B. up there." (A 'T.A.B.' [*Totalizator Agency Board outlet*] *is a betting shop*)
> (*Sun*, Melbourne, Thurs. March 9, 1989:64)

Every now and then the death columns reveal a delight. The following is not so much an example of exquisite euphemism, but stream of consciousness eulogy:

> MORGAN.—Theresa. The luggage van ahead, making plum pudding, bread and milk for breakfast, exploding spaghetti tins, you have gone, my childhood memories can never. Thank you.—Tracey, Geoff, and Kimberley.
> (*Sun*, Melbourne, Thurs. March 9, 1989:65)

Funerals and Such

In *The American Way of Death*, Jessica Mitford writes:

> 'The use of improper terminology by anyone affiliated with a mortuary should be strictly forbidden,' declares Edward A. Martin. He suggests a rather thorough overhauling of the language; his deathless words include: 'service, not funeral;

Mr., Mrs., Miss Blank, not corpse or body; preparation room, not morgue; casket
not coffin; funeral director or mortician, not undertaker; reposing room or slumber
room, not laying-out room; display room, not showroom; baby or infant, not still-
born; deceased, not dead; autopsy or post-mortem, not post; casket couch, not
hearse; shipping case, not shipping box; flower car, not flower truck; cremains or
cremated remains, not ashes; clothing, dress, suit, etc., not shroud; drawing room,
not parlour'.

   This rather basic list was refined in 1956 by Victor Landig in his *Basic
Principles of Funeral Service*. He enjoins the reader to avoid using the word 'death'
as much as possible, even sometimes when such avoidance may seem impossible;
for example, a death certificate should be referred to as a 'vital statistics form'. One
should speak not of a 'job' but rather of the 'call'. We do not 'haul' a dead person,
we 'transfer' or 'remove' him—and we do this in a 'service car', not a 'body car'.
We 'open and close' his grave rather than dig and fill it, and in it we 'inter' rather
than bury him. This is done not in a graveyard or cemetery but rather in a
'memorial park'. The deceased is beautified, not with makeup, but with 'cosmetics'. Anyway, he didn't die, he 'expired'. An important error to guard against,
cautions Mr Landig, is referring to 'cost of the casket'. The phrase 'amount of
investment in the service' is a wiser usage here.   (J. Mitford 1963:47f)

As is confirmed in a 1989 documentary film *Bodywork: Confessions from the
Funeral Trade*, Australian *funerals* are now usually *conducted* or *arranged* by a
funeral director from a *funeral parlor* or *funeral service*, not by a *mortician* and
rarely nowadays by an *undertaker*. People involved in the business of collecting the
body talk about *the removal* or *doing the contract*. We also come across words like
soup, a term by which those in the Australian death industry refer to "a badly
decomposing body," and where it may be necessary to carry out *the removal* using
buckets! Does the use of *soup* show a dysphemistic lack of respect, or is it a form of
gallows humor, a dysphemistic euphemism? In the documentary *Bodywork*, the
term was used without any hint of callousness or humor; it is a very apt description
of a body after 5–6 weeks of decomposition, and—provided it were not used to
bereaved family and friends—is a not inappropriate item of jargon for the trade.
Mortuary is a term restricted to medical and forensic use; if the corpse is laid out
for viewing in a funeral parlor, it will be in the *chapel*, not the *morgue* or
mortuary. The dead person is not referred to as a *corpse* in these circumstances,
nor as *the remains* (even after mutilation of the corpse), but as *loved one, dear
departed, the late Mr. X*, or the like; for example, a widow is much less likely to
refer to *my dead husband* than to *my late husband*. The newspapers and police
will refer to a dead person as *the deceased* (a nice euphemism); in militarese a
corpse is a *casualty*. As in America, the deceased will be put in a *casket* rather
than a *coffin* and then either be buried or maybe have the ashes scattered (or
otherwise located) in *a garden of rest* or *a garden of memories*. After the death
there may be *remembrances* in the form of donations for a commemorative
plaque, edifice, or the like; for instance,

   Johnson's family requests that any remembrances be made to the UA library, to be
   used to plant a tree with a plaque in his memory. Donations can be sent to . . .
                                   (*Arizona Daily Wildcat* Fri. January 12, 1990:3)

The Role of Flippancy in Downgrading Death and the Hospitalized

[*Two interns talking about a patient who expired*, 'post' = "postmortem"]
"But she came in healthy."
"Yeah, and then she boxed, and I get credit for the post."

(Shem *The House of God* 1978:153)

There are flippant euphemisms for death such as *call it quits, croak, check out, cock up one's toes, push up the daisies, buy the farm, bought it, kick the bucket, pop off, peg out, conk out,* and *cark it*. Why do we classify these slang expressions as euphemisms rather than dysphemisms? Once more it will depend entirely upon context. A jocular approach to death is only dysphemistic if Hearer can be expected to regard it as offensive. For instance, if a doctor were to inform close family that their loved one has *pegged out* during the night, it would normally be inappropriate, insensitive, and unprofessional (i.e., dysphemistic). Yet given another context with quite a different set of interlocutors, the same expression could just as well be described as cheerfully euphemistic. For instance, when JC was shortly to leave hospital after a colostomy she was telephoned by RR who said 'I just wanted to get you before it was too late.' It is thought that RR meant "before you leave hospital," though under the circumstances, the remark was rather dysphemistically ambiguous; when JC reported the event to KB, the latter commented 'It sounded as though he was expecting you to *peg out*.' Here the use of *peg out* was euphemistic: its flippant connotation distracts from the seriousness of the death threat in the prevailing circumstance, which makes it preferable to *die*.

Flippancy toward what is feared is widely used as means of coming to terms with fear, by downgrading it; and that is why euphemisms like those listed earlier are also used where Speaker is being hardnosed about the demise of the dead person, *the stiff*. Similar hardnosed jocularity is evinced in the criminal and law enforcement milieus with their euphemisms for capital punishment like *swing* for "hang" and *fry* for "electrocute"; and in describing the coffin as a *wooden overcoat*. For those involved in the industry of death, it is perhaps the only way to survive in such a profession. For example, author KB can report from first-hand experience that this sort of levity is extremely common among staff in nursing homes when talking to each other. For people who have to deal with the dying and with death everyday, this seeming irreverence for human life makes such work much easier to bear. It distances them from the sickness and death, and helps to blot out the awareness of their own vulnerability and that of their co-workers. In Australian hospitals, we find reference to *crumbles* "the frail and elderly at death's door"; *grots* "derelicts and alcoholics"; *vegetables* "unresponsive or comatose patients"; diagnoses like *F.L.K.* "Funny Looking Kid"; *G.O.K.* "God Only Knows"; and someone who has *passed through the valley of tears* is simply *cactus* or "dead". America has *gomers* and the feminine *gomeres* "Get Out of My Emergency Room", defined by Shem (1978:424) as 'a human being who has lost—often through age—what goes into being a human being'; and *LOL in NAD* "Little Old Lady in No Apparent Distress." This is just a small sample of hospital slang; and, depending on context, it can be described as euphemistic, dysphemistic, or simply descriptive. There are of course other important aspects to slang expressions of this kind. They identify activities, events, and objects that have become routine for those involved and have an important function in creating rapport in the work environment. Gordon (1983) claims that this is the only motive for the existence of such slang, but our (more exactly KB's) experience of hospital expressions does not support him. We will take up this question again in Ch. 8 when we examine the complex interaction between euphemism and jargon.

Euphemisms for diseases are discussed in Ch. 7, but germane to the present discussion is a defiantly flippant treatment of the plague in the well-known English nursery rhyme

> Ring-a-ring o' roses.
> A pocket full of posies.
> A-tishoo! A-tishoo!
> We all fall down.

The 'roses' are the pink rash associated with the plague; the 'posies' are the nosegays to counteract the 'corrupt vapours' that supposedly caused the disease; sneezing was brought on by pneumonic plague; and the almost inevitable consequence of the plague was to 'fall down' dead. What is now a nursery rhyme was originally nice example of flippant downgrading euphemism, and like most euphemisms of this type, it was used by people who were powerless against the disease they feared. Because such euphemisms are taunts against the metaphysical force that people fear, we could view them as dysphemisms against that force, but it is not clear to us that there is anything gained by so-doing.

Deceptive Euphemisms from the Killing Fields—and Hospitals

Therapeutic misadventure.

Substantive negative outcome.

The patient failed to fulfill his wellness potential.
          (From the hospital records of deceased patients, cf. *Medical Observer*
                                                              Jan 19, 1990)

Death while on military service is referred to by euphemisms that vary between the solemn patriotic (e.g., *do one's bit for one's country, make the ultimate sacrifice*) and the more flippant *come home in a box, X bought it,* and, for flyers, *be grounded for good*). Pompously heroic euphemisms like the former glamorize death and help to reinforce for those still surviving the illusory world of a 'glorious battlefield' (cf. Baruch's 1976:92 description of death in war). The humor that comes out of this sort of soldier's slang is a kind of release mechanism. Anttila (1972:139f) notes that during the two world wars weapons were sometimes given the names of household items; for example, a machine gun was called both *a sewing machine* and *a coffee mill*. Obversely, household items were given the names of weapons (e.g., *hand grenade* was used to mean "potato"). Lifton and Olson (1976:104) in their account of the Hiroshima explosion also describe how scientists of the time commonly used general-for-specific euphemisms like *the gadget, the device, the thing,* and even the all embracing *it* to refer to "the atomic bomb." There can be no doubt that an added motivation for this was secrecy; after all, that was the motivation for calling a tank *a tank* during the first world war. It is reputedly former U.S. President Ronald Reagan who dubbed the ten-warhead MX missile *Peacekeeper* (cf. *Time Australia* April 3, 1989:41). Clyne (1986) has identified comparable misleading euphemisms for nuclear weaponry and the like in German: *Atomkrieg* "atomic war" and *Atomwaffen* "atomic weapons" were exchanged for euphemisms based on *Kern* "core (of fruit), kernel (of nuts)"—*Kernkrieg, Kernwaffen*; a nuclear waste dump is *Entsorgungspark* "de-care park." The peace movement replaces *Kern-* by *Atom-* and calls the dump *Atommülldeponie* "atomic waste dump".

We live in a century during which states have sanctioned more killings than ever before. Eliot (1976:110–33) reckons that wars, executions, engineered famines, and such have led to more than 110 million state-sanctioned deaths since 1900. And of course advances in technology bring with them the threat of many more millions. Yet, paradoxically, in the face of such mayhem, enormous amounts of money and effort are spent on devices for saving and for prolonging the lives of individuals. There are many similar paradoxes in human behavior; another would be the energy and capital that is pumped into in vitro fertilization programs while millions of children around the world die daily from lack of food and proper health care. This almost schizoid trait in humans is supported linguistically by what Rawson (1981:3f) dubs the 'dishonest euphemism'—the euphemism which conceals, not so much to avoid offense, but rather to deceive. As George Orwell (1946) writes in his essay *Politics and the English Language* 'political language . . . is designed to make lies sound truthful and murder respectable, and to give an appearance of solidity to pure wind.' This is the world of circumlocution and political doublespeak. We recall a news report in which reference was made to the remains of those killed in the 1986 *Challenger* Space Shuttle disaster as *recovered components:* this horrible piece of dehumanizing bureaucratic terminology was intended to play down the loss of human life caused by an inattention to safety measures that was motivated by short-term political and financial gain.

Legal killings (as well as those carried out by terrorists) are called *executions: execute* is a euphemism whose literal sense is "do or perform." Otherwise it is *capital punishment*. In Latin a prisoner was *led away*—to be decapitated. In criminalese, Macbeth spoke of Duncan's *taking-off* (*Macbeth* I.ii.20), and more modern descriptions include *it's curtains for X, rub X out, take care of/terminate/waste/silence X* and from lynching *be/get strung up, to decorate a cottonwood*. We have heard that in a funeral oration for someone shot to death in a gangland killing, he was said to have died from *lead poisoning*. The U.S. CIA is reputed to have had a *Health Alteration Committee* to decide on *hits* (cf. May 1985:128). Although arguably no longer euphemistic, there is also the verb *to liquidate*, a metaphor which comes from the, by comparison, quite harmless practice of shutting down a business.

Militarese uses euphemisms that are loaded to the point of deception; perhaps *terminate with extreme prejudice* and *take out* [a target] are transparent; but *pacify* means "be killing (be at war with)"; *neutralize* means "kill selected targets, or conquer"; *mopping up operations* include killing off the remnants of *the resistance*. The Philippines army refers to the *salvaging*, not killing, of alleged communists (cf. *Time Australia* June 27, 1988:16). In nineteenth-century Australia, *disperse* used to be an officially sanctioned euphemism for "kill" in the context of Aboriginal killings by the 'native' police force set up in 1848. As Elder (1988:100) says, 'the aim was simple: set an Aborigine to kill an Aborigine.' Elder also cites an incident (p. 106, it was first reported by A.J. Vogan in his book *The Black Police: A Story of Modern Australia*) wherein a young sub, just new to the police force used the word 'killed' instead of the official *dispersed* in a report he had submitted. He was severely reprimanded for his carelessness and told to correct his error. The sub, described by Vogan as 'rather a wag,' rewrote the report thus: 'We successfully surrounded the said party of aborigines and dispersed fifteen, the remainder, some half dozen, succeeded in escaping.' (!) Another common euphemism used in reports on the activities of the force was *collision*; 'collisions' between Aborigines and the native police force invariably denoted a large scale massacre or, as they would describe it, 'dispersal.' Individual shootings were described as *self-defence* (as most police shootings are). Reports always recorded how the police, *exercising great restraint*, were *forced in the end to shoot*. What few reports were ever made, are nothing but euphemistic cover-ups to conceal the carnage that really took place.

In Nazi Germany *betreuen* "look after" was a euphemism for "commit to a concentration camp"; some prisoners sent there had their identity papers stamped *Rückkehr unerwünscht* "return unwanted"; these people were assigned *Sonderbehandlung* "special treatment"; and all this was part of *die Endlösung* "the final solution." The euphemistically entitled *Reichsausschuss zur wissenschaftlichen Erfassung von erb- und anlagebedingten schweren Leiden*, "Reich Committee for Scientific Research of Hereditary and Severe Constitutional Diseases" was just one of the many killing institutions of the Nazis. This one particular institution was responsible for killing an estimated 5,000 'racially valueless' children (cf. Dawidowicz 1975:173–78). It was part of a euthanasia program that, like *the final solution*, developed out of the Nazi's fanatical obsession with *racial health* and *positive eugenics*. The principal motivation for these Nazi euphemisms for killing was deception of public opinion rather than fear of death.

From the Vietnam War of 1966–1972 there came brutally dehumanizing expressions like *softskin target* and the infamous photo caption referring to *an oriental human being* depicted being dropped from a helicopter. One Vietnam War veteran writing of his combat training says:

> The death of an enemy was mentiond in aloof terms as if the enemies were stuntmen on a movie set who would get up and walk away after the take. Your own death was never dwelled upon; the training period was set in a world of illusions in which American soldiers seemed invulnerable to combat death.
>
> (Baruch 1976:92)

And Commager writes of euphemisms used during the Vietman war:

> Corruption of language is a special form of deception which [U.S. President Richard Nixon's] administration . . . has brought to a high level of perfection. Bombing is 'protective reaction,' precision bombing is 'surgical strikes,' concentration camps are 'pacification centers.' . . . Bombs dropped outside the target area are 'incontinent ordnance,' and those dropped on one of your own [i.e. allied] villages are excused as 'friendly fire.'
>
> (Henry S. Commager 'The defeat of America', *New York Review of Books* October 5, 1972:10)

When the United States went into Grenada in 1983, President Ronald Reagan at first referred to the act as an invasion; after it was all over, however, he objected to reporters about 'your frequent use of the word invasion. This was a rescue mission.' (*Time*, January 9, 1984:56). The word *invasion* is typically a dysphemism, with connotations of unwanted interference and enforced domination of others; whereas *rescue mission* is a euphemism, with connotations of humanitarian aid.

Sometimes those engaged in health-care make fatal mistakes, and when they do they sometimes seek to conceal the error (and save their tail) with a deceptive euphemism. In the *Medical Observer* of January 19, 1990, there is a report that Professor William Lutz of Rutgers University has collected the following examples of such deceptive euphemisms from hospital records: *therapeutic misadventure, diagnostic misadventure, diagnostic misadventure of the highest magnitude, negative patient-care outcome*, and *patient failed to fulfill his wellness potential*. Lutz also notes that an anesthesiologist recorded a *substantive negative outcome* after the patient had died as the result of his turning the wrong knob. Beware of medicos!

Summing up: whether the killers are civil, military, or criminal they seek to cloak in euphemisms the lethal act and the devices they use in carrying out the act. However morally justified a social group may feel they are when killing others, the language they use strongly suggests a shame, born perhaps of fear: fear of discovery, of retribution, of penury, or of death being inflicted on them in return.

Summary of Chapter 6

In this chapter we examined the different fears about death and dying—the fear of severed relationships; fear of the unknown that follows death; fear of the physical corruption of the body; fear of the souls of the dead; and finally, fear of an absurd and meaningless death—something which particularly touches those living in modern secular societies. These are what motivate the strong taboos that typically surround the subject of death.

We then briefly compared the medieval and contemporary attitudes to mortality in order to better understand the language of death in these two periods. Though there was certainly no less fear of it in the Middle Ages, people confronted death in a way which most of us now would not find acceptable. In the artworks, for example, death was given great exposure, and always with extraordinary literalness. This literalness distinguished also the language of the medieval Dutch texts examined. They were remarkably free of the elaborate euphemisms we normally associate with the topic of death. As discussed in this chapter, there were many reasons for this blunt approach, not the least important being simply that death during this time was almost certainly a commonplace experience for most people.

Such open acceptance of death contrasts sharply with what we find today (automobile accidents, plane crashes, and AIDS, notwithstanding). In both the original and modern senses of the word, death has become obscene—as Gorer so aptly describes it, pornographic. It has become the socially unspeakable of modern times (very much as sexuality was during the reign of Queen Victoria) and the seriousness of this taboo is reflected in the richly euphemistic language that is available. Using examples from local death notices, we examined the four broad categories into which the typical euphemistic formulae fell. These had to do with death as a loss, the anxiety surrounding the soul of the dead, death seen as some sort of journey, and as the beginning of a new life. But perhaps the rapidly growing number of publications by psychologists and sociologists on death and dying (e.g., Kubler-Ross 1969; Aries 1974; Becker 1973; Stephenson 1985) is a sign that death is emerging from being the great unmentionable in contemporary society.

We saw that people involved in the industry of death have another reason to use euphemisms, some of which are flippant, because flippancy is a way of coping with the seriousness of something. The euphemistic jargon of language varieties like funeralese, as well as blunting the reality of what is at hand, serve to create a certain professional status for the industry. From here we moved onto euphemisms to do with killing, which lead further into the realm of jargon, in this case political, military, and medical jargon. These euphemisms add the dimensions of deception and secrecy. We take up the question of the intersection of euphemism and jargon more closely in Ch. 8.

In Ch. 3 we discussed one way in which taboos surrounding the body and bodily processes can be linked to the fear of death and organic decay. As remote as this connection may at first appear, we believe there is yet another way in which they are linked. We return once more to the conflict of the mind versus the body.

The things that have to do with our bodies—our bodily functions, normal sexual urges, also sickness and disease—remind us that we too are a part of nature and therefore impress upon us the vulnerability which our bodies place us in. The simple fact is that we are contained within a fragile and temporal frame, and as much as we may resist, we can do nothing about this fact—death is simply another natural process. In this regard, we are no different from animals—there is no reasoning with the exigencies of the body! While our mind dreams of immortality, it is at the same time painfully aware of the inevitable extinction of both body and consciousness. Becker brilliantly captures the frightful dilemma of human beings:

> Man is literally split in two: he has an awareness of his own splendid uniqueness in that he sticks out of nature with a towering majesty, and yet he goes back into the ground a few feet in order blindly and dumbly to rot and disappear forever. It is a terrifying dilemma to be in and to have to live with.   (Becker 1973:26)

It is this dilemma—the paradoxical make up of human beings—that forms the basis of the nature–culture dichotomy described in ch. 3. While it may be farfetched to argue that the taboos surrounding our bodily effluvia exist because we associate them with the smell of putrefying corpses (cf. ch. 3 and Brain 1979), the connection between them and death is nonetheless justified. In that we do not want to be reminded of our natural mortal condition, such taboos are therefore motivated, at least in part, by our general denial of mortality and avoidance of death. T.S. Eliot comes close to capturing it all:

> Birth, and copulation, and death.
> That's all the facts when you come to brass tacks:
> Birth, and copulation, and death.
>   (*Sweeney Agonistes*)

CHAPTER

—7—

Unease in Talking About Disease

The very terms for "leprosy" and "tuberculosis" are restricted in use, because of the danger of "contagion". Especially in dwelling houses, euphemisms or alternative terms must be used; a sequence of these have been employed in recent decades, then discarded as they acquired dangerous power.

(Keesing & Fifiʔi 1969:160)

Diseases and Other Misfortunes: Matters Both
of Fear and Mere Distaste

Misfortune is taboo; we try to avoid *tempting fate* by not speaking of misfortune. On the other hand we try to protect good fortune by saying *touch wood*. The English word *accident* was once a euphemism because it had the much wider meaning of Latin *accidens* "happening"; but like many euphemisms it has now narrowed to "misfortune." There is a view that diseases are "accidents" of the body. In most societies such things are spoken of euphemistically either because they are thought to result from the actions of a malevolent spirit or person, or in consequence of their connexion with death, or just because of their intrinsic unpleasantness. Consider some euphemisms for *smallpox* in other languages. The literal *ndui*, in Kiswahili, gave way to *ugonjwa mbaya* "the bad disease" or *tete* "[cereal] grains," hence *tetwanga* literally "painful grains" is "chickenpox"; in Javanese, smallpox is called *lara bagus* "the pretty girl"; among the Sakarang Dyaks of Borneo it is known as "jungle leaves," "the fruit," or "the chief" (Frazer 1911:416); in India it was named after the goddess *sitala* "she who loves coldness." Expressions like these are motivated by deeply held fears and superstitions.

But the verbal taboos surrounding disease and sickness can have other motivations, too. For example, some euphemisms seem to reflect the cooperative desire not to impose one's trouble on others and not to be seen to whinge. For example, people talk about being *unwell* or *under the weather* rather than being *ill*. The English word *disease* was once a euphemism: its constituent morphemes are *dis-* "cease to" and *ease* "be comfortable." Doctors occasionally use a similar euphemism

172

today when they ask *Do you experience any discomfort?* instead of *Do you have any pain?*. The expression *to be in condition* is the contrary of to *have a condition*, which is used of some fairly serious illnesses, though usually with an appropriate qualifier such as *heart, liver,* or *kidney.* The expression *to be in a delicate condition* was once used of a pregnant woman, but because the social taboos surrounding pregnancy are no longer as strong, it is now less usual. Such euphemisms respond to our notions of decency and decorum and support the already strong taboos against mentioning one's bodily effluvia. For example, people talk about *being sick* to describe the condition of *vomiting*; they talk about having *an upset stomach* or *the runs* rather than *diarrhea*; an advertisement for travel sickness pills refers to *being uncomfortable* while traveling; a woman might say *I'm not feeling too good* rather than *I've got menstrual cramps.* Nicknames like *the barfs, the trots, Montezuma's revenge,* and *Bali Belly* are also commonly used of indecorous illnesses of this sort, which are more humiliating than life-threatening (see further examples in Johnson and Murray 1985:153). Humor is often used as a means of coming to terms with the less happy aspects of our existence. There are even jocular names for serious diseases like *Acquired Immune Deficiency Syndrome*—better known by its euphemistic acronym *AIDS*—such as *Anally Inserted Death Sentence* or *Toxic Cock Syndrome.* Here the motivation is primarily fear, as it is in the flippant downplaying of death and the hospitalized that we looked at in Ch.6.

Speaking of Diseases in the Middle Ages: Examples from Middle Dutch

The fear of disease is a happy restraint to men. If men were more healthy, 'tis a great chance they would be less righteous.
(Edmund Massey in a sermon, July 2, 1722, quoted by Black 1986:78)

There was tremendous fear and superstition attached to illness during the Middle Ages. Some of the treatments were frightful and there were very few effective remedies. The etiology of disease was shrouded in mystery, and explanations for sickness often connected the complaint with the workings of malevolent spirits or with divine punishment for sins committed. Epidemics were believed to be sent as retribution for the indiscretions of entire communities. As often as not, cures took the form of appeals for clemency to those higher powers believed responsible; people resorted to prayers, incantations, sacrifices, and sorcery. Medieval therapeutics in Europe show a curious mixture of Christian theology, and the superstitions of pagan antiquity: consequently, the most usual term for ill-health in Middle Dutch was *euel* "evil"; the cognate was also current in early English. Not surprisingly, the language of disease was rich in euphemism.

Pudendagra—*the Case of Syphilis*

In the world of euphemism, it seems that names and expressions for something multiply in direct accordance with the severity of the taboo—whether the taboo is motivated by fear, disgust, unease, or whatever. Syphilis is a good example. Although syphilographers disagree as to when the disease first appeared, by the Middle Ages it was already endemic and probably one of the most dreaded diseases of the time. This was not on account of its mortalities, however, because there were many diseases that were just as deadly. What was most significant about this disease was its sexual origins: the moral depravity that was believed to cause it was made obvious in the physical symptoms of the disease.

The linguistic taboo was severe, and the disease masqueraded under literally dozens of different names, many of which are euphemistic or ironic. To the English it was known as *cupid's measles*, to the Dutch *Venusziekte* (cp. Neo-Latin *lues venerea* and English *venereal disease*). Its cause may well not have been known precisely, but its connection with sexual conduct certainly was; and none of the countries where the disease raged would admit to its origin. Each blamed it on another, exemplifying the common dysphemistic practice among human groups of blaming outsiders for importing vice and immorality. The Italians attributed the epidemic to the French and called it *Mal Francesse* or *Morbus Gallicus*. The French retaliated and called it *Mal de Naples*. The Germans also blamed it on the French and described it as *Franzosen böse Blattern* "French bad blisters." The English followed their example with *French Pox/Marbles/Aches/Fever/Malady/Gout*; Shakespeare, and presumbly others, referred to it as 'Neapolitan bone-ache' (*Troilus and Cressida* II.iii.19). The Russians blamed it on the Poles while they in turn called it the *German Disease*. To the Dutch, it was *Spaensche Pokken* "Spanish Pox." In the sixteenth century, it finally received the name *syphilis* a euphemism that, it seems, had universal appeal. The name is figurative in origin. It derives from an imaginary shepherd who was suffering from the disease, the protagonist in a poem entitled *Syphilis, sive Morbus Gallicus*. The poem was first published in 1530 by Girolamo Fracastoro of Verona, a doctor who also wrote a treatise on the disease. Sadly, the name's original pastoral associations have long been lost.

Even with most of the mystery of cause and cure now gone, *social diseases* like syphilis are scarcely more freely named today: general-for-specific abbreviations like *S.T.D.* and *V.D.* are preferred. And doctors often avoid using the word *syphilis* with their patients, resorting instead to labels like *treponemal disease, luetic disease* (from Latin *lues* meaning "contagion, plague"), *spirachoetal disease*, and even *special disease!* It would seem that the linguistic taboo is, after all, still strong.

Mental Illness in the Middle Ages

The superstition that attached to disease was, as in many places it still is, great and extremely primitive in nature; for instance, many people believed that naming the disease would cause it to strike (cf. Ch. 2). In these circumstances, it is not surprising then that the most dreaded of diseases typically go by a number of different names. Many were favorable appellations, obviously aimed at somehow placating the unseen powers responsible. The quotation from Keesing and Fifi?i (1969) at the beginning of this chapter refers to twentieth-century practices among the Kwaio, an Austronesian people: "The very terms for 'leprosy' and 'tuberculosis' are restricted in use, because of the danger of 'contagion'. Especially in dwelling houses, euphemisms or alternative terms must be used;" the attitudes revealed here would have been true of medieval Europe and very likely of our own community for some occasions, as we shall explain shortly.

A demonological concept of mental illness prevailed in the Middle Ages, which meant that people suffering from mental disorders were especially feared. Usually, their symptoms were thought to be the work of malicious devils; therefore, cruel, and often drastic, measures were used to dislodge the demons. In the case of mental illness, scourging of evil spirits was practiced well into the nineteenth century. People whose madness manifested itself in the form of religious fervor were more fortunate. Their madness was usually seen as a gift that set them above ordinary human beings, and they were able to escape the brutal treatments normally recommended. Whether touched by holiness or by evil, however, madness was linked with the supernatural world and the insane were feared and shunned much like lepers.

Epilepsy was viewed by many as some sort of divine afflatus, and is a good example of an especially feared disease that went by a multitude of different names. In ancient times it was euphemistically referred to as *morbus sacre/divus* "the sacred disease." Caelius Aurilianus, a Roman physician at the time of Galen explains the use of *sacre*, which he implies was a kind of omnibus euphemism for diseases of this sort:

> It is known as 'sacred disease' either because it is thought to be sent by a divine power or because it is centered in the head which in the judgment of many philosophers is the sacred abode of that part of the soul which originates in the body, or because of the great power of the disease, for people generally call what is powerful 'sacred'.   (Quoted in Gordon 1959:552)

This name also appeared in the Middle Dutch medical texts but there was a plethora of alternatives also available. Many are general terms, suggesting more a discomfort (like the original of our own word 'dis-ease') or a simple weakening; some may be prompted by compassion for the sufferer. Consider the following.

> *epilencie* "epilepsy"—from the Greek *epi* "upon" + *lepsy* "to seize"
> *caducum morbum/morbus caducus* Latin—"falling disease" (it was known in
>   English as *the falling sickness*, cf. Shakespeare *Julius Caesar* I.ii.254)
> *vallende euel* "falling evil"
> *tfallende fledersijn* "falling flapping sickness"—'fledersijn' means literally "flap-
>   ping" and must refer to the thrashing movements of the patient during a fit.
>   Oddly, this term alone is also used for diseases like gout and rheumatism.
> *tgroet onghemac* "the great discomfort"
> *tgroet euel* "the great evil"
> *tonghemac* "the discomfort"
> *Sente Cornelys euel* "St Cornelius' evil"
> *Cornelys onghemake* "Cornelius' discomfort"

The last two illustrate the practice of labeling the most feared of disorders with the names of saints, which we turn to next.

The Cult of Saints

Some 130 saints were invoked to serve as protectors and comforters of the sick (Gordon 1959:33–40; Huizinga 1924:173–177); and, strange to our eyes, their names appear in the names for diseases. *Sente Loys euel*, *Sente Quintijns euel*, and *Sente Vencentius euel*, for example, are all names of various ulcerous skin complaints. So great was the terror surrounding the plague, that it masqueraded under a number of different saints' names. For example, *St Adrian's/St Christopher's/St Valentine's/St Giles'/St Roch's Disease* were just some of the many protectors named. Another epidemic, also of enormous proportion, that swept through medieval Europe was *St Vitus' Dance* or *the dancing mania*. It caused thousands of people, believed to be possessed by demons, to take to the streets. They could be seen leaping about in mad frenzy for hours, sometimes for days, until they dropped foaming at the mouth. The name *St. Vitus' Dance* suggests more a joyous romp and conveys nothing of the horror and suffering that attended this particular mass psychotic disorder. *St Antonius' Fire* (*Sente Antonius vier* as it was known in Dutch) referred to another epidemic which raged successively throughout the Middle Ages, killing and deforming on a huge scale. Descriptions of the disease (cf. Gordon 1959:469ff) suggest it was a variety of herpes, most probably *herpes zoster*; but because of the general confusion surrounding disease, the name of this particular patron saint may also have been used for other diseases. It is worth noting that *herpes* was originally a euphemism from Greek; it literally means "a creeping." In the Dutch texts, this was just one of the many different euphemistic names for this particular disease. Among them was *ommeloep* "walk around," *ommeleghe* "lie around," and *sprengh vier/springet vier* "jumping fire." Like *herpes* these descriptions make reference to the effects of the rapidly spreading sores characteristic of the complaint. The term 'fire' which recurs in many of the disease-names presumably describes the effect of the burning pain that accompanied them. These labels could perhaps be called metonymic or part-for-whole euphemisms in that they figuratively describe one of the symptoms of the complaint (cp. *the runs* for "diarrhea"). But from a present-day perspective, it is extremely difficult to assess the euphemistic quality of terms like this.

If ever a euphemistic practice backfired it was this use of saints' names. The diseases became so associated with the names of the saints that the saints themselves came to be seen, not as comforters and protectors of the faithful, but rather as wrathful tyrants to be feared as perpetrators of the disease; in the minds of sufferers of herpes zoster, for instance, it was now St. Antonius who was stoking up the fires in their burning blisters. Much to the horror of the church, this change in perspective inspired a rather dramatic return to pagan worship, and because of the return to paganism, the cult of saints was for a time tabooed by the church and the practice largely disappeared. This is a striking illustration of the pejorative path that euphemism often takes.

Of King, Fire, Rose, and Wolf

There are many other examples of euphemistic names for diseases, some almost reverential. Most reflect real fear and superstition; and many, like the use of saints' names discussed earlier, seem like attempts at mollifying the forces of evil. One such example is *sconincseuel/coninx onghemac/morbus regnis* "King's evil/disease" as it was also known in English; a sickness that was thought to be curable by the touch of a king, perhaps because a king was invested with God's power on earth. This was understandably a more palliative expression than *scrofula*, a term that was known then but it seems to have been little used (cp. Middle Dutch *die lopende scrofulen* "loping scrofules"). The name *dat heyliche vier/ignis sacer* "sacred fire," presumably also because it was believed to be despatched by God, was euphemistically applied to many of the nasty cutaneous eruptions (including St Antonius' Fire) that afflicted people during these times. Erysipelas was another one. This unpleasant skin inflammation was variously called *wilde vier* "wild fire" and *roos* "rose" (Modern Dutch *belroos*). It is certainly euphemistic to associate roses with the ugly red blotches characteristic of this disease; yet the scratch of the rose thorn may suggest something of the pain the disease causes. The term 'fire' may be similarly descriptive.

Wolf was the name for a highly erosive skin complaint known also by the Latin equivalent, *lupus*. The name may originate from the practice of smearing wolf's blood on the skin to remove ugly marks, or the wolflike face which resulted from the disease tautening the skin. In those days, there were also ulcerous sores that erupted all over the body as the disease spread: 'hungry like unto a wolf,' in one seventeenth-century English account of the disease. A fourteenth-century Dutch description also suggests this etymology is the correct one: *es en herde quade aposteeme ende es te segghen dat hem selven eet* "is a very malignant ulcer and this is to say which eats him [the patient] himself." Nowadays *lupus erythematosus* and the rarer *lupus vulgaris* cause a red facial rash, but not the ulcerous sores described of medieval lupus; given the imprecision of medical labels in the middle ages, it is possible that *lupus* denoted a host of cutaneous disorders. For instance, the disease closely resembles (and may well be confused with) another, known as *noli me tangere* "don't touch me," which is also an erosive skin disease that specifically attacks the face. The name almost mockingly makes reference to the extremely contagious nature of the disease as well as the agony caused to the patient whenever s/he has to be handled—because to touch the patient also worsens the sores. Both *wolf* and *noli me tangere* represent perhaps a callously jocular means of coping with what at this time must have been extremely unpleasant diseases, unless, like the animal names denoting leprosy (see the following), these names are distressingly accurate in describing their dreadful effects.

Leprosy—Disease as Punishment

> 44He is a leprous man, he is unclean: the priest shall pronounce him utterly unclean; his plague is in his head. 45 And the leper in whom the plague is, his clothes shall be rent, and his head bare, and he shall put a covering upon his upper lip, and shall cry, Unclean, unclean. 46 All the days wherein the plague shall be in him he shall be defiled; he is unclean: he shall dwell alone; without the camp shall his habitation be. (*Leviticus* 13:44–46)

One of the most striking examples of disease as punishment is leprosy, which inspired great fear and prejudice; it is therefore very curious that there are so few euphemisms for it. In Dutch it was known variously as *lepora/leprosia* from the Greek *lepra* "scaly" describing the scalelike lesions symptomatic of the disease. It was also called *laserscap* and *lazerie* derived from the name of the biblical character Lazarus (*Luke* 16:20; this is not Lazarus of Bethany whom Jesus resurrected from the dead). Lazarus was the victim of a skin disease that may well have been leprosy although this is uncertain. As a description of leprosy, the name undoubtedly began as a general-for-specific euphemism for any deforming disease (cp. early English *lazar* "a poor diseased person") but most commonly applied to leprosy cases, whence it was quick to narrow to its modern sense. However, its use may also reflect the great ignorance surrounding the disease in these early times, when many different complaints, including psoriasis and syphilis, were confused with true leprosy (cf. Shakespeare's *Timon of Athens* IV.i.30 and Partridge 1955:75, 141). The proper name Lazarus also gave rise to the term *lazaret(to)*, used of a medieval hospital established and run by the Order of Saint Lazarus to care for the diseased and poor, and more particularly lepers (these hospices may well have been named after Lazarus of Bethany). The term later came to be used exclusively of leprosariums and therefore shows the same pejorative narrowing as a word like *asylum*, discussed shortly.

Medieval physicians generally distinguished four different types of leprosy and the labels for each of these are all derived from animal names; though, once again, it is likely that other cutaneous complaints are being included here. The following descriptions come from the fourteenth-century surgery text written by the Brabant physician Meester Thomaes van Scellinck. He lists the four types and briefly explains why each is so called:

> *olifancia na een dier dat heit olifant ende het es dat alrestercste dier dat men vint. Ende alsoe is dit onghemac.*

> " 'elephantiasis' after an animal which is called elephant; and it is the strongest animal which one finds. And so is this discomfort."

> *leoninana na eenen leeuwe omdat hij is anxteliic ende rampende ende die lede vanden vingheren ende vanden teen vallen gheerne af.*

> " 'leontiasis' after a lion, because he is frightening and rampant; and the limbs of the fingers and of the toes often fall off."

> *tijria(na) na een serpent ende is gheheeten tijrus. Ende dat serpent doet dicke siin sluijf af alsoe doet die sieke hem scellet dicke siin huiit . . . ende hi stinct herde seer.*

> " 'tijriana' after a snake which is called Tijrus. And the snake often loses its skin so the patient often sheds his skin . . . and he stinks very badly."

> *allopicia na den wolf allopes . . . ende haer aensicht is root ende ghedronten ende die wiinbrauwen vallen uut.*

> " 'alopecia' after the wolf allopes . . . and their face is red and swollen and the eyebrows fall out."   (*Boeck van Surgien* p.198f)

Most of the medieval examples of disease names we have considered up until now have involved euphemism. Yet to suggest that someone has a lionlike appearance, or sheds skin like Tijrus the snake, seems a curiously offensive way to describe the clinical features of a disease. These names are really almost dysphemistic in the way that they depict in a most graphic manner the repulsive effects of leprosy on its victims.

In order to fully appreciate the animal nomenclature here, we have to take into account the perception of lepers in a medieval community. It is obvious that the names are not born out of any compassion for the leprosy sufferer, but reflect real loathing and contempt. In the eyes of the rest of the community, leprosy was the reward for sin and heresy; lepers were viewed as both physically and morally dangerous. In line with early Roman and Greek beliefs, diseases like leprosy were believed to result from contact with a menstruating woman because menstrual blood was widely perceived as a source of evil and disease (see Ch.3). So, within society, lepers were branded as *tanquam mortuus* "as though dead" (i.e., the living dead). Fear of contagion meant they were deprived of all normal community rights, including rights of inheritance; and in some places they were even expected to undergo a sort of ritual burial before entering the leprosarium—a ritual presumably marking the civic death of these individuals (cf. Richards 1977 and Gottfried 1983:13–15). What was perceived as their animal-like physical appearance was the fitting punishment for their transgressions against God. Along with other despised members of the community, lepers were commonly blamed for disasters and epidemics like the plague. As you might expect, an effective means to bring contumely upon someone one didn't like was to spread a rumor that s/he had leprosy.

The horrible physical disfigurement of lepers was believed to reflect an inner corruption and mental derangement. Meester Thomaes van Scellinck explains this according to contemporary theories of the 'humors'.

> *Lepora is corrupcio corporis exterius et interius . . . haer complexie is quaet ende ghecorrumpeert also siin haer gheduchten ende haer ghepens is quaet ende ghevenieint. Ende daerom salmense sceiden uut den ghesonden luijden.*

> "Leprosy is corruption of the body externally and internally . . . their complexion [i.e., the lepers' combination of humors] is bad and corrupting and so are their thoughts and their mind is bad and poisoned. And therefore one should separate them from healthy people."

Unfortunately, the enormous social upheaval that frequently went hand in hand with epidemic diseases like leprosy encouraged moral judgment of this sort and strengthened the connection between physical and mental corruption in the medieval mind. The upheaval was directly due to the terror of contagion when an epidemic hit a community, and understandably put extraordinary strain on family and social ties. Mass forms of psychosis like the dancing mania and flagellation, for example, often coincided with the outbreak of disease; furthermore, diseases like bubonic plague caused physical intoxication of the nervous system which then resulted in complex psychological and neurological illnesses; the resulting chaos was taken as further evidence of the evil influence of diseased people like lepers. As we will see in the next section of this chapter, there are certain diseases today that inspire similar attitudes because the disease is linked with moral depravity. Typically, these are diseases for which there are abundant euphemisms, and it is striking therefore that leprosy does not seem to have inspired the same linguistic treatment. While the despised status of the leper might explain the instances of dysphemism we have found, it still does not explain a conspicuous poverty of euphemism.

Despite the fact that there is no longer any mystery surrounding the cause and cure of leprosy, much of the frightful stigma that was attached to this disease still exists today. It is apparent in the idiom *to be treated like a leper* and in other powerful images which *leper* and *leprosy* retain in their metaphorical use. Examples like the following taken from among the *O.E.D.* entries for leprosy demonstrate the persistent dysphemism of the epithet: *Idleness is a moral leprosy which soon eats its way into the heart* (H. Smith); *When nations are to perish of their sins, 'tis in the church the leprosy begins* (Cowper); *that leperous stain Nobility* (Coleridge); *the leprous humour of Popery* (Sanderson). (Compare also the modern French *lépreuse* to refer to an eroded stone edifice.) These examples show clearly that leprosy is still strongly linked with sin and heresy. Given this stigma, it is not surprising that in Hawaii the word *leprosy* is banned, and it is called *Hansen's Disease* after the physician who in 1871 discovered the bacillus responsible for it. This is also the favored term in medical circles.

Disease and the Twentieth Century

As the study of disease in the Middle Ages shows, where there is fear and ignorance, there is generally euphemism. The then dearth of knowledge concerning bodily organs, their processes, and their pathological changes produced exotic medical doctrines that were built for the most part upon imagination and superstition. To be fair, in their own context these doctrines did not appear as absurd as they do from today's perspective; and in many cases simple faith in the ritual may well have been enough to effect a cure. Nonetheless, disease was generally thought of as something mysterious and supernatural and this gave rise to a language of disease generally rich in euphemism. From the secure position afforded us by twentieth-century medicine, it is difficult to see any connection whatsoever between these medical events of the past and those of today. Yet despite the achievements of technology and the fantastic advances made in modern medicine, there are still many diseases of mystery. Today diseases like AIDS and cancer are met with very much the same fear and superstition and we see once again, the terrifying becoming the unspeakable.

In Ch.2 we examined some of the beliefs that center around the power of naming: the belief, for example, that knowing a name gives one power over the named person or object; alternatively, that not uttering a name can avert the incursion of something feared—for example, if a more favorable appellation were substituted, as was exemplified in the first paragraph of this chapter. Such practices are by no means confined to the past or so-called primitive human groups today—there is still plenty of evidence for the powerful magic of names in contemporary society, particularly in the vocabulary of disease.

The Vocabulary of Disease

Attaching some sort of diagnostic label to someone's suffering is a two-edged sword. Though we do not intend to enter into a full-blown discussion of the labelling theory here, some relevant points should be made clear. Many of us have experienced the enormous sense of relief that comes from hearing an ailment identified and named. The name legitimizes our status as a patient (see Field 1976:355); it stamps on our collection of symptoms the seal of a genuine illness. Rightly or wrongly this seems to remove much of the mystery for us and consequently much of the anxiety. We feel more in control somehow, for the name has identified the disease, which allows for a plan of action, a possible treatment, a likely prognosis. In this way, a simple label can make a world of difference to the rate of a patient's recovery. This is a very positive side to labelling.

But there are also negative aspects. There is a great deal of evidence to suggest that attaching some sort of medical label to someone can sometimes be enough to induce the appropriate physical symptoms (cf. Field 1976:358, Sontag 1979:6). For example, after RSI (Repetitive Strain Injury, tenosynovitis) received a lot of media attention in Australia during the 1980s the number of cases appeared to mushroom, and one discovered astonishingly large numbers of people suddenly suffering from an affliction that almost no member of the lay public had heard of ten years before. We assume that the overwhelming majority of those Australians who took sick leave, early retirement, and so on did have a genuine complaint (given prevailing beliefs); but what is puzzling is why one had not heard of RSI before, and why it did not have such a large incidence in comparable countries where it was given little or no media attention. The conclusion must be that when the condition was virtually unknown, people suffering from what would later be known as RSI attributed it to arthritis, or dismissed it as a nonspecific ache or pain. Once the condition is known, people are found to have it. It may be that AIDS is not as new as we have assumed it to be, and that people had it long before it was recognized. Lliam Harrison has told us that samples of blood serum kept for research purposes since the 1960s have shown a number of cases that are HIV positive. Today these people would be added to the AIDS statistics: what statistics did their deaths inflate all those years ago? We draw the following conclusion from these observations. It used to be said that in primitive societies people would become ill and die if the witch doctor pointed the bone at them: A lot of baloney, you say? It looks to us as though you can persuade some of our people that they have a disease simply by talking about it hard enough.

The drawbacks of medical labelling become particularly acute where the illness is a stigmatizing one. Such illnesses tend to be those of mystery; typically, diseases of uncertain biological nature with no ready cure. As we will see, these illnesses are quick to become tainted with attitudes of shame and disgrace. In some cases, patients may find themselves branded with a label that they then find impossible to lose, even after treatment is complete. Mental illness carries such a stigma. The lay public does not readily accept that it is curable. A negative stigma can destroy the possibility of normal social interaction for patients who, fearing the withdrawal of others, find themselves retreating from contact. This of course can seriously affect a patient's receptiveness to treatment. We have already discussed this behavior among patients whom the medical profession euphemistically describe as 'patients in a terminal situation.'

A contrast of the vocabulary relating to cancer and heart disease will illustrate the very different linguistic treatments that stigmatizing and nonstigmatizing illnesses receive. Even though cancer and heart disease are both life-threatening, from the lay point of view there is a world of difference between them. As we will see, this can have unfortunate consequences for the way in which patients are then received by society. We wish to make it clear that the discussion that follows is concerned with the emotional and social aspects of illness. The medical profession recognizes something like 200 different varieties of cancerous diseases, each with its own biology; the lay person, on the other hand, uses the word *cancer* as if there were a single, clearly defined condition. This is not a medical textbook, it is about the way people use language; so it is the response of the lay public that is of concern to us.

The Contrast Between Having Cancer and Having 'A Cardiac Condition'

> Cancer the Crab lies so still that you might think he was asleep if you did not see the ceaseless play and winnowing motion of the feathery branches round his mouth. That movement never ceases. It is like the eating of a smothering fire into rotten timber in that is is noiseless and without haste.
>
> (Rudyard Kipling, quoted in Patterson 1987:31)

> 'I heard the word cancer, and I didn't hear another word the doctor said to me.'
>
> (Patient interviewed in 'Breaking the News' Australian Broadcasting Commission's *The Health Report*, January 22, 1990)

The euphemisms for *cancer* are many (e.g., *The Big C, CA, a long/prolonged/ incurable illness*); one author (KB) recently heard a seriously ill cancer patient described as someone who had *a touch of the c's!* These sorts of euphemism typify mention of cancer and they are usually also accompanied by a characteristic set of paralinguistic phenomena such as lowered eyes and hushed tones. All this suggests a superstitious reluctance to pronounce the word *cancer*, as if perhaps there were some hidden supernatural power responsible. For example, in an obituary for someone who did in fact die of cancer, the disease is never mentioned; here are the relevant paragraphs.

> Penny Fisher died after a long illness on Tuesday, 31 January 1989; she was 35.
>   Tragically her health began to deteriorate in 1985. She left Monash in March 1986 but continued to work as a consultant for more than a year. . . .
>
> (*Monash Reporter* March 1989:8)

The euphemism *died after a long illness* is a fairly common one when someone has died from cancer. Sontag (1979:6) writes 'The very names of such diseases are felt to have a magic power'; despite the fact that *died suddenly* should often be interpreted as "died from a heart attack," the taboos on mentioning heart disease are nothing like those on naming cancer. (It has been suggested to us that the use of expressions such as *a long illness* in death notices may be prompted by a fear of burglary, as the homes of cancer patients are believed to contain narcotic drugs. Although this could occasionally be true, we do not believe it to be the usual motivation for this mode of expression.) The reason for the different linguistic treatment of cancer and heart disease cannot lie in the mortality rate because heart disease is the most common cause of death in so-called advanced nations today. By contrast, most cases of cancer, despite its reputation, are not in fact terminal. The reason for the very different linguistic treatment must lie elsewhere.

With cancer, the lay public cannot usually see any obvious cause for the disease, even though they may have heard of invisible (and therefore mysterious) carcinogens. There is also no straightforward cure. Cancer is often a hidden disease with no symptoms and no visible signs of progress. One could be dying and not even know it. The image of cancer is very much that of a latent malignancy that is ready to strike again even after treatment. Of course, there are also cancers that do not remain hidden. Lumps can appear on the body, tissue wasting can become apparent, and there may be unpleasant discharges; additionally, some cancers distort normal excretory functions, and therapy can be mutilative. Chemotherapy is toxic and has side effects such as nausea, vomiting, diarrhea, hair loss, and even sepsis and bleeding. As one doctor put it to us, 'Cancer can be a dirty disease.' All these frightful symptoms and effects serve to strengthen the lay public's view of cancer as a contamination, and perhaps to some minds the wages of sin. Just the kind of evaluation that the medievals gave leprosy. We may no longer believe, like the medievals did, that the cancer patient is invaded by a deadly crablike demon, yet all this is strikingly reminiscent of the sort of fear and hysteria that attended disease during the Middle Ages.

Concealing a cancer diagnosis, particularly if it involves incurable cancer, is a common practice, or at least has been (cf. Hinton 1976:303–14). It is still a contentious issue in current medical spheres as to whether or not doctors should inform patients that they have a fatal condition—do they conceal or tell all, or just say what they think the patient wants to hear? As Stephenson (1986:103–9) describes, it is only since the early 1970s that it has become usual for patients to be told the truth. Up until then avoidance of unpleasant truths was the common practice—ostensibly to spare the feelings of the patient, but most likely also to spare the doctor, who—as we have already observed—often finds these situations more perturbing than the patients themselves. For example, it was not until after the death of Emperor Hirohito that the Japanese nation learned the full truth about their Emperor's 'condition.' A false diagnosis of chronic pancreatitis was made public after a tissue biopsy had been carried out; even the Emperor himself was not informed of the fact that he had cancer. 'Cloaking unpleasant truths' is how *Time Australia* (January 23, 1989) described it; however, it is not such a peculiarly Japanese phenomenon as the *Time* article makes out. The cancer diagnoses of American Presidents Grover Cleveland and Lyndon Johnson, for example, were not disclosed to the public; and the mooted malignancies of both George Washington and Franklin D. Roosevelt have never been confirmed nor disconfirmed publically. It is also only relatively recently that authorities have overcome their reticence to use the word *cancer* in the names of hospitals, clinics, and special units for cancer patients, preferring something like *oncology* instead (from the Greek *onkos* "mass"). In 1906 Charles Childe, the leading British expert in the disease at the time, was prevented from using the word *cancer* in a book title that then became *The Control of a Scourge*. Even as recently as 1987, Patterson (p. 329, footnote 7) writes that publishers tried to persuade him to drop *cancer* from the title of his history of the disease, *The Dread Disease: Cancer and Modern American Culture*, which is a nice irony in view of the fact that the book documents the reticence and fear inspired by cancer, and he gives many instances of people's unwillingness to use the word!

We now see much more frequent use of the word *cancer* in the media, names of hospitals, and so on. And there was wide and quite explicit coverage of former U.S. President Ronald Reagan's malignancies. (Significantly, however, Reagan denied that he himself had cancer: 'I didn't have cancer. I had something inside of me that had cancer in it and it was removed.' Quoted in Sontag 1989:66.) But although there may be greater public acceptance today, there are still many reports of the reluctance among doctors to use the word *cancer* with their patients because of its demoralizing effect on them. For example, Karl Menninger writes that 'the very word cancer is said to kill some patients who would not have succumbed (so quickly) to the malignancy from which they suffer' (quoted in Sontag 1979:6). The label used by many Australian doctors in place of *cancer* is *mitotic disease*. This is a nonspecific term that refers generally to multiple cell division and includes both *malignant* and *benign* cancers. As we already mentioned, patients themselves prefer to use instead of *cancer* words like *tumor* or *growth* that do not evoke the same unpleasant imagery of decay and corruption. Whereas a cancer diagnosis is equated with malignancy and death, tumors and growths have a much more benign image. Growths can be removed but a cancer continues to lurk in the system. Such a strategy of avoidance may not always be on the part of the patient, however. There is plenty of evidence to suggest that the medical profession, as much as the patients themselves, require the soothing ministrations of euphemism. The following are instructions issued by a doctor to a nurse in charge:

> There is nothing we can do; that thing has attacked everything lungs, spleen, everything. Go and help them [the family and patient], go talk to them. You're her nurse today, you go talk to them. You should be able to help them; I've done all I can.   (Quoted in Stephenson 1985:108f)

Because there is no obvious reason for cancer, there is sometimes a perception that the disease can come about through some defect in character, and therefore the feeling is that cancer patients are somehow responsible for the fact they are sick. This has given rise to popular articles and books with titles like 'Can your personality kill you?' (cf. Sontag 1979:49–51). Even well into this century, cancer was believed to afflict those people, particularly women, who were nervous and stressed. There was even a strong feeling that it was a disease caused by "luxurious living". This is not so far removed from the medieval humoralist concept of disease; for example, according to this theory melancholic women were believed to be more prone to breast cancer than others.

By contrast, there is no mystery to heart disease but an obvious reason for a person's condition. The attendant imagery is not that of an invisible growing evil as in the case of cancer, but one of a simple 'nuts and bolts' breakdown in the machinery (hence the use of mechanical terms like *ticker* and *pump* to describe the heart). As one medical student from the University of Pennsylvania put it 'other serious illnesses are similar to a machine breaking down, whereas cancer is macabre' (Quoted in Cassileth and Cassileth 1979:305). There is no shameful stigma attached to having a 'heart condition,' no more than there is to having the flu. Accordingly, people don't usually modify their attitudes toward heart patients just because they are known to have a weak heart; and heart patients are rarely denied the truth of the seriousness of their condition. This does not of course mean that heart patients never find the need to resort to euphemism; for a variety of reasons they may want make light of their complaint and use language like *there's something wrong with my ticker* or *the old pump's not working too well*, rather than using expressions like *heart disease* or *heart attack*. As we have already seen, this sort of euphemistic dilution of the medical facts helps to alleviate the threat of death, but the motivation may also be more mundane: it might simply reflect the desire not to impose one's own troubles on others; not to be seen to complain. This sort of euphemistic behavior is governed by general principles of social interaction. It is not motivated by the same superstitious dread that we find associated with the cancer euphemisms. We conclude with an anecdote that nicely illustrates what we have been discussing here. A close relative of WM's died not so long ago of heart disease. In the subsequent death notice, it was requested that instead of flowers, donations be sent to *Cancer Research*, which was in error for *the Australian Heart Foundation*. WM reports that the deceased's widow was considerably distressed that people would mistakenly believe her husband had died of cancer rather than heart disease.

It is notable that cancer has always had a strong association with women. In light of the strong pollution taboos that surround women's bodies (Ch. 3), this association may contribute to the stigma that is attached to the disease. One early twentieth-century writer (F.B. Smith *The People's Health 1830–1910*, quoted in Patterson 1987:43) observed that 'want of proper exercise and changed surroundings' was emasculating men and making them prone to 'women's diseases,' and he believed this explained the rising incidence of mortality from cancer among men. Even today we find cancer described as a woman's disease whereas heart disease is typically associated with men; the implication being that 'cancer is a "female" and a "dirty" disease, whereas heart disease is a "male" and a "clean" one' (Meigs 1978:318). The cancer mortality rates have always been higher for women than for men; and even in antiquity, physicians noted that the disease was particularly prevalent in women; no doubt they were swayed by the fact that it commonly strikes breasts, the cervix, and other "secret" or "private" organs of females. Through fear and shame, many women kept their illness secret, which only made the aura of mystery and fear surrounding it that much more devastating. In the early twentieth century it was not so very different. Greta Palmer in 'I Had Cancer' *Ladies Home Journal* July 1947 (quoted by Patterson 1987:150) wrote, 'So many women of the last generation speak about cancer as if it were one of the social diseases—a taboo word never to be mentioned in polite company.' In consequence of this attitude, many such women failed to seek professional help, to the distress of the medical profession at that time.

We will return once more to this link between women and disease in Ch. 10 when we address the wider issues of taboo and euphemism.

Tuberculosis: The Cancer of the Nineteenth Century

To step back in time for a while: roughly speaking, tuberculosis was both linguistically and medically the nineteenth-century counterpart of cancer today. It was then an 'unmentionable' disease whose production and cure were steeped in mystery. As the killer disease of the nineteenth century, tuberculosis went by a number of different euphemistic labels, among which the abbreviation *TB* was, and perhaps still is, the most common. The term *consumption* began as a euphemism but soon ceased to be one. Its literal meaning is "an eating up," and in the context of disease could refer to any general wasting of the body by illness. The most common of these wasting diseases was tuberculosis. Today, *consumption* is no longer used, having been replaced by *tuberculosis*, which is the term for a more precise medical concept. In fact, many people today would not be familiar with a disease called 'consumption,' nor would they be aware of the fact that euphemistic expressions like *raising color* or *spilling rubies* refer to "spitting blood," one of the characteristic symptoms of tuberculosis.

It is curious when you consider the hideous reality of TB that it came to be associated with such romantic imagery. Sontag (1979 Ch. 4) remarks that it was once even fashionable to appear consumptive: perhaps the closest thing today would be the 'anorexic look.' Its connection with artistic creativity meant that many lyric poets of the nineteenth century aspired to it, because a consumptive individual was believed to be passionate, sensitive, and intellectual. In contrast, good health and a ruddy complexion was mundane, even vulgar. All this, however, is somewhat at odds with the true symptoms of the disease, which arise from the destruction and wasting away of tissue, leaving a severe cough and foul-smelling sputum, and finally lead to the expectoration of frothy blood—the signal of imminent death. The reality of this debilitating disease suggests that the romance of 'consumption' may well have been an imaginative fiction of writers and musicians of the time, whose romantic heroes and heroines languished in long lyrical consumptive death scenes. Accounts by those with first-hand experience of the disease paint a different picture. Franz Kafka writing from the sanatorium where he was to die two months later, described his experience as follows:

> *In Worten erfährt man freilich nichts Bestimmtes, da bei Besprechung der Kehlkopftuberkulose jeder in eine schüchterne ausweichende starräugige Redeweise verfällt.*
>
> "Verbally one really doesn't learn anything definite, since in discussing tuberculosis of the larynx everybody falls into a shy, evasive, glassy-eyed manner of speech."
>   (Postcard to Robert Klopstock April 7, 1924)

It is notable that the name *sanatorium* is a euphemism: although etymologically it implies that the patient will be cured of the illness and return to a normal life, in reality many people left a TB sanatorium in a box.

Mental Illness

In the twentieth century, mental illness remains an area of abundant euphemism. Once again, in order to explain why this might be so, we need to consider something of the nature of the "disease." Although Western society no longer believes in a demonological concept of mental illness, its behavior toward the mentally afflicted generally expresses the same attitudes of compounded fear and contempt. One reason for this is that mental illness still has a great deal of mystery to it. For one, it is nowhere near as easy to define as physical illness. The term covers an enormous assortment of conditions, ranging from mildly eccentric or neurotic behavior, to severe psychotic disorders where a patient might lose total contact with reality (as in the case of severe schizophrenia, for example). To the layperson who lumps all these together as *insanity*, the picture is indeed a confusing one. When is nonnormal behavior to be considered an illness? When is behavioral deviance considered problematic?

As in most cases of stigmatizing illnesses, the origins of mental illness are usually mysterious. Because of this, there is great shame attached to having the disease. Again, the lack of an obvious cause means that much of the burden of responsibility falls on the patients themselves. Mental illness is viewed not so much as a disease, but more as a moral failure! While it is of course perfectly acceptable to be physically ill, it is not acceptable to be mentally ill. More than any other disease, mental illness suggests a deficiency in the person, some sort of weakness of character, and it seems to make no difference if the disease is the result of a tumor or an accident of some sort. The role of patient is not one that is readily accepted in the context of mental illness. Onlookers are much more judgmental than they are with other illnesses. This is partly the reason that, until quite recently, people used to visit an insane asylum for entertainment, much as they now visit the monkey house in a zoo. (With an entry fee of one penny, Bedlam had revenue of about £400 per annum in the early eighteenth century!) This is the reason that many terms for madness are associated with the funny: *funny* (*in the head*), *funny farm, wacky, wacko, mad, crazy, bats, nuts,* and so on. And the loss of control that is a feature of both slapstick humor and insanity is evident in the different meanings of the terms *mad* and *crazy* in normal non-clinical usage, such as we shall discuss very shortly. There is a long history to the perceived link between madness and funny behavior.

Because the behavior of mental patients does not conform to morally and socially accepted norms, it is usually viewed as threatening and strange and is often believed to result from maliciousness of character, particularly if the patient some-how 'looks different.' This perception of mental patients is still firmly tied to the old fashioned notion linking internal and external corruption. It shows also a lingering fear of 'moral contagion'—the belief that the madness might somehow transmit itself to others (cf. Gillis 1972:175–79). In summary, the stereotype of the mental patient is someone who is morally deficient, incurable, and potentially dangerous; someone who is best *locked up* or *put away.* Paradoxically, the word *idiot* has the same origin as *idiom* or *idiosyncratic,* deriving from the Greek *idios* meaning "pecu-liar to oneself, private"; in ancient Greek an *idiōtēs* was "a private person," so an *idiot* was perhaps perceived as someone locked up in their own private world; compare this with current connotations of *hermit* or *recluse.* Nevertheless, for a long time it was thought (and a lot of people still think) that such people ought to be locked up by us sane ones. And because of the vagueness of the term *mental illness,* even very minor anxious or depressive disorders tend to carry the same negative stigma as the severe psychotic cases. The result of all this is a stigmatizing illness with a long chain of euphemistic labels as impressive as any that might have existed in the Middle Ages. The strength of the stigma is also evident in the proliferation of maledictions invoking mental abnormality—for example, *retard, moron, jerk, spas*(*tic*)—and also in the rapid degeneration of the euphemisms. As the following few examples illustrate, these are quick to narrow to their taboo sense alone.

Lunatic (hence *loon, loony*) is an early example of euphemism. It was originally used to refer to a certain type of madness that was believed to be caused by the changing phases of the moon. The name derives from Latin *Luna*, the name for the goddess of the moon. It therefore reflects an early supernatural concept of the disease; but its euphemistic motivation is no longer apparent to many people.

Describing someone as *touched* suggests they have been touched by the hand of God. If so there was possibly a similar motivation behind the Swiss French euphemism *chrétin* "christian," which has somehow given rise to the current English dysphemism *cretin*.

Insane, from Latin *in-* "not" + *sanus* "healthy," is now confined to "mentally unsound" but originally had a much broader domain encompassing all bodily organs and their functions. Today, even the form *sane* (without the negative prefix) has narrowed under the influence of *insanity* to denote only a mental condition.

Crazy (and hence *crazed* and *cracked*) originally meant "cracked, flawed, damaged" (cp. *crazy paving*) and was applicable to all manner of illness; but it has now narrowed to "mental illness." It captures the stereotypical mental patient as someone "flawed, deficient" (cf. *mentally deficient*), and is the basis for many euphemistic expressions for madness: *crack-brained, scatter-brained, shatter-brained; head-case, nutcase, bonkers, wacko, wacky; falling to pieces; have a (nervous) breakdown; unhinged; having a screw/tile/slate loose; one brick short of a load, not a full load; not playing with a full deck, three cards short of a full deck; one sandwich short of a picnic; two bob short of a quid, not the full quid; his elevator doesn't go to the top floor; a shingle short*; and perhaps *he's lost his marbles*. This is a kind of euphemistic formula that appears in a number of languages; compare for example, German *nicht alle Tassen im Schrank* "not all the cups in the cupboard"; Dutch *hij heeft ze niet allemaal op een rijtje zitten* "he doesn't have them sitting all in a row"; *hij is er niet helemaal* "he's not altogether there" (cp. English *not all there*).

The history of the word *deranged* illustrates a typical path of development for euphemisms. Originally from a verb meaning "to disturb, disarrange, disorder," it could be qualified with the modifier *mental* to be used of people who were "disturbed in the mind." Once used in the context of "madness," however, the word soon became contaminated; and now, without the modifier, it has narrowed to the "mad" sense alone (cf. *be/go mental* and become a *mental patient*). The word has recently given rise to a string of expressions that may once have been euphemistic, but are now probably neutral: *mental health/depression/fatigue* and also *mental hospital/home/health/center/institution*. It may well be that the terms *disordered* and *afflicted* are moving in the same direction; though still requiring a modifier like *mentally* to refer specifically to psychiatric illness, this is now becoming their normal context of use. Similarly the general-for-specific euphemism *sick* is frequently used to describe someone who is *mentally challenged* or in other words *of an unsound mental condition* (that's two more euphemisms); but it remains to be seen whether it will narrow its meaning.

The names for establishments holding mental patients show rapid pejoration. A striking example is *asylum*: Latin *asylum* originally meant "place of refuge, retreat," the sense still retained in *political asylum*. At first it appears to have been used of an institution for debtors and criminals; but was later broadened to include the orphaned, the blind, the deaf, and also the mad—when it required some sort of qualifier (e.g., *lunatic asylum*). *Sanatorium/sanatarium* as either an institution for mental or TB patients has shown similar deterioration to *asylum* (cp. also *home*). In recent years attempts have been made to demystify mental illness, and people are now treated in *hospitals* as they are for physical illnesses, sometimes even in general hospitals that have special units. This practice seeks to eradicate the dysphemistic stigma of labels like *loony bin*, *funny farm*, *insane asylum*, *mental institution*, and *lunatic/mad house*, which are still very much tainted by the earlier brutal treatment of the mentally insane. In the same spirit, the *Matteawan State Hospital for the Criminally Insane* was rechristened the *Correction Center for Medical Services*; and other institutions, too, have been renamed (cf. Rawson 1981:265f).

As we have said, many euphemisms seem to develop out of an acute fear that humans feel when they lose control over their thoughts and actions. Madness is perceived as a lack of control; hence there are expressions like *out of/losing one's mind*. The loss of control that is a feature of slapstick humor links it with insanity and with historical treatments of the mad as witless clowns. Loss of control is the motivation in many of our uses of the terms *mad* and *crazy*; for example, *I'm going mad/crazy* (indicating forgetfulness, confusion, losing one's grip [on sanity]); *I must be mad/crazy* (when said of undertaking something beyond one's capabilities, doing something excessively foolish); *He was mad/crazy* (when said of someone in frenzied rage); *He was quite mad/crazy* (when said of someone carried away by excessive enthusiasm). All these suggest extravagant actions and emotions that are out of control, as do many compounds formed on the words *mad* and *crazy*: for example, *boy-mad/boy-crazy*, *music-mad/music-crazy*, *car-mad/car-crazy*; and *madman/crazyman*; *mad dog/bull*.

According to Gillis (1972:177) 'the fear of becoming insane is one of the most common of fears felt by normal people, taking equal place with those of cancer and death.' This fear has inspired some of the strongest linguistic taboos to be found in the general area of illness and disease.

AIDS—Acquired Immune-Deficiency Syndrome

It is widely—but probably inaccurately—believed that with AIDS the world is once more facing an epidemic as terrifying as both leprosy and syphilis were in times past. In fact both linguistically and socially, AIDS has much in common with these two epidemic diseases. (Strictly speaking AIDS is 'a medical condition whose consequences are a spectrum of illnesses.' Sontag 1989:16.) For example, there are those who genuinely believe that AIDS has been sent as awful retribution for an epoch of excesses and impiety; hence, we find it being referred to by the acronym WOGS, which stands for *Wrath of God Syndrome*.

The vast and open media coverage that AIDS has been receiving might give the impression that we have finally come to terms with such diseases; but the real experience of handling the disease on a day-to-day basis suggests that this is far from the truth. There has been much publicity recently about discrimination against individuals with AIDS, those thought to have it, and even those with the job of caring for them. The question of testing for the disease highlights the problem. As British authors Steve Connor and Sharon Kingman describe, the AIDS test 'has become a threatening finger of guilt, in many cases pointing out those who have transgressed what some see as the normal rules of society.' ('AIDS The Disease that Changed the World' *The Age*, Melbourne, Mon. December 5, 1988:8) Aroused by fear and ignorance, the popular prejudice against victims of the disease is strong. (AIDS activists prefer the abbreviations PWA / PLA / PLWA "Person Living With AIDS" as more optimistic labels than *AIDS victim*—the message is, of course, that AIDS is something people LIVE with.) For many people, its connexion with both homosexuals and injected-drug abusers links the disease with deviant (and therefore unnatural) behavior, enforcing a correlation between moral and physical corruption. Not surprisingly, the prohibitions and social taboos surrounding AIDS are severe. However, the term has already been extended to metaphorical use by an environmentally conscious population: the phenomenon of *land degradation* in Australia which results from overfarming (specifically, the erosion of topsoil and salination caused by irrigation after deforestation) is now regularly known as *AIDS of the earth*; the same metaphor is also used in America.

Since AIDS is a relatively new disease, it remains to be seen what clusters of euphemistic synonyms will emerge. Before it received the name AIDS, the disease went by a number of different labels; for example, *Gay Cancer, Gay Plague* (it is doubtless appropriate that it should be associated with such deadly diseases), and *GRID* (Gay Related Immuno-Deficiency). These were perhaps abandoned because they linked the disease directly with the gay community, thus upsetting heterosexual hemophiliacs and other AIDS-afflicted non-gays. The acronym *AIDS* was probably chosen because of its optimistic ring—as though it were 'a disease that wanted to help, not hurt. Just a plague whose intentions were good.' (Black 1986:40) But it happens to be homophonous with the name of an American brand of diet-candy (*AYDS*), which seems to have been withdrawn from the market (at least under that name). It is also homophonous with the suffix *-ade* used of sweetened soft drinks such as lemonade or orangeade. This homophony has given rise to a number of puns and riddles like the following:

Question   'What do you call homosexuals camping out at the North Pole?'
Answer   'Koolaids.'
Question   'What do you call homosexuals on roller skates?'
Answer   'Rolaids'.

(CF. Schmidt 1985/6:70)

It will be interesting to see whether these products will get renamed.

There are a number of examples of verbal flippancy in reference to *AIDS*: two were mentioned earlier, *Anally Inserted Death Sentence* and *Toxic Cock Syndrome*. There is also the new acronym *GAY* for *Got AIDS Yet?*; which is a sick joke, playing on the name homosexuals, one of the groups at greatest risk of contracting AIDS, use for themselves. The high-risk groups have been labelled the *4-H Club*. Like many clever acronyms this is a double-whammy: first, it recognizes the fact that

AIDS sufferers are HIV positive (where the abbreviation for *Human Immuno-deficiency Virus* is reinterpreted as "H—roman 4"); and second, it captures the four groups at greatest risk—Homosexuals, Haitians, Hemophiliacs, and Heroine-users. Johnson and Murray (1985:156) point out '4-H clubs are also organizations of wholesome young American future farmers.' A nice irony.

As we have seen on many other occasions, omission can prove to be an effective euphemistic strategy. This strategy can only be effective if Speaker gets the message across by hinting so strongly we are led to read between the lines. For instance, suppose we were to read in the obituary of an unmarried man in his early 40s, 'Associates said the cause of death was complications from Burkitt's cancer.' Here the cause of death is not given by a hospital, doctor, family member, friend, or care-giver; but a spokesperson disguised by the word *associates*. Does the plural marking on 'Associates' indicate that the information came from more than one source? If so, why should a consensus have been necessary? This may not be reporting a coverup, but it smacks of one; and leads us to ask: What is being concealed? What is not being said? We may guess what from the fact that Burkitt's lymphoma is taken to be an indicator of presumptive AIDS; it is associated with both cytomegalo virus and Epstein–Barr virus, and someone with AIDS would be open to virulent attack from either of these complications. So we conclude that such a report looks like a euphe-mistic obituary for someone who died because he had AIDS; and in the current climate of opinion, such evasion is understandable. Now take another example, this time from a recent obituary in *Australian Aboriginal Studies* written for a victim of AIDS (Rose 1988). We are never told what the deceased died of, only that he was 'a participant in what is becoming one of the key scenarios of life in the late twentieth century, young people dying of incurable diseases.' The entire obituary highlights the nonconventionality and controversy that characterized his life; and it seems we are to infer the cause of his death from this. For example, we are told how 'the declining physical person defied time, decay, and conventional expectations' and challenged 'current social taboos.' Descriptions of the recently dead often use these veiled and guarded phrases that invite one to read between the lines (cf. Ch. 6). The obituary is also interesting in another respect. It is clear that its subject was a difficult person, who would appear to have trodden on a number of toes; consequently, a stream of unqualified eulogy would certainly not ring true. Yet the obituary writer has made the effort not to speak ill of the dead; and in consequence, the whole obituary is a nice example of the subtle art of euphemism that invites us to interpret the phrase 'a brief, brilliant, and controverisal career' with focus on 'controversial.' Under the circumstances, this sort of ritual euphemism makes everyone feel a good deal more comfortable than the naked truth would have done.

Summary of Chapter 7

In this chapter we examined another of the less happy aspects of human existence; namely, disease. We discussed both medieval and present day perspectives on it. As expected, we found the language of disease in both eras to be profoundly affected by strong taboos.

As in other cases of euphemism, a mild or indirect term for a disease will generally serve to make the harsh reality more tolerable; for instance, *long illness* and even *malignant condition* are more palliative expressions than *cancer*. The euphemistic language of disease and illness is sometimes motivated by politeness conventions. We prefer to hint at an indecorous and embarrassing illness rather than to name it directly; hence, the general-for-specific description *be sick* in place of *vomit*, or the use of *blood disease, catch a dose, social disease, VD, STD*, or even *sexually transmitted disease* in reference to specific venereal diseases; this correlates with the strong taboos against mentioning certain body parts and their functions, which we discussed in Chs. 3 and 4. Yet on other occasions we may want to play down an illness for the simple reason of not wanting to appear to complain: talking about one's illness has the reputation for being boring to others—although in fact such a judgment may itself be a euphemism disguising Hearer's embarrassment.

Fear and superstition also figure strongly in euphemisms for diseases. More serious illnesses are referred to euphemistically because of their connection with death. Much of this fear and superstition must also be due simply to the inexplicable nature of disease. What is a disease after all? There are symptoms and there are sick patients, but there is often nothing tangible in disease itself. To the lay person it seems to arrive out of the blue and then just as mysteriously seems able transmit itself from person to person affecting some while leaving others curiously untouched. The mystery would have been even greater during the Middle Ages when physicians had a much inferior knowledge of physiology and fewer sophisticated instruments to guide them; and recall that at this time it was illegal to open up a human body. It is hardly surprising, then, that people arrived at fantastic conclusions about the causes of disease. The most common assumption at this time linked it to demonological influences or the wrath of celestial powers. Most of the examples of euphemism were therefore favorable appellations, undoubtedly with the aim of somehow placating the malevolent supernatural powers.

It is clear, however, that such practices are by no means confined to the past. Even given the remarkable achievements of modern medicine, there is still much of an aura of mystery around illness. This is particularly true in the case of stigmatizing illnesses like cancer, AIDS (when will the name for this disease become well-enough accepted to become lower-case *aids* like the French translation equivalent *sida?*) and the plethora of disorders included in the description *mental illness*. The current social attitudes toward these diseases still reflect the medieval equation of good and evil with illness and disease.

CHAPTER
—8—

Distinguishing Between Jargon, Style, and X-Phemism

EXPLANATORY NOTE

Regulation 3 of the Local Government (Allowances) Regulations 1974 ('the 1974 regulations') (S.I. 1974/447) made provision prescribing the amounts of attendance and financial loss allowance to members of local authorities. Regulation 3 of the Local Government (Allowances) (Amendment) Regulations 1981 ('the 1981 regulations') (S.I. 1981/180) substituted a new regulation for regulation 3 of the 1974 regulations. Regulation 3 of the Local Government (Allowances) (Amendment) Regulations 1982 ('the 1982 regulations') (S.I. 1982/125) further amends regulation 3 of the 1974 regulations, with effect from 8 March 1982, by increasing the maximum rates of attendance and financial loss allowances. Regulation 7 of the 1982 regulations would have revoked both regulations 3 and 5 of the 1981 regulations (regulation 5 being a regulation revoking earlier spent regulations) with effect from 1st April 1982. These regulations preserve regulations 3 and 5 of the 1981 regulations by revoking regulation 7 of the 1982 regulations.

(Quoted in Cutts and Maher *Gobbledygook* 1984:57)

jargon *n* 1. the language peculiar to a trade, profession or other group.

(*Macquarie Dictionary*)

What We Mean by Jargon

He was ... ful of jargon as a flekked pye [= magpie]
(Chaucer *Merchant's Tale* 1386, 1.604)

The term jargon probably derives via the Old French *jargon/gargon* from the same source as *gargle*: Indo-European **garg-* "throat"; it originally denoted any noise made in the throat. In Middle English it was generally used to describe the chattering of birds, or human speech that sounded as meaningless as the chattering of birds—what the dictionary calls unintelligible gibberish (*O.E.D.* 1,3). For reasons that will become apparent, if they are not already evident, it came to be 'applied

193

contemptuously to any mode of speech abounding in unfamiliar terms, or peculiar to a particular set of persons, as the language of scholars or philosophers, the terminology of a science or art, or the cant of a class, sect, trade, or profession' (*O.E.D.* 6). This has now become the primary sense, as witness the *Macquarie Dictionary* entry for **jargon**:

> 1. the language peculiar to a trade, profession or other group: *medical jargon*. 2. speech abounding in uncommon or unfamiliar words. 3. (*derog.*) any talk or writing which one does not understand. 4. unintelligible or meaningless talk or writing; gibberish. 5. debased, outlandish or barbarous language.

Sense 1 from the *Macquarie Dictionary* is the one we ascribe to 'jargon,' using this term in such a way as to include what some scholars call 'technical language.' We will soon show why it is that senses 2–5 are consistent with this.

Jargon has not been given precise definition within linguistics; however, as far as we can tell, jargons are what some people call registers—language varieties determined by the occasions of use. Wardaugh, in his introductory text book on sociolinguistics, has this to say about register:

> Register is another complicating factor in any study of language varieties. Registers are sets of vocabulary items associated with discrete occupational or social groups. Surgeons, airline pilots, bank managers, sales clerks, jazz fans, and pimps use different vocabularies.   (Wardaugh 1986:48)

One notable difference between *jargon* and *register* is that only the former is used pejoratively. And, whatever may be the case with registers, jargons involve more than simply lexical differences; they may also differ from one another grammatically and sometimes phonologically, as we shall demonstrate. We distinguish different jargons on the basis of subject-matter and also on the basis of domain; that is, the activity in which the language plays a part: the law, government, linguistics, religion, and so on are all domains. Although we use a label like legalese as if it were a discrete entity, there is no one-to-one correspondence between the contextual factors on the one hand and the features of jargon on the other. Jargons are by no means fixed. They vary continuously in response to a complex of different situational factors; if any one factor is changed, the language may change accordingly. These factors are: (1) the relationship between Speaker, Hearer, and any other participants in the speech act; (2) the setting of the speech act; (3) whether the spoken or written medium of transmission is used.

Within any one jargon, then, there are individual styles that respond to different arrangements of situational factors like the preceding. For example, take the different varieties of legalese. Four of the five basic styles of English identified by Joos (1961) (cf. Ch. 2) can occur in the legal setting. 'Frozen' style typically occurs in written documents like wills, but can also be spoken in the case of jury instructions or in a witness's pledge to *tell the truth, the whole truth, and nothing but the truth.* Written 'formal' style is found in statutes and briefs and can be spoken in the court room in the arguments of counsel and the examination of witnesses. 'Consultative' style is spoken in lawyer–client interaction and lay witness testimonies; while 'casual' style can usually only occur during in-group interaction: for example in conversations between lawyers (cf. Danat 1980:470–72 for a comprehensive discussion of legal style). All these are varieties of legal jargon. They all show some features of legalese (discussed shortly), but differ in the relative frequency of these features.

It is because lexical features are the most obvious and usually the most accessible, that jargons are often characterized solely in terms of their vocabularies, as we see from the *Macquarie Dictionary* sense 2 and Wardaugh's definition of register (both quoted earlier). And there are some scholars who restrict the notion even further. For example, Charrow (1982:84), in her discussion of legalese, uses the term to refer specifically to 'terms of art' in the legal profession (i.e., terms whose meanings are conventional among lawyers). These include expressions like *month-to-month tenancy, negotiable instrument,* and *eminent domain,* but curiously she does not include technical legal vocabulary like Latin *mens rea* and French *tort* or even *action* for "lawsuit"—which is a commonly used word that has an additional specialized legal meaning. Charrow's use represents a strangely narrow application of the term *jargon.* For Hudson (1978), an expression is only considered to be jargon if it can be translated into something simple. No matter how highly technical the terminology is, he does not deem it jargon if it is the only viable means of communicating the information; unfortunately, however, he gives us no criterion for determining such uniqueness!

The view adopted here encompasses a much broader understanding of the term. While it is certainly true that specialist vocabularies constitute an important facet of jargon, we feel that jargons are more than simply esoteric vocabularies. They are characterized by inventories of style features that draw from all levels of the grammar; when compared to "normal" adult conversational discourse, they may show grammatical and phonological differences like tone of voice, as well as paralinguistic differences like gesture (although we shall have nothing to say on this last matter). Individuals will choose from these bundles of style features in response to the demands of a particular speech event.

Our discussion so far implies that features of jargon are necessarily a matter of stylistic choice; that is, the form of a jargon expression represents a choice on Speaker's part from a number of possible alternatives for the sole purpose of maintaining a particular style. But this is, of course, not always the case. When lawyers use the term *plaintiff* or when cricketers refer to *a man at deep fine leg* they do so not because of stylistic appropriateness, but because no viable alternative exists. Such expressions are therefore motivated by the characteristics of the denotatum, (see Allan 1986a §2.9). It is not always clear, however, whether a particular usage has stylistic implications or not. For example, the characteristic lack of anaphoric pronouns in legalese supposedly ensures clear unambiguous identity of a denotatum. An example is given in the following:

> 'Mixed hereditament' means a hereditament which is not a dwelling house but in the case of which the proportion of the rateable value of the hereditament attributable to that part of the hereditament used for the purposes of private dwelling is greater than the part used for other purposes.   (Cutts and Maher 1984:14)

Such a feature is motivated by the degree of precision demanded by the legal profession and is therefore a necessary part of legal jargon and not a purely stylistic choice within that jargon. However, there are many occasions where a denotatum is quite apparent and the use of a pronoun instead of lexical repetition would be unambiguous. In these circumstances to choose lexical repetition is to succumb to the needless extension of a conventionalized jargon marker of legalese; such usage is the linguistic equivalent of the gowns, wigs, and ceremonial procedures of the courtroom.

Individual jargons will differ in this matter of stylistic modification. We will argue later, for example, that the linguistic features of bureaucratese (official language), which are often parodies of legalese, are typically not in response to functional needs but for the most part represent stylistically motivated choice. It is this kind of jargon that Hearer is the most likely to perceive as dysphemistic gobbledygook. In American English the general term for bureaucratese is *gobbledygook*, whereas in British and Australian English the sense of this word has generalized to mean "(any type of) incomprehensible language." The word was coined by Maury Maverick (presumably his pseudonym) originally to describe the language of Washington, D.C., official documents. He is quoted as saying 'People ask me where I got gobbledygook. I do not know. It must have come in a vision. Perhaps I was thinking of the old bearded turkey gobbler back in Texas, who was always gobbledy-gobbling and strutting with ludicrous pomposity. At the end of this gobble there was a sort of gook.' (cf. Partridge in *Vigilans* 1952:16). Notice that gobbledygook is the mouthing of *a turkey*, which is the term used for someone who in Australian is *a dill* or *a dag* and in Britain *a wally* or *a stupid prat*.

The Interface Between Jargon and X-Phemism

Jargon has two functions: the primary, and orthodox, function is to serve as a technical or specialist language; the other is to promote in-group solidarity, to exclude those people who do not use the jargon as out-groupers. Hence Hudson's (1978:1) nice description of jargon as 'a kind of masonic glue between different members of the same profession.' In the case of criminal jargon and other argots, there is the added motivation of supposed secrecy. This may well be true also of legal English: legalese has been described by many as a kind of secret language, designed deliberately to assert the superiority of the legal profession and keep all the rest of us in the dark, (cf. Benson 1985:530ff). It is, of course, out-groupers who find jargon 'abounding in uncommon or unfamiliar words', and therefore 'unintelligible or meaningless talk or writing; gibberish'—to quote definitions 2 and 4 from the *Macquarie Dictionary*. So while jargons aim to facilitate communication on the one hand, they also erect communication barriers on the other. As we will see, it is these conflicting roles that render jargon both euphemistic and dysphemistic.

Obviously, subject matter and setting will have an important bearing on our evaluation of jargon as X-phemistic. Funeralese, for instance, is overwhelming euphemistic precisely because it deals with death. Legalese, certainly in Britain and Australia (which is what we shall discuss here) is euphemistic when it deals with the distasteful; for example, in matters like *indecent exposure of the person*. There is euphemistic jargon in the title of the 'British Mental Health Act' 1959 which replaced the by then dysphemistically named 'Lunacy and Mental Treatment Acts' 1890–1930 and the 'Mental Deficiency Acts' 1913–1938. Similarly, the Queensland 'Intellectually Handicapped Citizens Bill' 1983 replaced the 'Backward Persons Act' 1938 (cf. Pannick 1985:144). Prisoners are euphemistically given *life sentences* or may be *detained during Her Majesty's pleasure* rather than *for an indefinite period of time*. And euphemisms also have a part to play in the everyday running of the legal system. By convention, members of the legal profession are excessively polite to one other and this manifests itself in euphemism. Barristers will address a judge *My Lord/Your Honor/Your Worship; May it please your Lordship/If your Lordship pleases*, and so on. They will then proceed to *make their submissions* rather than *present their arguments*; and this is always carried out *with (the greatest of) respect*. One Melbourne lawyer claimed to us that the more uncooperative the judge was, the greater the verbal respect accorded to him. A barrister briefed by the state is an *amicus curiae* "a friend of the court." The opposition barrister is referred to as *my learned friend*, and the retort *In answer to my learned friend's erudite submission* will be understood to mean "you are wrong." These are just a few examples of the sort of euphemistic doublespeak that goes on in a court room. It enables one to sink the legal slipper at the same time as maintaining an aura of good will and harmony by using jargon as masonic mortar.

In this era of specious egalitarianism, euphemistically known as *equal opportunity*, some professional jargons have another motive to be euphemistic. Educationalese, for example, has sought to abandon the simple value-laden terms like *lazy, idle, stupid, clever*, and *poor* and replace them with expressions like *educationally and socially disadvantaged groups, underachievers, those on the lower end of the ability scale, high verbal-ability subjects, disadvantaged home environments*, and *underprivileged children*. Both the clever child and the below-average child become *the exceptional child*, not, note, *the abnormal child*. Such euphemisms seem to be aimed at obscuring the differences between educationally successful and educationally unsuccessful children.

There are some occasions when the public would probably prefer to hear semicomprehensible jargon; for example, in the medical sphere. We have already discussed in Ch.7 the role of euphemistic technical jargon in blunting the edge of unpleasant reality for both patient and doctor; and so the terms *spirochetal/treponemal/luetic disease* are preferred to terms like *syphilis*. Patients also have certain expectations of their physicians, and probably prefer the jargon expression *patellar tendon reflex* to everyday *knee-jerk*; and whereas *pityriasis rosea* sounds like an expert diagnosis, *rash* does not; indeed, after the latter diagnosis a patient might well wonder why s/he has bothered to pay for a medical opinion at all! Full command of medical jargon is viewed as part of the competence of a medical expert.

But what about those jargons where situational factors are apparently not the motivation for euphemism: after all, most of legalese does not deal with delicate matters like *indecent exposure of the person*. To what extent does X-phemism overlap with jargon in legal nomenclature like *action* "lawsuit", *of course* "as a matter of right", *instrument* "document"? To the initiated, the in-groupers, these expressions are preferred to others when used within their normal legal contexts: therefore they are euphemistic. They are a true jargon and promote in-group solidarity, but they would be dysphemistic if used to the uninitiated because to out-groupers they would be inappropriate. However, jargon and X-phemism are not one and the same thing, even though some writers (e.g., Enright 1985:12) do seem to use the terms as if they were. To gain a clearer picture of what is involved we will first examine one kind of jargon to establish some of the characteristic lexical, grammatical, and discourse features. By way of illustration, we take the case of "legalese" in Britain and Australia; presumably, many of the features we will discuss are also to be found in American legalese as well.

Aspects of Legalese—Vocabulary

> THE TELEVISION EQUIPMENT (herein referred to as the Equipment) shall remain the sole and absolute property of the Owners and the Renter shall not sell, assign, pledge, underlet, lend or otherwise deal with or part with possession of the Equipment and shall not without the Owners' consent remove the Equipment from the installation address set out overleaf or from such other address as the Owners may from time to time consent to and will protect the Equipment against distress, execution, poinding, landlord's sequestration and any manner of seizure and indemnify the Owners against all losses, costs, charges, damages and expenses incurred by the Owners by reason of or in respect thereof.
>
> (Quoted in Cutts and Maher *Gobbledygook* 1984:16)

Legalese is the variety of English used by members of the legal profession when speaking of legal matters. It is primarily a written language used in the preparation of legal documents like wills, statutes, forms, contracts, leases, laws, and briefs; but is also spoken in court proceedings in the examination of witnesses, in putting submissions, in expert-witness testimonies, and so on. It is well known that jargons like legalese are full of terms unfamiliar to the layperson; yet it is not generally accepted that jargon serves a necessary function. One reason that jargons like legalese are unfamiliar is because they are full of technical terms that are not needed outside their speciality: *affidavit* "a written statement sworn or affirmed"; *habeas corpus* "writ to literally 'have the body' of a person before the court"; *garnishment* "the attachment of a debt to a judgment debt as a means of enforcing a judgment"; *plaintiff* "one who brings an action to law"; *lessor* "one who grants a lease"; *hereditaments* "a real property right capable of devolving an heir"; among others. Another reason the terminology is unfamiliar is because some common words are used with meanings that have been specially narrowed and circumscribed: *contract* "an agreement enforceable at law"; *domicile* "country of one's permanent residence"; *action* "civil proceeding"; *good faith* "honesty"; and so forth. Specialized languages like legalese and religiousese have been very successful in resisting change; and for that reason the vocabulary used is very often archaic. Words like *chattels* "personal property"; *hearken* "listen"; *witnesseth* (with the archaic *-eth* ending), and the adverbials *hereinafter*, *hereto*, and *herebefore*, for instance, have more or less disappeared from ordinary language. Others like *instrument* in the sense of "document" retain a meaning that has long since ceased to exist outside the law. As we will see, legalese is also characterized by grammatical fossilization. It seems that the narrower or more specialized the context in which a jargon is used, the more rigid the style becomes along with a greater likelihood of archaisms surviving in marginal and special functions (cf. Burridge [forthcoming] Ch. 1).

There are many more aspects of legal vocabulary that distinguish it from ordinary usage. One is the preponderance of borrowings, largely from French and Latin; for example, *ex aequo et bono* "on the basis of what is fair and good"; *mens rea* "guilty mind"; *chose in action* "incorporeal personal property right enforceable in court of law"; *voir dire* "to say truly"; *de novo* "from the beginning; anew." Another is the common practice of stringing together two, sometimes three synonyms; the so-called doublets or triplets: *act and deed; goods and chattels; in my stead and place; cease and desist; remise, release, and forever discharge; rest, residue, and remainder*. These derive from an early literary practice of conjoining one noun of Germanic origin with a synonym of Romance origin, as in the case of *false and untrue; will and testament*. It is usually assumed that the practice became a necessity when English replaced French in the courts (cf. Charrow 1982:91), but the same sorts of synonym strings are extremely common in early Dutch and German literature, for which no such explanation is tenable; we suggest that, even in legal documents, they have always been a matter of stylistic ornamentation and have little to do with the need for greater clarity in the law. A similar explanation presumably holds for the practice of constantly varying vocabulary (e.g., alternation of *person* with *defendant, author* with *writer*, etc.).

It is not difficult to see that many of these aspects of legal vocabulary could cause problems for readers outside of the law. With the synonyms we have just been discussing, for example, a reader may well wonder if in fact the same denotatum was intended on each occasion. But the most befuddling aspect of legalese to the layperson is probably its grammatical structure.

Aspects of Legalese—Grammatical Structure

Long and extremely complex sentences can render a piece of legalese almost incomprehensible to the general reader. Consider the following two examples. The first (A) is subsection 27(12) from the Companies (Acquisition of Shares) (Victoria) Code. The second (B) is taken from the Sales Act (State of Victoria):

(A)   Where an offeree who has accepted a take-over offer that is subject to a prescribed condition receives a copy of a notice under sub-section (10) in relation to a variation of offers under the relevant take-over scheme, being a variation the effect of which is to postpone for a period exceeding one month the time when the offeror's obligations under the take-over scheme are to be satisfied, the offeree may, by notice in writing given to the offeror within one month after receipt of the first-mentioned notice and accompanied by any consideration that has been received by the offeree (together with any necessary documents of transfer), withdraw his acceptance of the offer and, where such a notice is given by the offeree to the offeror and is accompanied by any such consideration and any necessary documents of transfer, the offeror shall return to the offeree within 14 days after receipt of the notice, any documents that were sent by the offeree to the offeror with the acceptance of the offer.

(B)   95. (1) A term of a sale (including a term that is not set out in the sale but is incorporated in the sale by another term of sale) that purports to exclude, restrict or modify or purports to have the effect of excluding, restricting or modifying—

(a) the application in relation to that sale of all or any of the provisions of this Part [of the act];

(b) the exercise of a right conferred by such a provision; or

(c) any liability of the seller for breach of a condition or warranty implied by such a provision—

is void.

(2) A term of a sale shall not be taken to exclude, restrict or modify the application of this Part unless the term does so expressly or is inconsistent with that provision.

The average legal sentence contains fifty-five words, which is twice the number for 'scientific English' and eight times the number found in dramatic texts (cf. Danat 1980:479). Quotation (A) consists entirely of one sentence, comprising 174 words. Notice also the characteristic dearth of punctuation. It begins with a conditional clause 'Where . . .,' which contains within it no less than four relative clauses. Not until half way down the passage do we encounter the main clause—beginning 'the offeree may'—which is itself long and extremely complex. Notice the huge gap between the auxiliary verb 'may' and the main verb 'withdraw,' which are separated by thirty-six words and three propositions, and the somewhat narrower gap of eleven words identifying an indirect object and a durative adverbial between the verb 'return' and its object 'any documents.' One general feature of legal syntax is a consistent violation of what has become known as Behagel's first law; namely, 'that which belongs together mentally is arranged close together' (1923:4, our translation). In the case of quote (B), the subject 'a term of sale' is separated from the rest of the sentence 'is void' by a complex of embedded material containing as many as eleven propositions expressed in eighty-eight words. It is the constant interruption of the significant proposition expressed in the main clause which makes it so hard for the reader to extract the information. The constant shifting of topics blurs the main focus of the discourse, with the consequence that the whole point of the passage gets lost.

Both quotes (A) and (B) contain a very high frequency of passive constructions. This is a striking feature of legalese that distinguishes it from 'normal' discourse. Another is the absence of anaphoric pronouns; as Crystal and Davy (1969:202) state, in legalese 'they seem to be eschewed as a species.' The repetition of nouns (cf. the constant repetition of *offeree* and *offeror* in [A]) is presumably to remove any doubt as to the intended denotatum. However, as previously mentioned, omission of pronouns does not always seem to be with the aim of facilitating easier communication.

It is well established in the psycholinguistic literature that multiple negatives pose difficulties for cognitive processing, (cf. references in Danat 1980:486). But multiple negatives are rampant in legalese. This includes not only overtly negative markers like *not, never,* or *un-,* but also semantically negative words like *default* and the connectives *except, unless, provided that,* and *however.* The second section of the Sales Act quoted includes a string of negatives 'not,' 'exclude,' 'restrict,' 'unless,' 'inconsistent,' and this significantly complicates the comprehensibility of the passage.

Probably, the most distinctive feature of legalese is its heavily nominal style, which originated in imitation of Latin style, and may have been influenced by the heavily nominal style of nineteenth-century German scientific prose. For example, verb forms like *assign* and *consider*, are replaced by nominalizations *make assignment* and *give consideration*. A piece of prose that is heavily nominal in style is much more general and abstract than one that is not. Doing away with verbs allows for the omission of subjects and objects, and Speaker can be noncommittal on who is doing what and to whom. This is an extremely useful device and one that is frequently used in government offices for those occasions where it is precisely desirable to conceal this sort of information (see the following discussion). The nominalizations of legalese have been shown in the psycholinguistic research of Charrow and Charrow (1979) to present significant processing difficulties for Hearer. Compare the nominal versions in (1–2 a.) with their more natural counterparts in (1–2 b.):

(1) a. *in the event of default in the payment*
    b. if you don't pay what you owe
(2) a. *unless a written demand has been made on the supplier for satisfaction of the judgment*
    b. unless the buyer has written to the supplier asking him or her to comply with the judgment

Of course all these features occur in ordinary language, but as with any style feature it is a matter of relative frequency. In legalese, and in other jargons including bureaucratese/officialese (gobbledygook), medicalese, educationalese/ educanto/pedaguese, and our own jargon, linguisticalese, some peculiar lexical and grammatical features occur with much greater frequency than they would in other jargons and in everyday language—if there is such a thing.

Jargon and X-Phemism Revisited

To the initiated then, jargon is efficient, economical, and even crucial in that it can capture distinctions not made in the ordinary language. For instance, legalese is (at least sometimes) necessary because ordinary language cannot adequately capture all the precision that legalese can capture (though see Pannick 1985:136). As linguists, we would claim the same for linguisticalese; for example, the bolded items in the following passage more or less randomly chosen from an introductory book on linguistics:

> **Syntagmatic relations** are characteristically based on the **co-occurrence of elements** in the **speech chain,** while **paradigmatic oppositions** only obtain within the total **system,** all elements of each network of relations but one being absent from the actual **string** of **phonemes** or words through which *langue* manifests itself in *parole.*   (Atkinson, Kilby and Rocca 1988:106)

Although this quotation comes from an introductory book, it necessarily uses jargon in order to achieve its purpose. Such professional jargons do not simply generate cross-varietal synonyms, they generate vocabulary and phraseology that can only be paraphrased in ordinary language by extensive and discombobulating circumlocution.

Many jargons take time to learn, and a certain prestige is awarded to those who command them. It would be dysphemistic, for example, for a lawyer not to use jargon when s/he is contextually expected to do so, for example when creating a legal document: this is exactly what legalese is for. However, the combination of esoteric vocabulary, grammatical complexity (abnormally long sentences, large number of passives, nominalizations, multiple embeddings, intrusive phrases, multiple negatives, etc.), and unconventional information structure (presentation and maintenance of topics, etc.) typically leads the outsider to perceive a legal document as having poor (i.e., hard to follow) discourse structure. So when the public, who are by definition out-groupers, are presented with documents written in legalese, they will often feel that the cooperative maxim of manner ('Speaker's meaning should be presented in a clear, concise manner that avoids ambiguity, and avoids misleading or confusing Hearer') is being violated; consequently, they are offended by the perception that the writer requires them to expend unreasonable effort in order to understand what the document means. As a result they may feel incapable of understanding the implications of what is said without help from a lawyer. Being supported by the sanctions of the state, legalese is an extremely powerful weapon with which the legal profession is able to intimidate and dominate the public (cf. Benson 1985:530).

But the jargon of disciplines like linguistics and sociology can be even more alienating. The everyday nature of their subject matter renders these jargons even more baffling and offensive to out-groupers who feel they ought to understand. Take the example then of linguistics, the study of language. Outside the discipline there already exists an extensive nontechnical vocabulary used by the lay public when talking about language; unfortunately, this terminology is too imprecise and vague to be of use in the discipline of linguistics. Linguists are therefore faced with having to narrow and redefine everyday terms like *sentence*, *word*, *syllable*, and *grammar*, as well as add a number of new terms to overcome the imprecisions and to distinguish denotata that ordinary language ignores. For example, linguists find the term *word* insufficiently precise for all their purposes and so occasionally need to distinguish between *grammatical*, *orthographic*, and *phonological words* as well as introducing new terms like *lexeme* and *morpheme* to capture additional distinctions. If this were quantum mechanics and not linguistics, no one would question the right to furnish the discipline with a technical vocabulary all of its own. But to the nonlinguist such practice seems unnecessarily pedantic. The technical language is perceived as intellectual hocus-pocus, and all the more offensive precisely because it deals with familiar subject matter. Sociologists and psychologists are faced with the same problem. In a serious piece of academic writing, a child psychologist could hardly refer to *hitting a naughty child*—but does s/he really have to go as far as to say *punitive external control* (to which the child then provides *sensory feedback* rather than *yells*)? When the lay public comes across something like the following:

> the objective self-identity as the behavioural and evaluative expectations which the
> person anticipates others having about himself
>   (Smith *British Journal of Sociology* 1968:80)

there is a feeling that the writer is deliberately mystifying the topic: why does Smith not use *self-image*? Is this just puffed-up jargon to augment Smith's own 'objective

self-identity as the behavioural and evaluative expectations which the person antici-
pates others having about himself'? To be fair, this particular gem of jargon does not
look so bad in the context of the whole paper from which it is selected; and we must
bear in mind that it is easy to tilt at the jargon of others.

Where the cause of offense is jargon, jargon is dysphemistic. This accounts for
the many social and political movements currently pushing for clear and simple
English, particularly in laws, legal documents like contracts, and government documents of all kinds. Such movements seem to have sprung up throughout Europe;
for example, Cutts and Maher's *Gobbledygook* (1984), from which we have quoted
several times in this chapter, is the work of a British society for plain English. There
are also plain English movements in America and Australia (cf. publications by the
Law Reform Commission of Victoria on 'Plain English and the Law'). In Sweden
there is currently a strong push for comprehensible Swedish officialese in place of
krångelsvenska "muddled Swedish," and in Norway there are courses on 'Plain
Norwegian for Bureaucrats.' Parallel cries can be heard in Germany and France (cf.
Danat 1980:451f).

These movements therefore represent a current push to dejargonize legal register, a notion that effectively means "dejargonize legal jargon." This gloss exposes the
inherent contradiction, which arises for the following reason. There is a presumption that lawmakers should be permitted to use only the kind of language used in
National Geographic magazine or on the front page of the Melbourne or London
Sun—say, sixth-grade English. But it is hard to see how a legal document could
properly incorporate the appropriate legal definitions and exclusions using just sixth-
grade English. Supporters of legalese, like Gowers (1962:18–25), argue that it is not
so much complexity of language but complexity of law that is the problem. The
special needs—namely, exact and precise phraseology—of the legal profession simply cannot be met by ordinary language. But isn't this showing jargon in a very
favorable light? Even given the need for precision and the conceptual difficulty of
the material, the elimination of some of the lexical and syntactic sources of difficulty mentioned earlier must improve comprehension and make legalese more
accessible to the general public and even to lawyers. For example, contracts should
be understandable to those who are required to sign them! Nevertheless, precision
should not be carelessly exchanged for something readable.

The difficulty of translating jargon into ordinary English suggests there is a real
distinction to be made between jargon and X-phemism, namely:

1. since X-phemisms by definition have cross-varietal synonyms, there are alternatives to X-phemisms available; but
2. jargon is necessary to frame a message because there really is no viable alternative; (i.e., there are no reasonably simple cross-varietal synonyms available).

Yet, in fact, such a conclusion is incorrect on both counts.

X-phemism is always a matter of contextually motivated choice, but we saw in
Ch. 1 that X-phemisms cannot normally be freely exchanged with their alternatives:
for the most part, Speaker simply cannot exchange *feces* for *shit* or *terrorist* for
freedom-fighter, and so forth, without changing the connotations of the message s/he
intends to convey. To change the connotations is to change the message. Hence
hypothesis (1) is unacceptable.

And (2) fares no better. Some legal terms, for example, have no convenient substitute: they either have no equivalent in ordinary language (e.g., *mandamus* "we command"—a writ to force someone to do something); or they have a special legal meaning that differs from ordinary usage (e.g., *defendant* "a person against whom civil proceedings are brought"). But many do have acceptable equivalents in general usage (e.g., *action* "court proceedings" and *instrument/presents* "document"). Many legal terms are simply stylistically more formal than their everyday counterparts: for example, *equitable* instead of "fair," *prioritise* instead of "rank"; or *expiration* instead of "end." And there are numerous loan words that have perfectly acceptable English equivalents: *ab initio* "from the beginning"; *corpus delicti* "facts constituting an offense," and so on. Most of the archaic words like *indenture, chattels, situate,* or *witnesseth* have modern substitutes. Vocabulary items like these are not chosen in the interests of precise unambiguous identification of concepts and objects; they are fossilized relics of times past, and wholly contextually bound. And while the *aforesaids, hereinbefores, hereons, hereunders, hereuntos, hereins, thereofs,* and *whereofs* are helpful in clear referencing and textual cohesion, more often than not they are quite redundant. Almost incantatory in their repetitiveness, they become, like the black gowns, the wigs and the fanciful ceremonies, part and parcel of the rituals of legal process.

Still other jargons consist almost entirely of vocabulary with cross-varietal synonyms; for instance the funeralese quite properly does so, just because it must deal with the strong taboos on death. Street jargon, druggie jargon, criminalese, and the like, are full of terms with cross-varietal synonyms that serve to maintain in-group recognition devices that also (purportedly) disguise meanings from out-groupers; consider these examples from several regions: *work* for "work as a prostitute," *a trick* or *a john* for "a prostitute's client"; *tea* or *grass* for "marijuana," *shit* for "cannabis resin," *speed* for "amphetamines," *horse* or *brown sugar* for "heroin," *snow* for "cocaine," *the works* or *a fit* for a mainliner's "hypodermic syringe and spoon"; *a grass* for "a police informer"; *doing porridge* or *at college* for "being in prison"; and so on. And although the hamburger industry's use of the term *autocondimentation* as opposed to *precondimentation* is an economical way of distinguishing a client's right to sauce his or her own hamburger, it is certainly not necessary to use *autocondimentation* in order to get the meaning across. So why use it? The answer is, of course, that it confers on the hamburger industry a certain dignity. The dignity comes from the Greco-Latinate lexicon used because it is reminiscent of such prestigious jargons as legalese and medicalese.

Dressing Up the Goods—Euphemism and the Special Case of Bureaucratese

[*Amendation to the traffic plan for a London borough*] Line 5. Delete 'Bottle-necks', insert 'Localised Capacity Deficiencies'.
(Quoted in Cutts & Maher 1984:45)

Both in the past and in certain contemporary societies we have seen how people avoid words thought to be ominous or evil or which they believe are somehow offensive to hidden powers. For example, they may substitute the original name of the feared powers with some sort of euphemistic appellation in the hopes of somehow being able to appease them and win over their favor. For this reason, the elves and fairies of folklore were once referred to as *good neighbors*. The ancient Greeks called the *Furies* the *Eumenides* "the well-minded ones." Now, in the world of politics and government, we find the very same practice at work—only this time it is the electorate who represent the all-powerful beings whose favors are sought.

Politicians must accordingly be cautiously considerate and tender of their electorate's feelings—they must of course avoid words that have unpleasant associations and they must be excessively polite! Small wonder that the language of the civil/public/government service, known as *bureaucratese, gobbledygook*, or *officialese*, is so overwhelmingly euphemistic. Understandably, a politician will prefer a term like *special developmental areas* to describe "marginal constituencies," *categorial inaccuracy* for "lie," or *scheduled for discontinuance* instead of "due to be closed down." Their language is full of caution and circumlocution. 'Cautionary clichés', as Gowers (1962:74) describes them, are the earmark of bureaucratic language: *Inclined to think; at present advised; prima facie; it is advisable to; will you be good enough to; it would serve no useful purpose; you would be well advised; in respect of*; and so on. In structure and vocabulary, bureaucratese has been strongly influenced by legalese. It is, therefore, characterized by an excessively nominal style. Here the abstract nominal serves as least three essentially euphemistic functions for Speaker, although Hearers may find them dysphemistic: (1) As we have already said when discussing legalese, nominalizations are notoriously difficult for readers to process. Bureaucrats seem to believe that the greater complexity displays greater intelligence on their part. (2) Nominalizations also have the great advantage that they avoid the need for explicit subjects, which enables Speaker to evade responsibility for an action or event. For example, a number of years ago the governor of California, when asked why he had allowed a man to die in the gas chamber (a highly unpopular act at the time) replied: 'There was insufficient evidence on which to base a change of decision.' He could hardly have said *I couldn't find enough evidence to make me change my mind and decide to save the man's life*. As Williams (1981:162) points out, the nominalizations of the verbs *change* and *decide* enabled the governor to avoid reference to both the dead man and himself, thereby concealing his own responsibility in the matter. It is easy to see that the preference for nominalizations leads to the deceptive prose style which cynics would say is eminently suited for the jargon of government. (3) Because of the omitted subjects, nominalizations are characteristic of an impersonal style; they give bureaucratic prose an air of being objectively scientific, and concomitantly profound and reliable. This is reinforced by the fact that such nominals are typically built up from Greco-Latinate morphology.

A striking feature of bureaucratese is the use of long and complex strings of nouns such as *young driver risk-taking research; prototype crisis shelter development plans*; or *the FEMA-sponsored host area crisis shelter production planning work-book* (cf. Charrow 1982:88f). Telescoped noun strings like these characterize other jargons, too (e.g., in educationalese *teacher behavior* replaces "behavior of teachers"; *teacher satisfaction* replaces "satisfaction among teachers"; and *teacher effectiveness* "effectiveness of teachers"). Consider the following extract from the *Journal of*

Educational Research 61, 8, 1968 (quoted in Hudson 1978:94, our emphasis): 'Darley and Hagenah point to prestige drives among youth as an important source of OCCUPATIONAL CHOICE–VOCATIONAL INTEREST CONGRUENCY.' What is it about 'occupational choice–vocational interest congruency' that makes the writer prefer it to the more comprehensible *source of congruency between occupational choice and vocational interest?* One could perhaps argue that compounds like these serve the interests of economy, but certainly not intelligibility! Yet surely its real function is to give the passage a specious air of academic profundity.

Jargons like bureaucratese have two motivations. The first, the exclusion of out-groupers, is shared with criminal jargon; the second, and more prevalent, it shares with the hamburger industry. The matters with which bureaucrats deal are so mundane that they can be fully described and discussed in sixth-grade English such as one finds in the tabloid press. In order to augment their self-image, therefore, bureaucrats create cross-varietal synonyms using a Greco-Latinate lexicon, seeking to obfuscate the mundane (and often trivial) and endow it with gravity. For example, the following was reported to us as occurring among instructions to counter clerks at a British employment office:

> the occupational incident of the demand change is unlikely to coincide with the occupational profile of those registered at the employment office

which apparently means "people may not be suited to the job they want" (though one cynic, whom we won't name, disagrees with us and interprets it to mean "the unemployed are unemployable"). Based on the view that such jargon upgrades not only the language but also the status of the bureaucrat, it should be euphemistic to in-groupers. But to other people (out-groupers), the cooperative maxims of manner and perhaps also quantity are violated and the effect is dysphemistic. Such jargon is not used for efficiency, economy, or because it is crucial in making distinctions not captured in the ordinary language; however, it is technical and it is certainly used to exclude out-groupers.

Bureaucratese or civil service language has been beautifully parodied by the characters in the BBC series 'Yes Minister' and its sequel 'Yes Prime Minister.' Consider the following (?parody of) bureaucratese from *Yes Prime Minister*, the purported memoirs of the Right Honourable James Hacker, Prime Minister of Great Britain; it comes in a letter from Sir Humphrey Appleby, the Cabinet Secretary, complaining because the locks to the PM's office had been changed (to keep Sir Humphrey at bay):

Cabinet Office

Dear Prime Minister,

I must express in the strongest possible terms my profound opposition to the newly instituted practice which imposes severe and intolerable restrictions on the ingress and the egress of senior members of the hierarchy and will, in all probability, precipitate a progressive constriction of the channels of communication, culminating in a condition of organizational atrophy and administrative paralysis which will render effectively impossible the coherent and coordinated discharge of the function of government within Her Majesty's United Kingdom of Great Britain and Northern Ireland.

> Your obedient and humble servant,
> Humphrey Appleby

I read it carefully. Then looked up at Humphrey. 'You mean you've lost your key?' I asked. 'Prime Minister,' he said desperately, 'I must insist on having a new one.'   (Lynn and Jay 1986:136f)

It would spoil the fun to say more than that Jim Hacker got it in one. Bureaucratese is noted for puffed up circumlocutions; for instance, by 'imposes severe and intolerable restrictions on the ingress and the egress of senior members of the hierarchy' all that Sir Humphrey meant is "stops me coming in when I want to." And in his diary entry for Sunday, May 15, Sir Humphrey writes the following—a splendid example of the excesses of bureaucratic euphemism:

> Sometimes one is forced to consider the possibility that affairs are being conducted in a manner which, all things being considered and making all possible allowances is, not to put too fine a point on it, perhaps not entirely straightforward.
>   (Lynn and Jay 1987)

He could of course have simply said *You are lying*.

The following extract, however, does not come from Sir Humphrey's diary. It is a genuine piece of bureaucratese, a Government Department's letter to its advisory council:

> In transmitting this matter to the Council the Minister feels that it may be of assistance to them to learn that, as at present advised, he is inclined to the view that, in existing circumstances, there is, prima facie, a case for . . .
>   (quoted by Gowers 1962:74)

The translation would be simply *The Minister wants the Council to know that he thinks there is a case for.* . . . The parody in 'Yes (Prime) Minister' is not much of a hyperbole on the real bureaucratese it derides.

To end this section, consider whether or not the following version of Luke (2:8–10) is euphemistic, dysphemistic, or just puffed up jargon:

> Additionally, the identical geopolitical hegemonic system contained ovine supervision operatives in a nocturnal holding posture vis-à-vis a carcass and fleece production pasture unit. And, of limited stature, they were interfaced with a paranormal extraterrestrial android manifestation in a hyper-illuminated adulatory authoritarian environment and they experienced a deleterious mood change motivated by intrapersonal insecurity. The paranormal manifestation previously identified verbalized antithetically to their self-induced negative confidence syndrome. "Become visually cognitive of the informational benefits I present. Which will orientate your peer group within a meaningful celebration situation."
>   (Richard Stilgoe, BBC broadcast 'Potted Tongues')

We regret that we don't know the original author of this gem, but whoever wrote it deserves our congratulations and a well-paid sinecure in a government office. For the reader whose Bible is out of reach, there is a version of *Luke* (2:8–10) at the end of this chapter.

Summary of Chapter 8

A fast-medium right arm inswing bowler needs two or three slips, a deep third man,
a gully, a deepish mid-off, a man at deep fine leg and another at wide mid-on.
(Cricketing jargon)

As we have said, the setting and the nature of the subject matter for which a jargon is developed has important bearing on our evaluation of it as X-phemistic. Restaurants not only upgrade food by, for example, presenting what the supermarket can calls *Spaghetti in Tomato Sauce* as *Pasta in Marinara Sauce* (perhaps justified by waving a crabstick over the cooking pot); they also use, and more often misuse, French as a jargon when describing their *cuisine* on the menu, so a soup is a *potage de* whatever and French fries are *pommes frites* (with variable spelling). Among some nice examples cited in Lehrer [forthcoming] are 'Le Coupe aux Marrons Sundae' and our favorite 'Stuffed Tomato aux Herbs, Shoreham Style'. Restaurantese (perhaps we should call this 'menuese') in gourmet restaurants might sometimes distinguish in-groupers from out-groupers, but restaurants can get by without it; indeed, as Lehrer points out, many menus explain the jargonized names of dishes, and she quotes as an example 'ENTRECOTE AU POIVRE MADAGASCAR, sirloin steak topped with green peppercorn, served with cream sauce and cognac'. Some other jargons seem better motivated. Funeralese is excessively euphemistic precisely because it deals with the taboo topic of death. Legalese is euphemistic when it deals with indecorous matters like 'indecent exposure of the person.' On the other hand, much of legalese, cricket talk, and the jargons of horsebreeders, mountain climbers, interior decorators, and so on are preferred by in-groupers, but can only be said to have more pleasing connotations than alternative ways of speaking, by virtue of their being jargon. With these we seem to have reached the outer bounds of euphemism. In summary then, many uses of jargon are X-phemistic. Jargon may conceal the fearful and the shameful, sweeten the distasteful, and even elevate the trivial. But all too often, jargon is, in the words of the *Macquarie Dictionary*, 'speech abounding in uncommon or unfamiliar words,' which people suspect are used primarily to disguise the ordinary and to befuddle out-groupers rather than to serve as a precise and economical expressive tool. Seen from this perspective, the jargon becomes 'unintelligible or meaningless talk or writing; gibberish.' On all counts, this renders such jargon dysphemistic to out-groupers, and gives the public a low opinion of 'jargon' despite the fact that every member of the public will themselves necessarily use jargon for some purpose or other.

One of the best examples of the intersection of jargon and euphemism must surely be found in the cosmetics of advertising, and with this topic, we will conclude the discussion. Real estate agents (like used-car salesmen) use euphemistic hyperbolic jargon in order to upgrade. Consider the following:

MIDDLE PARK
150 Neville St.
Inspect Today 12.15–1.00
Exciting Opportunity to Renovate!

A wonderful chance for the creative renovator to restore this WEATHER-BOARD VICTORIAN VILLA in a premier location by the bay. Accommodation comprises: 2 well-proportioned bedrms., both with OFP's, spacious lounge, kitchen leading to bathrm., sep. WC & sunny paved courtyard. NOTE: CAR ACCESS FROM RIGHT OF WAY AT REAR.

(*The Age*, Melbourne, Wed. 6 July 1988:43)

What more could one ask of someone trying to sell your dilapidated old artisan's cottage (only a five-minute drive from the sea)?

Fulfilling the Promise Made Earlier

To fulfill our earlier promise, here is *Luke* 2:8–10 from the *New International Version of the Holy Bible* (1980):

And there were shepherds living out in the fields near by, keeping watch over their flocks at night. An angel of the Lord appeared to them, and the glory of the Lord shone around them, and they were terrified. But the angel said to them, "Do not be afraid. I bring you good news of great joy that will be for all the people."

There is a certain amount of Christian religious jargon here (e.g., 'an angel of the Lord'), but it is not too distinct from modern standard English.

CHAPTER
—9—

Language as a Veil: Artful Euphemism

Poetic diction is the systematic application of euphemism to maintaining eleva-
tion of style, and since nobody to speak of tries to maintain an elevated style any
more, this sort of euphemism would seem to be on the verge of obsolescence.

(Adams 1985:46)

Although Adams is right about allegorical poetry, the artful use of euphemism is far
from being obsolescent. In this chapter we examine some exploitations of euphemism through which the author protects him or herself when talking about taboo
topics by artfully trading on metaphor and figurative interpretations of the locutions
used. There is a cline among such euphemisms that stretches from street slang to
poetic allegories like the thirteenth century 'Le Roman de la Rose' an allegory of the
pursuit of the flower of womanhood (the rose is a euphemism and symbol for the
blood from a newly split hymen; or, it is sometimes said by romantics, a figure for
the vagina whose fleshy red folds resemble the open petals of a red rose). Works such
as Jonathan Swift's *Gulliver's Travels* and *Tale of a Tub* or George Orwell's *Animal
Farm* and *Nineteen Eighty-Four* are political allegories. Shakespeare's sonnets are,
mostly, lyrical allegories, but his low comedy is closer (at least in spirit) to today's
street language.

We will begin with a modern allegory, the 'Camera Song.' In this, as in other
examples of euphemism as art, the author makes the pretense of adhering to the
middle class politeness criterion (see Ch. 1), even though the doublespeak text has
tabooed denotations. It is a bawdy allegory, and like a diaphanous nightshift, such
artful euphemisms are designed to titillate; and the best of them succeed.

Camera Song by Grit Laskin [*to the tune of 'Three drunken maidens'*]
[1] Early Saturday morning whilst strolling in the wood,
I chanced upon a lady who by the wayside stood.
'And what pray tell would such a lass as you be doing here?'
'I've come to take some photographs,' she said, as she drew near.

210

[2] Says I to her: 'I do declare, this is a fateful day;
For I have come to photograph, the same as you did say.'
So I pulled out my Nikkon F and placed it in her hand;
Says she: 'That's quite a camera you've got at your command.'

[3] My camera so delighted her that with no more delay
She let me see her camera case wherein her accessories lay.
Says I to her: 'You've got most everything that can be bought!
Just help me spread my tripod before I take some shots.'

[4] We photographed in hay-lofts, and up against the wall.
(If you've not photographed on Saturday night you've not photographed at all)
She had her shutter open wide, the daylight was all gone;
Likewise my naked camera lens had its filter on.

[5] This lady had experience with cameras, yes she did indeed;
I thought that her exposures were the best I'd ever seen.
Although she seemed to tire not, as on and on we went,
Says I: 'We'll have to finish soon; my film supply is spent!'

[6] Says she: 'I've had Minoltas, Yashicas and Rollei,
Hasselblad and Pentax, likewise a Polaroid;
Miranda, Leica, Nikkomat, a Kodak and the rest—
But now I've tried your Nikkon F, it surely is the best.'

Stanza [1] presents a traditional boy meets girl beginning such as is found in dozens of folk songs. Although there is no explicit evidence anywhere in the lyrics that 'I' is male, the story convention is so strong that we immediately presume this to be the case; and our presumption is confirmed circumstantially in the final line of stanza [3]—of which more in due course. The first couplet of stanza [2] continues in the same vein, but the second couplet begins to allow of a double-meaning. It arises from the man reporting *I pulled out my* X and gets a boost from the conjoined proposition 'and placed it in her hand.' While this line can certainly be interpreted literally, it nonetheless hints to the imaginative Hearer that 'Nikkon F' may be a euphemism. For two reasons this implication tends to be confirmed by the lady's response in line 4. First, the locution *That's quite an* X *you've got at your command* expresses admiration for the awesome X: this trades on the folklore that men are given to boasting about their genitals; and the male protagonist is doing just that in reporting the lady's compliment. She, like many women, is playing up to male vanity. Second, this locution implies that the object to hand, 'X,' is something over which Hearer ('you') has command. It would not normally be said that *one has command over one's own camera* (where 'camera' is understood literally) because this locution implies that 'X' itself has some power, that 'X' (in this case his 'camera') is animate, or a body-part, or a weapon. By the end of stanza [2], therefore, there is good reason to infer that 'Nikkon F' is a euphemism for the protagonist's penis. Why should *Nikkon F* be chosen for the euphemism rather than any of the other camera names listed in stanza [6]? From this list, the only alternative for the author's purpose would perhaps be *Rollei* with its phonetic echo of *roll* "fornicate". However, *Nikkon F* was chosen instead. That final 'F' certainly allows for associations with *fuck, fucking, fucker*; and the first syllable of

Nikkon rhymes with *dick*, and the whole word offers a partial consonant rhyme with *nookie* (see Chs. 1, 4, and 5). A camera is, of course, used for photographing, and the song frequently refers to this activity. It is no accident that the initial phoneme of *photograph* is /f/ the same as that for *fuck*, and in the bawdy interpretation of the song it means "fuck". Furthermore, the pronunciation of *Nikkon F*, namely /nɪkʌnlef/, suggests a metathesis between first and last phonemes and the first two vowels to give /fʌkɪnlen/, *fuckin N*, interpretable as "fucking nuts". Such speculation is highly imaginative, but no more imaginative than many of the etymologies seriously proposed by Socrates in Plato's *Cratylus*, Varro in *De Lingua Latina*, or by Isidore of Seville (see Allan 1986a §2.8 for discussion). We do not pretend to have unravelled the real reason for author Laskin's use of *Nikkon F* as the euphemism for "male genitalia," but we do suggest that the choice of this particular camera name is very probably motivated by the coalescence of some such set of associations as are given here (see Ch. 4 for similar examples of motivation for the use of forms like *cock* and *pussy*). At least we have shown that, in the context of the song, 'Nikkon F' is suggestive of these associations; and that, of course, is all that is necessary for bawdy allusion to succeed.

Stanza [3] of the 'Camera Song' exemplifies the best aspects of euphemism as art: where the purported literal meaning is perfectly viable as an interpretation, but at the same time the double-meaning (the smutty interpretation) is equally clear. Text understanding is the process of constructing a mental model of the world spoken of; in a text such as the one being described here, which has a double-meaning, Hearer is led to construct dizygotic-twin worlds. Hearer might be led up a garden path by 'She let me see her camera case' in the second line, by assuming that it is a metaphor of the 'vagina as camera case.' Such an image would be consistent with a number of terms which recognize that the physical characteristics of the female genital organ are determined by its function as a container for the penis during copulation (see Ch. 4 and **case** in Partridge 1955). But such a metaphor would be infelicitous because, whereas in the literal world a camera is taken out of the case when photographing, in this song it would have to be put into it! Instead what this first couplet of stanza [3] suggests is that the lass whips up her skirt, revealing her underpants 'wherein her accessories lay'; and thus it plays on male fascination with females in underwear.

In the final line of stanza [3] the use of 'tripod' is original: it denotes the tripartite male genitalia of penis and twin testicles. At this stage in the song's proceedings a little help in setting up the tripod (i.e., foreplay) is not inappropriate. The final clause in the stanza uses the standard weapon metaphor for the penis, in which ejaculation is *shooting*.

The first couplet of stanza [4] invokes two classic locations for casual fornication, and surely leaves no doubt as to the double-meaning of 'photograph' in this song. This is partially confirmed by the rather lame echo of the well known chorus from 'Four and Twenty Virgins': *Balls to your partner, and backs against the wall, / If you've never been fucked on Saturday night you've never been fucked at all*. The second couplet begins by exploiting the metaphor of the vagina (strictly speaking the vulva) as a winking eye (cf. Farmer & Henley 1890–1904, Vol. 4, Partridge 1955:109), a metaphor shared with the camera's shutter—which is what the author is trading on here. The final line presents the image of the erect penis as a camera lens protruding, as high focal length lenses do, from the body; and, in this era of safe-sex, it is sporting a condom.

Stanza [5] develops the action of a spectacular romp. The use of the word 'exposures' is nicely ambiguous: it trades on the at least partial nakedness of the 'photographers', and on its phonetic and etymological association with *positions* that augments the degree of sexual experience attributed to the lady—and therefore the amount of sexual enjoyment to be had with her; and finally, it suggests the lady was baring her soul in orgasm. The last line of stanza [5] also has one of the best analogies in the song: that between (1) shooting with a camera, and so using up its supply of film; and (2) shooting semen, and so using up the supply of that precious sign of sexual energy (cf. Ch. 4 under 'orgasm').

Literal meaning dominates stanza [6] of the 'Camera Song,' which is comparatively weak in euphemism, even though in the context of the preceding stanzas it must suggest that the lady is admitting to very extensive and varied sexual experience. Nonetheless, this stanza very cleverly strongly implies "Whatever you, the audience, might have thought this song is about, it is really just about photography." A clever artist exploits euphemism to portray a bawdy topic while shielding him or herself from the shocked wowser by establishing the defense that the text is innocent, and any bawdy interpretation is a construction in the wowser's mind. This is, of course, a prevarication—or in blunter terms, a downright lie.

The 'Camera Song' uses modern motifs, but 'The Blacksmith' uses traditional folksy ones. It is also less subtle than the 'Camera Song.' The smith is a standard symbol of masculinity, partly because of the power to form artifacts out of metal, but largely because of the sexual imagery of heating the furnace, putting a hard rod into the fire, softening it, and creating new forms by banging away with his tool on the anvil. (An alternative version, which we would have used if we had discovered it earlier, is to be found among D'Urfey's songs, in D'Urfey 1719 Vol. 4: 194f.)

The Blacksmith

A lusty young smith at his vice stood a-filing,
His hammer laid by but his forge still aglow.
When to him a buxom young damsel came smiling,
And asked if to work at her forge he would go.
CHORUS: With a jingle-bang, jingle-bang, jingle-bang, jingle
        With a jingle-bang, jingle-bang, jingle-hi-ho.

"I will," said the smith. And they went off together;
Along to the young damsel's forge they did go.
They stripped to go to it—'twas hot work and hot weather;
She kindled a fire, and she soon made him glow.
   Chorus.

Her husband, she said, no good work could afford her;
His strength and his tools were worn out long ago.
The smith said, "Well, mine are in very good order; and
Now I am ready my skills for to show."
   Chorus.

Red hot grew his iron, as did their desire;
And he was too wise not to strike while 'twas so.
Quoth she, "What I get is aglow from the fire—
I prithee strike ho, and redouble the blow."
  Chorus.

Six times in her forge by vigorous heating
His iron grew soft in a minute or so;
And then it would harden, still beating and heating,
But the more it was softened it hardened more slow.
  Chorus.

"Good smith," said the dame, "you are excellent skilful.
Oh what I would give could my husband do so!
Pray won't you come hither with your hammer tomorrow?
But please work your iron once more ere you go."
  Chorus.

In 'The Blacksmith' the sexual innuendo works through euphemistic metaphors. It is the tale of a smith, whose *hammer* identifies him as one who "bangs," and a dame dissatisfied that her husband is no longer capable of *working* her *forge* with his worn out *tools*. The pair strip off to get down to *work*, and she soon *kindles a fire* in the smith, and has him *aglow* with sexual arousal. The smith's *iron* rod softens on *heating* in the dame's *forge*, but soon *hardens* once it is taken out. But the more often it re-enters the dame's forge, the softer it gets, and the more difficult it is to harden it. Nevertheless, the dame seems well satisfied and expresses a wish to return on the morrow; but begs for just one more fling before she leaves. As we said, this is much more obviously sexual than the 'Camera Song.'

Perhaps the all-time master of euphemistic bawdy in English is William Shakespeare, particularly in his low comedy. There are examples in nearly all his plays (see Eric Partridge's *Shakespeare's Bawdy* 1955); here is one. (Line numbers in the quotations BEGIN from the line numbers in the Arden edition of 1966, and line numbers from outside the text quoted here also refer to that edition of the play; but, of course, line numbers from within the quoted text accord with those given.)

	In a tavern: A Drawer, the Hostess, Doll Tearsheet (a whore).	
HOSTESS	I'faith, sweetheart, methinks now you are in an excellent good	
	temperality. Your pulsidge beats as extraordinarily as heart	
	would desire, and your colour I warrant you is as red as any	
	rose, in good truth, la! But i'faith you have drunk too much	25
	canaries, and that's a marvellous searching wine, and it per-	
	fumes the blood ere one can say, 'What's this?' How do you	
	now?	
DOLL	Better than I was—hem!	
HOSTESS	Why, that's well said—a good heart's worth gold. Lo, here	30
	comes Sir John.	
	Enter Sir John Falstaff, singing.	
FALSTAFF	'When Arthur first in court'—[*to Drawer*] Empty the jordan	
	[*exit Drawer*]—'And was a worthy king'—How now, Mistress	
	Doll?	

HOSTESS	Sick of a calm, yea, good faith.	35
FALSTAFF	So is all her sect; and they be once in a calm they are sick.	
DOLL	A pox damn you, you muddy rascal, is that all the comfort you give me?	
FALSTAFF	You make fat rascals, Mistress Doll.	
DOLL	I make them? Gluttony and diseases make them, I make them not.	40
FALSTAFF	If the cook help make the gluttony, you help to make the diseases, Doll; we catch of you, Doll, we catch of you; grant that, my poor virtue, grant that.	
DOLL	Yea, joy, our chains and our jewels.	45
FALSTAFF	'Your brooches, pearls, and ouches'—for to serve bravely is to come halting off, you know; to come off the breach, with his pike bent bravely; and to surgery bravely; to venture upon the charged chambers bravely;—	
DOLL	Hang yourself you muddy conger, hang yourself!	50
HOSTESS	By my troth, this is the old fashion; you two never meet but you fall to some discord. You are both i'good truth as rheumatic as two dry toasts, you cannot one bear with another confirmities. What the goodyear! one must bear, [*to Doll*] and that must be you—you are the weaker vessel, as they say, the emptier vessel.	55
DOLL	Can a weak empty vessel bear such a huge full hogshead? There's a whole merchant's venture of Bordeaux stuff in him; you have not seen a hulk better stuffed in the hold. Come, I'll be friends with thee, Jack, thou art going to the wars, and whether shall ever see thee again or no there is nobody cares. *Enter Drawer.*	60
DRAWER	Sir, Ancient Pistol's below, and would speak with you.	
DOLL	Hang him, swaggering rascal, let him not come hither: it is the foul-mouth'dst rogue in England. [. . . *The Hostess has no wish to allow a 'swaggerer' ("hooligan") into her tavern, but Falstaff persuades her that Pistol is a harmless 'tame cheater' ("decoy duck") and so . . .*] *Enter Ancient Pistol, Bardolph, and Page.*	65
PISTOL	God save you, Sir John!	
FALSTAFF	Welcome, Ancient Pistol. Here Pistol, I charge you with a cup of sack. Do you discharge upon mine hostess.	110
PISTOL	I will discharge upon her, Sir John, with two bullets.	
FALSTAFF	She is pistol proof sir; you shall hardly offend her.	
HOSTESS	Come I'll drink no proofs nor no bullets: I'll drink no more than will do me good, for no man's pleasure I.	
PISTOL	Then to you, Mistress Dorothy; I will charge you.	115
DOLL	Charge me! I scorn you, scurvy companion. What! You poor, base, rascally, cheating, lacklinen mate! Away you mouldy rogue. Away! I am meat for your master.	
PISTOL	I know you, Mistress Dorothy.	
DOLL	Away you cutpurse rascal! You filthy bung, away! By this wine, I'll thrust my knife into your mouldy chaps, an you play the saucy cuttle with me. Away, you bottle-ale rascal! You	120

basket-hilt stale juggler, you! Since when, I pray you sir? God's
light, with two points on your shoulder? Much!

PISTOL   God let me not live, but I will murder your ruff for this.   125

(Shakespeare, *Henry IV Pt2*, II.iv.22ff)

The nine opening lines set the scene for the events that occur in the body of the quotation. The Hostess is revealed as something of a precursor to Sheridan's Mrs Malaprop, since she uses a mixture of neologisms and malapropisms: 'temperality' (23) "temper," 'pulsidge' (23) "pulse," 'extraordinarily' (23) "ordinarily, normally," 'canaries' (26) "canary, sweet wine," 'perfumes' (26) "perfuses, pervades," 'calm' (35) "qualm, illness," 'rheumatic' (52) "choleric," 'confirmities' (53) "infirmities." In line 30 she blends the proverb A *good heart conquers ill fortunes* with A *good name is of greater value than gold*. In short, she is a pleasant old fool. *Doll Tearsheet*'s name, like the name of many personae in Shakespeare's low comedy, is suggestive of her character: she is a doll, a plaything, and her profession is hinted at in *Tearsheet*; it is unclear whether *tearsheet* is also intended to indicate her prowess as a whore.

The first few lines quoted reveal that Doll has been drinking; and the boozy theme is continued with Falstaff garbling the ballad 'Sir Lancelot du Lac' (cf. Arden edn 1966:64) and ordering the drawer to empty the *jordan*—a euphemism for "chamber-pot." All this is very appropriate to a tavern scene, and forms a backdrop to Falstaff's remarks at line 36, where he plays with the meanings of *sect* "profession" or "sex/gender" and *calm* "calm" or "qualm, illness," so that he is able to imply all the following: (1) the prejudiced observation that "women are always ill, even in calm waters (to use a maritime image)"; (2) "prostitutes don't like it when things are quiet; if there is no trade they have no income"; (3) "a prostitute is only off the job if she is sick (either menstruating or with venereal disease)." In line 37, Doll describes Falstaff as a *muddy rascal*: she uses *muddy* much as we would use *dirty, rotten,* or *lousy. Rascal* is ambiguous: it seems to have had a similar meaning then to its current meaning, roughly "bad guy"; but in addition it meant "an out of season, immature, or undernourished deer." Falstaff's reply allows of three plausible interpretations. Ostensibly the meaning is "you make rascals fat" by encouraging them in loose living; but there is a *double entendre* "you give a man an erection"; and last there is the possibility that *make* is used in the sense "copulate with," so that *you make fat rascals* means "you copulate with fat rascals like me"—Falstaff is notoriously fat (cf. Doll's remarks in 11.57–59).

Falstaff's next rejoinder (42–44) undoubtedly refers to sexually transmitted diseases. *Catching* and *angling* were contemporary terms for copulation (see Ch.4), which is obviously a pertinent notion when addressing a whore. Note the playfully dysphemistic *my poor virtue* addressed to Doll. Doll returns in line 45 with *joy* "pet," although it has been suggested that the original version in performance may have been *Iesu* "Jesus" (cf. Arden edn 1966:66), which would certainly be appropriate in making this a sharp sarcastic reply. The euphemistic interpretation here is that her clients steal her jewelry; but we suspect a bawdy reading would have been available to an Elizabethan audience. *Chain* in Shakespeare's time was used of 'a barrier to obstruct the passage of a bridge, street, river, the entrance to a harbor, etc.' (*O.E.D.* 6) and *jewel* can be interpreted as "vagina": in which case, the bawdy sense of *catch our chains and jewels* would be roughly "snatch at our clothing and bodies"—which makes very good sense in this context. The opening words of Falstaff's reaction to this in line 46 are also intended to be bawdy. Ostensibly, *brooches, pearls, and ouches* mockingly echoes Doll's complaint of stolen jewelry; an *ouch* is a "a brooch or clasp to hold the sides of a garment together," but also "carbuncle or sore." Furthermore, *brooch* is 'the same word as BROACH' (*O.E.D.*) and *broach* means "A pointed rod of wood or iron; a lance, spear, bodkin, pricker, skewer, awl, stout pin" (*O.E.D.* 1) and so a classic euphemism for *penis* (see Ch.4). The verb, of course, meant "to pierce, penetrate, etc." Given this interpretation, along with the fact that *pearls* are "the whitish seed in an oyster" (= semen in the vagina), and *ouches* are "the painful effects of venereal disease", we can see that Falstaff has conjured an aphoristic caveat against consorting with a prostitute. This coheres with the obvious bawdiness of the rest of this speech and, furthermore, sets the pattern for its interpretation. Falstaff uses metaphors from warfare to imply a sexual skirmish in terms of a military one. Thus *serve bravely* suggests "copulate vigorously"; *come halting off* "come away exhausted", and *come* certainly evokes "orgasm." *The breach* is "the vagina"; *with his pike bent bravely* "with the penis no longer erect after such brave action"; *to surgery bravely* "to get some medicine for the pox"; *to venture upon the charged chambers bravely* "to brave the discomfort of using the discharging pox ridden organ". At this point he is cut off by Doll's bantering reply (50), which surely shows that she has understood Falstaff to be speaking of a sexual skirmish; her *conger* describes him as a "conger eel" with all the sexual connotations and fishy associations that has of a penis wriggling in a hole. *Conger*, for which an alternative spelling is *cunger*, perhaps evokes *cunnie* or Shakespeare's *coun* (from French *con* "cunnie, cunt") which he uses in *Henry V* (III.iv.48f).

Doll's next speech (57–61) is equally full of innuendo. She describes herself an an *empty vessel*, presumably waiting to be filled by Falstaff's bodily fluid, since she immediately introduces an image of herself *bearing* (i.e., lying underneath) the liquor-filled barrel that is Falstaff). And she seems to be flattering him via the double meaning of *you have not seen a hulk better stuffed in the hold*, where the marine image is surely secondary to the compliment "he's a good screw (for such a fat man)."

At this point *Ancient Pistol* is introduced to the scene, though he does not put in an appearance until after his hooliganish character and Doll's obvious distaste for him have been established, presumably in order to explain the ensuing trouble— very little of which is quoted here. The title *Ancient* is a form of "ensign" (i.e., the lowest grade of commissioned officer), which Doll will later claim puts him beneath her consideration; but she dislikes him anyway—and with good cause—because of his bad character. The weapon metaphor is active in the play on Pistol's name, and it is exploited as soon as he comes on stage. Indeed, there is more than one play on his name, because a few lines after those quoted here (158) the Hostess addresses him 'Good Captain Peesel,' that is, *pizzle* "piss organ, penis" (and, typically, she gets his title wrong).

In line 109 there is a play on *charge* in the senses "fill a glass" and "drink a toast," and later when Pistol uses it to Doll (115) "have intercourse with a woman." *Discharge* means "pay," "propose a toast in turn," and also "ejaculate semen." *Bullets* are "little balls." The Hostess, being a bit simple, probably makes the ambiguous statement *for no man's pleasure I* inadvertently; it can mean both "I will do it for no man" and "I am not a whore." Doll possibly mistakes Pistol's *I will charge you* (115) to mean "demand money"—and since she is a whore, she would understandably be outraged at this. But, it is more likely that she is rejecting Pistol's offer to have sex with her, or even—to take the superficial euphemistic interpretation—the implications of Pistol's proposing a toast to her, because she (1) loathes him and (2) has her eye on Sir John Falstaff. Consequently, Doll's *I am meat for your master* (118) means both "I am good for your master and too good for you" and "I am your master's lover, not yours." Pistol's *know you* in the reply 'I know you, Mistress Dorothy' (119) means both "am acquainted with you" and "(already) have carnal knowledge of you."

Doll is angered by what she claims is presumption from a lowly ensign with only 'two points' on his shoulder, and she ostensibly accuses him of being (1) *cutpurse* "a thief" who would use a *cuttle* "knife"; (2) *bung* "a lazy bum" or perhaps "a purse-snatcher" deriving from *bung* meaning "purse"; (3) *bottle-ale rascal* "a drunk" or perhaps "small-beer"; and (4) *stale juggler* "a boring buffoon." The *O.E.D.* says of **juggler** 'Often used with implied contempt or reprobation'. Of course, Shakespeare intended his audience to figure out some other insults as well. In using *cutpurse* Doll suggests that Pistol is emasculated: his *purse* "balls" has been cut off. The two sentences 'Away, you cutpurse rascal! You filthy bung, away!' (120) can be interpreted as jibes against his would-be penetration of her (i.e., taking his *cuttle* "weapon" to her *purse* "vagina" and *bunging* it up).

Her next remark (120–122) is a threat to turn tables on him by thrusting her weapon into his *chaps* "lips, mouth" before he gets his *saucy* "impertinent, lascivious" ?"spermy" *cuttle* "weapon" or "fishy bone(r)" into her. Shakespeare uses *saucy* as a term of more serious condemnation than it is today; nevertheless, in this context, it is difficult to imagine him overlooking a figure based on the fact that sauce adds moisture to meat or fish. He, and his audience, probably knew that the common cuttlefish is 'also called the *inkfish* from its power of ejecting a black fluid from a bag or sac' (*O.E.D.*): black fluid rather than white, but that would suit Doll's malicious purpose just fine, given her hearty dislike for this foul-mouthed 'swaggerer'!

In the context of 'you bottle-ale rascal', and the association of Pistol's name with piss, the 'stale' of *you basket-hilt stale juggler* evokes "pissing" (cp. a horse staling); and Doll might here be suggesting that Pistol is such a drunkard that he has to juggle with his balls in order to piss, so there is no way he could manage to screw anyone. This, however, is not the most obvious covert meaning. The association of *juggler* with *balls* is certainly intended: *basket-hilt* denotes a "sword-hilt"—the wrong end of the sword, one should note—and hence implies the uselessness of Pistol's *basket* "genitalia." The most immediate interpretation of *basket-hilt stale juggler* is therefore a worn-out ('stale') practitioner of sword-tricks; and once again Shakespeare invokes the weapon metaphor, *sword* being a well-established figure for the penis. In short, what Doll is saying is that Pistol is such a piss-artist he couldn't be any good in bed (in contrast with her stated opinion of Falstaff). Therefore, it is just possible that the *two points on your shoulder* are a cuckold's horns; however, this is probably not intended because a cuckold's horns are normally located on the head, cf. 'There the devil will meet me, like an old cuckold, with horns on his head' *Much Ado About Nothing* (II.i.42f). Pistol's angry response to Doll's insults is to threaten to tear off one of the marks of her profession: the large ruff (125); at least, this is how Doll interprets him in line 141 (not quoted). But we might, along with Partridge (1955), perceive a another meaning in *ruff*; by alluding to the salience of the woman's pubic hair, Pistol refers to her vagina (cf. Ch.4). On this interpretation, he threatens to cut up Doll's vagina and thus destroy her source of income—to say nothing of her health!

The scene from which this passage has been quoted continues in much the same vein. The running gag uses a warfare metaphor for a sexual skirmish, in which sexual organs are weapons. Shakespeare is a master of double meaning—to the extent that one is occasionally tempted to read too much into the text; on this occasion most of what we have drawn attention to is indicated in good editions of Shakespeare's plays. We hope to have demonstrated to the readers' satisfaction that, as is so common in Shakespeare's low comedy, the double meanings exploit euphemism to the full for an uproariously ribald comic purpose.

We remarked in Ch.4, and elsewhere, that many of the terms for tabooed body-parts show a keen inventiveness, and some of them are delightfully poetic—for example *the miraculous pitcher, that holds water with the mouth downwards* or *the one-eyed trouser snake*. Further examples of folk art are found in street jargon and criminalese (e.g., druggie jargon like *tea* or *grass* for "marijuana," *shit* for "cannabis resin," *speed* for "amphetamines," *horse* or *brown sugar* for "heroin," *snow* for "cocaine," or *the works* or *a fit* for a mainliner's "hypodermic syringe and spoon" are all metaphorical, and even poetic). Most of these euphemisms barely veil their denotata; but occasionally they do. For instance, rhyming slang and acronyms function as euphemisms (mostly euphemistic dysphemisms), often disguising the original taboo term to such an extent that for most speakers it has disappeared completely. For example, British rhyming slang like *berk* (*Berkeley Hunt*, "cunt") is widely used by speakers who have no notion of the implications of the unabbreviated version; *cobblers* (*cobbler's awls*, "balls") is another example. *Time* magazine and other U.S. media frequently use the mildly dysphemistic acronym *snafu*, presumably without intending to imply the unexpurgated original meaning of *Situation Normal All Fucked Up* (sometimes 'fucked' is euphemized to 'fouled'). Such innocence probably does not accompany *SFA* or *Sweet Fanny Adams*, where 'Fanny Adams' or 'FA' are usually taken to be euphemistic remodellings of *Fuck All*. According to McDonald (1988), however,

> *sweet F.A.* is not an abbreviation of *sweet fuck all*, but rather that *sweet fuck all* was invented to fit the older *sweet F.A.*
>
> The original *F.A.* was *Frances Adams*, a little girl who was brutally murdered in 1867. This was around the time that tinned mutton was first bought by the Royal Navy, and the sailors humourously pretended that sweet Fanny Adams' dismembered body had found its way into their new tinned meat. They started to call the meat *sweet Fanny Adams*, and then, because they did not have a very high opinion of it, they started to refer to anything regarded as worthless by the same name. From meaning "of no value" it was a short step for the expression to acquire the meaning "nothing".

Euphemisms like *china*, "vagina" or the derived metaphor *her crockery* (originally a "vagina with intact hymen"), along with many of the other euphemisms for genitalia that we saw in Ch. 4, and even the remodellings that serve as euphemisms for "God," "Jesus," and "Christ" (cf. Ch. 2) are instances of folk art—as indeed are all the metaphors that we use as euphemisms.

The reason that some such euphemisms survive without being recognized as euphemistic is that their original denotations have been lost to the general public. Thus the euphemism as a work of art falls into three categories: there are the artful euphemisms, like many of those used in street language, which make a striking figure, but which are the everyday vocabulary of a particular jargon; there are the artful euphemisms which mask their original taboo denotations to such an extent that the latter are not generally recognized; and finally there are the artful euphemisms which are meant to be as revealing—and in their own way as provoking—as diaphanous lingerie. The reason that texts like *Gulliver's Travels*, the 'Camera Song,' 'The Blacksmith,' 'Le Roman de la Rose,' and Shakespeare's bawdy are—in their different ways—so successful is that they exploit euphemism to publically expound taboo topics, yet at the same time pretend to disguise that purpose. Like any tease, such disguise may itself be titillating because, as Epstein says, 'the best pornographer is the mind of the reader' (1985:64). And as political satirists like Swift and Orwell know, titillation of Hearer's mind is the best way to draw Hearer's attention to Speaker's message, whatever it may be.

CHAPTER
—10—

Language Used as Shield and Weapon

If Bob Haldeman or John Ehrlichman or even Richard Nixon had said to me,
"John, I want you to do a little crime for me. I want you to obstruct justice," I
would have told him he was crazy and disappeared from sight. No one thought
about the Watergate coverup in those terms—at first, anyway. Rather, it was
"containing" Watergate or keeping the defendants "on the reservation" or coming up with the right public relations "scenario" and the like.

(John W. Dean III *New York Times* April 6, 1975:D25)

A Very Brief Review

Our book has surveyed euphemism and dysphemism in English, with some excur-
sions into euphemism in Middle Dutch and occasional remarks on X-phemisms in
other languages for the purposes of comparison. (You will recall that X-phemism
was the term we coined to name the union set of euphemisms and dysphemisms.)
We defined euphemism and dysphemism in such a way as to suggest that every
utterance is either euphemistic or dysphemistic:

> A euphemism is used as an alternative to a dispreferred expression, in order to
> avoid possible loss of face: either one's own face or, through giving offense, that of
> the audience, or of some third party.

> A dysphemism is an expression with connotations that are offensive either about
> the denotatum or to the audience, or both, and it is substituted for a neutral or
> euphemistic expression for just that reason.

There is no room between them: either one gives offense and is dysphemistic or,
should this effect be dispreferred, the utterance will be euphemistic. If our definition of euphemism seems to diverge from the lay usage (we doubt that there is a lay
usage of dysphemism), the divergence is more apparent than real; witness the
definition from the *Macquarie Dictionary*: 'The substitution of a mild, indirect, or
vague expression for a harsh or blunt one.' A 'harsh or blunt' expression is nothing
more than an expression that is dispreferred for the reasons we give. To speak

euphemistically is to use language like a shield against the feared, the disliked, the unpleasant; euphemisms are motivated by the desire not to be offensive, and so they have positive connotations; in the least euphemisms seek to avoid too many negative connotations. They are used to upgrade the denotatum (as a shield against scorn); they are used deceptively to conceal the unpleasant aspects of the denotatum (as a shield against anger); and they are used to display in-group identity (as a shield against the intrusion of out-groupers). We found that there are many, often antithetical, sources for euphemisms: figurative imagery; circumlocution, abbreviation, omission; synecdoche and metonymy; hyperbole and understatement; other languages; different varieties of the same language. Dysphemisms have similar sources. To speak dysphemistically is to use language as a weapon to assault another or perhaps just to exclude them. Or, if one utters an expletive, to vent one's anger, frustration or anguish at malevolent fate. People use euphemisms and dysphemisms in order to communicate an attitude both to Hearer and also to what is spoken about. In the typical cases of language usage, having chosen his or her topic, Speaker will have to choose the vocabulary, syntax, and prosody through which to articulate the topic: these choices will reflect politeness phenomena and aspects of X-phemism.

In the remainder of this chapter we will take up some thoughts evoked by matters we have raised in the preceding pages, and also tie up some of the loose ends left dangling.

Cases of Nonverbal X-Phemism

> This Nicholas was risen for to pisse,
> And thoghte he wolde amenden al the jape;
> He [Absolon] sholde kisse his ers er that he scape.
> And up the wyndowe dide he hastily,
> And out his ers he putteth pryvely
> Over the buttok, to the haunche-bon;
> And therwith spak this clerk, this Absolon,
> "Spek, sweete bryd, I noot nat where thou art."
> This Nicholas anoon leet fle a fart
> As greet as it hadde been a thonder-dent,
> That with the strook he was almoost yblent;
> And he was redy with his iren hoot,
> And Nicholas amydde the ers he smoot.
> Of gooth the skyn an hand-brede aboute,
> The hoote cultour brende so his toute,
> And for smert he wende for to dye.
>   (Geoffrey Chaucer 'The Miller's Tale'
>   11.3798–3813)

[Key: *ers* "arse, ass"; *yblent* "blinded"; *toute* "behind"; *smert* "pain"; *wende* "wanted"]

Perhaps because we are linguists, we have concentrated in this book on X-phemisms in language; but they occur as part of our nonverbal semiotic environment as well.

Society has many instances of visual euphemism. Bald men wear toupées. Both sexes wear contact lenses. Fig leaves hide the genitals of statues. Pubic hair was airbrushed out of soft-porn photographs until the 1960s. The Society for Indecency to Naked Animals designed boxer shorts, knickers, and petticoats to cover the sex organs of animals during the 1960s (cf. Fryer 1963:19). Frilled pantalettes modestly hid the *limbs* (*legs* could not properly be mentioned, especially in America, see Read 1934:265) of the table and the pianoforte during the Victorian era. A red rose on the prostitute's dress once indicated that *the shop was closed to business*. The *Beautiful Memory Picture* created by the modern-day embalmer vivifies the cold corpse. Flowers planted around the outhouse (dunny), pink floral toilet tissue, and flower-shaped air-fresheners within; the shapely, slender-waisted bottles of low calorie salad dressing, *diet*-this and *diet*-that—these are all commonplace nonverbal euphemisms that have just the same irrational underpinnings as their verbal counterparts. There have always been religious taboos on food: kosher food for Jews, halal meat for muslims, no meat on Fridays for Roman Catholics, and so forth. Today, many people, not just secular yuppies, taboo foods high in calories, carbohydrates, cholesterol and other animal fats, sugar, salt, and caffeine, so that foods are often advertised as having low quantities of these dreadful things. Lehrer [forthcoming] quotes an advertisement for ham saying 'only twenty-five calories a slice' and '95% fat free', and even seltzer water emphasizes it contains 0 calories, 0 carbohydrate, 0 fat, and 0 sodium. On the other hand, vitamins are added, particularly to drinks. High-fiber foods are also thought be healthy, so the current euphemistic food descriptives are *low fat/cholesterol, lite, diet, added fiber, vitamin enriched,* and so on. Lehrer mentions a chocolate torte called *decadence*, whose name exploits the notion that such rich food is sinful, and trades on the thrill people experience when transgressing taboos.

The British Museum Library relegates certain items to the so-called Private Case, fondly referred to unofficially as The Dirty Book Collection; in a striking example of nonverbal euphemistic omission, these books are not listed in the general catalogue. Euphemism motivates retail prices like $89.99 for a pair of shoes, so that they cost less than $90, and a hotel owner's decision to number floors and rooms 12, 12a, 14, to ensure that no one gets unlucky 13 (in Japan they omit unlucky 4, one of the words for which, /ši/ is a homophone of the word for "death"). An American friend recently remarked on the fact that, though her dimensions and weight have remained pretty stable over the years, her dress size has gradually reduced from a size 12 down to a size 8. We believe that this is a nonverbal euphemism on the part of the rag-trade; and our hypothesis would seem to be confirmed by the experience of one of the authors in Australia, whose bra cup size increased without concomitant increase in the size of the breasts they cup. We conclude that nonverbal euphemisms in the rag-trade reflect the current ideal shape for a woman: slim with big breasts. A similar reflection of current ideals is to be found disguising the dysphemistic reduction in the weight or volume of food items. Readers have probably noticed the considerable decrease in size of their favorite candy bar over the years. The nonverbal euphemism is to retain the same price and develop a *new fun size* or *handy-size*—which are euphemisms for "smaller than before"; or else the familiar packaging is retained with the generously increased content of *two candy bars for the price of one*, although the two together don't quite add up to the weight of the old one. Attractive packaging itself is a kind of euphemism: emphasis on appearance instead of the product contrasts strikingly with the old-time grocer who displayed items in bulk. Lighting effects that redden meat, the waxing of fruit, and the attractive packaging are cosmetic; and like verbal euphemism, they create a positive illusion. Still photography, film, and television are superb media for deceptive euphemisms. Short men like Michael Dukakis stand on concealed raised platforms to seem as tall as tall men like George Bush. The ugly are made up to look marginally attractive and the camera positioned so that the paint-job won't show. You know the story. These media present a world of perfected forms in which there is romance in the toilet bowl cleaner, poetry in the sanitary napkin, temptation in the tampon, and beauty in a glass of dentures.

Then there is the ancient practice of *bundling*. It is a mode of courtship wherein young men and women "sleep" in the same bed fully clothed, with or without a bundling board to separate the players. According to custom, the young man is expected to pay his visit late, well after the girl's parents have gone to bed, and stay until the early hours of the morning. Curiously, bundling is practiced by intensely religious groups with otherwise very rigid sex-codes, such as the Amish and Old Order Mennonites in North America. It was presumably meant to be a test of will power such as Mahatma Gandhi exercised himself with; to what extent premarital intercourse or even surreptitious groping occur, it is impossible to know (and, really, who cares). No less euphemistic than the ancient custom itself is the explanation that it represents nothing more than a charming custom of convenience, made necessary by the extreme cold and inadequate supplies of fuel. (Was there never any bundling in summer?) But recall that one person's euphemism can be another person's dysphemism: the practice is now disappearing among both the Amish and Mennonites because it is 'misinterpreted' by outsiders like ourselves. There is also the pretence that electric heating and other modern conveniences make it redundant (cf. Hostetler 1980:150). The mainstream equivalent of bundling is the drive-in, where rows of vehicles, which Australians appropriately call *sin-bins* (a.k.a. *panel-vans*), rock in disunison on their shock absorbers to horizontal dances within. Parents of the kids inside would probably be just as adamant as the early supporters of bundling that 'innocent endearments should not be exceeded.' (Stiles 1871:14)

Nonverbal offensive behavior has many forms, most of them readily imagined by our acute readers, we have no doubt. One nasty nonverbal parallel to *Shit on you!* is the *flying pasty*, where shit wrapped in paper is thrown at the target (see Spears 1982). The Greeks have an insulting gesture known as the *moutza*, in which the palm is thrust toward the target's face; the tradition is said to derive from an ancient Byzantine custom of smearing feces in criminals' faces. There are also gestures like *hanging a rat, mooning, pressing hams, brown-eying*, and others that expose genitals or buttocks (as did the foolish Nicholas in Chaucer's 'The Miller's Tale'). Then there are such things as *giving the finger/the V-sign, poking out the tongue*, and other such dysphemistic gestures. Of course, one must always be wary of cross-cultural differences: the American A-OK ring sign appears as an obscene gesture in a number of other nations. Some women (and perhaps even a few men) find the use of scantily clad girls in advertising to be dysphemistic; and many people find pornographic pictures, posters, movies, and videos to be dysphemistic. Blasphemous and profane depictions of religious icons are also occasionally complained of.

Earlier in the book, when we introduced the notions of dysphemistic euphemism and euphemistic dysphemism, we explained the existence of such paradoxical phenomena by appealing to the fact that human behavior is complex: Speaker may feel the inner urge to swear but at the same time not wish to appear overly coarse in his/her behavior. By saying *Shoot!* rather than *Shit!*, Speaker accomplishes a dysphemistic illocutionary act with a euphemistic locution. Intuitively, there are nonverbal manifestations of this sort of behavior, though we have no criteria comparable to 'locution contra illocution' for determining the matter. For example, *bustles*, once popular, aimed to draw attention to women's buttocks, while at the same time ensuring they were concealed under copious quantities of cloth. *Padded bras* serve a comparable purpose. *Codpieces* were highly conspicuous and decorative bagged appendages attached to the front of close fitting breeches in the fifteenth and sixteenth centuries to highlight a man's cods (genitals). All these, and perhaps the tinier bikinis, serve to emphasize tabooed body-parts while at the same time keeping them under wraps. Crutchless panties don't even do that!

In this section we have suggested that X-phemisms are not confined to language, but are also manifest in other semiotic systems used by humans. They undoubtedly play a part in nature too, in the appearance and behaviors of animals and insects; perhaps even of plants. Such systems would, however, differ very greatly from X-phemism in human language, where the use of a euphemism or dysphemism is an intentional communicative act on the part of the individual.

There Are Taboos and Taboos

Sweet words dance hand in hand with dreadful facts.
(Enright 1985:1)

In Ch. 1 we introduced the notion of taboo into the discussion of euphemisms, and examined the effects of taboo and euphemism on general vocabulary. The word *taboo* was originally borrowed from Tongan to denote "prohibited behavior"; in particular, behavior believed dangerous to certain individuals or to the society as a whole. Tabooed expressions in Austronesia are avoided because they are thought to be ominous or evil or somehow offensive to supernatural powers. In this context, the safest way to talk about these powers is to use euphemism—it is believed that dysphemism can be literally a matter of life or death. Austronesian type taboos are said to hold for all occasions except when there is ritual taboo reversal. But societies like Austronesia are by no means closed to innovation; the taboos are not immutable in the face of intrinsic forces or of cultural attitudes imported from outside the region. Austronesia also has counterparts to the taboos of contemporary Western society, which typically rest on traditions of etiquette and are intimately linked with social parameters like the age, sex, education, and social status of the interlocutors, and on the setting for the particular speech event. Here euphemism is the polite thing to do, and dysphemism is little more than the breaking of a social convention.

The differences between Polynesian-style and Western-style taboo are more apparent than real. Taboos, and attitudes toward taboo violation, change over time. It was not so very long ago that some transgressions against our own Western taboos were severely punished—by such as imprisonment, hanging, or burning at the stake. There are still people who would take such biblical commandments as *Leviticus* (24:16) literally: 'He that blasphemeth the name of the LORD, he shall surely be put to death, and all the congregation shall certainly stone him.' Even though few in our technically advanced and secular twentieth century would admit to the sort of fear and superstition that we associate with the taboos of exotic and unenlightened peoples, there are many in it who carry talismans when they travel, avoid walking beneath ladders, knock on wood, and believe in lucky and unlucky numbers; for some people, talking about a disease is enough to bring on specific symptoms; and medical students are notorious for discovering in themselves the symptoms of whatever disease they happen to have been studying. But if words can make us sick, they also have curative power; thus, doctors admit that many cited cancer cures have nonrational explanations.

If nothing else, all this, together with the comparative data we have presented on Middle Dutch and present-day English, show the similarity of human beings of different eras and different cultures to one another. At any rate, the similarities are more significant than the differences. Looked at it this way, diachronic and cross-cultural differences in the use of euphemism and dysphemism are superficial; the underlying motivations are similar and, consequently, so are the definitions of euphemism and dysphemism. We have shown that all verbal taboos, even those imposed by purely social conventions, serve direct human interests. Euphemism creates harmony and strengthens the social fabric by avoiding those things which threaten to cause offense and distress. Private dysphemism in the form of swearing relieves the frustration and distress of the swearer; but other forms of dysphemism make for discord. Avoidance speech styles, such as the 'mother-in-law languages' once widely used in Australia, prevent conflict where relationships are sensitive and threaten to disorganize the family unit. The fact that humans are not confined to mating cycles and, unlike most other animals, can copulate at any time, means that sex poses a real danger to social harmony. Perhaps this is why many subcultures place inordinate value on celibacy, a value that Nature has determined shall be more often preached about than practiced. The love, jealousy, possessiveness, and hate that can arise from our sexual proclivities all too often interfere with easy-going social interaction. So it is not surprising that human sexuality has been one of the principal drivers of verbal and nonverbal taboos. Taboos on the language denoting sexual acts and body-parts, especially in gender-mixed company, are partly motivated by a wish to avoid possible disharmony.

For reasons of fear and or distaste, which we have discussed at length in Chs. 3, 4, and 5, all human groups experience anxieties over certain body parts and bodily effluvia. For the same reasons, all also have taboos of one kind or another on some diseases and some aspects of death, as we showed in Chs. 6 and 7. In Ch. 2 we showed that supernatural beings and powerful persons are also the cause of euphemistic expressions for addressing and naming them; the practice has extended through the community because of politeness conventions, and the fact that euphemism and dysphemism are defined on face effects. Euphemism has probably been around ever since recognizably human language developed; and dysphemism must have been around for at least as long, to judge from animal behavior. No matter which population group one looks at, euphemism and dysphemism are powerful forces within it. Where populations differ is in the degree of tolerance with which dysphemisms are treated, and this depends on the values and belief systems of individual societies.

Dirty Words, Dirty Looks, Dirty Deeds

> Ma's out, Pa's out—let's talk rude:
> Pee—Po—Belly—Bum—Drawers!
> Dance in the garden in the nude:
> Pee—Po—Belly—Bum—Drawers!
> Let's write rude words all down our street,
> Stick out our tongues at the people we meet,
> Let's have an intellectual treat:
> Pee—Po—Belly—Bum—Drawers!
> (From Flanders & Swann
> *P** P* B**** B** D******* 1977)

[Key: *po* "potty"; *bum* "bottom, fanny"; *drawers* "panties"]

In this section we discuss the reasons that so many dysphemistic terms and acts are described as *dirty* and also what the connection is between death, disease, shit, and sex, that they should all figure in a discussion of euphemism and dysphemism.

One characteristic of euphemism is that it involves doublethink: in a given context, something, often tabooed, can be acceptably spoken of using a euphemism, but not by using a dysphemism. It is as if the denotatum were viewed from two opposing points of view. The ability of individual human beings to hold contradictory points of view on a common entity or phenomenon is something that will be very difficult to incorporate into artificial intelligence. We guess that such doublethink is necessary to permit an intelligent organism to pragmatically adapt to its environment. Most human groups have forms of ritual ceremony symbolizing death followed by resurrection or new birth, thus embodying the paradox that out of life comes death and out of death comes life. In Ch. 3 we showed how the prohibitions and taboos surrounding sex and bodily effluvia are linked to the same paradox and appear to derive from our acute fears about and revulsion from death and organic decay. This is how we can account for the fact that, throughout history, and doubtless in prehistoric times too, there have been social groups who, in certain contexts, perceive otherwise polluting substances to be agents of health and well being. Take ourselves: decaying corpses, rotting meat, tomato paste encrusted with mold, forgotten vegetables in the refrigerator crisper, shit—we generally perceive all these to be revolting and odious to the sensory organs. But we also recognize their creative benefits when we use them to enrich the soil and promote growth. Similar regenerative and creative powers are believed to be inherent in all body substances. In traditional folk medicine, dead human matter and the vermin associated with its decay have strong health-promoting aspects; relics of dead humans, earth from their graves, sites of murdered people, and executed criminals, by repute have all effected cures. It is consistent with this *weltanschauung* to view the womb as containing feculent and rotting organic matter (some of which is discharged in catamenia), and therefore to impose the same kind of sanctions on women's bodies as on rotting corpses. For reasons explained in Ch. 3, menstrual blood is overall the most lethal and the most strongly tabooed of bodily effluvia and consequently under some special conditions has exceptional curative and protective powers.

We have explained why bodily effluvia are tabooed and revolting, but not why decaying matter is revolting. Perhaps it simply is so by association with bodily effluvia. It seems more likely, however, that decaying matter has long been recognized as a source of disease. The motivation for homeopathic cures such as those discussed earlier was quite possibly a recognition of the inoculative power of substances that give rise to disease, which we now know to cause the development of pathogenic antibodies; doubtless this observation was extended by analogy to cases where it was completely unwarranted. In short, sympathetic medicine using urine, feces, menstrual blood, and material from corpses probably has a rational basis. Of course, it is also in the interests of the medicine men and shamans to perpetuate these homeopathic practices to maintain their own power and standing as controllers of such dangerous substances; how like they are to today's M.D.s.

The modern links with these ancient and prejudicial pollution taboos are preserved in the dirt metaphor. Epithets for taboo topics and the words naming them include *dirty, filthy, impure, grotty, unclean, bad*, and *foul*. Tabooed and offensive behavior is described as *off*, like it was rotting food; tabooed language is said to be *off-color*, like a sick person or, again, like rotting food. There is no need for us to reiterate the connection between bodily effluvia, rotting organic material, and dirt: today they all require that we 'wash our hands' after contact with any of them (certainly any with a r[evoltingness]-rating greater than R) before we eat, or shake hands with others. In former times, foul language was occasionally sanitized in a similar way:

> When my brother Roy was a little boy—on this occasion I was undressing him to put him to bed—and I said, "I'm a duffer, I didn't take your bodice off before I put your pyjamas top on," and he said, "Well, take my bloody bodice off." And mother went into the bathroom and she came out with a cake of soap, and pulled his tongue out and slithered it up and down his tongue. The poor kid. He was like someone having a fit. The soap was foaming all over the place.
>
> (Miss A.M. Miller b. 1906, interviewed 1978
> by Moya Gunn of La Trobe University)

Thus *dirt* and other words within the same semantic field generally have dysphemistic connotations. So dysphemistic behaviors are drawn into the net and get described as *dirty looks, dirty deeds, dirty business, do the dirty on, foul play*, and so on. One *gets and gives the dirt on* someone to *blacken* their name (rotting substances go black, though this use of the verb is doubtless connected to the darkness of night, too). A dysphemistic act is said to *stink*, and so forth. Diseases *contaminate* people and so do other people with *bad* characters. People without an S.T.D. might describe themselves to a prospective sexual partner as *clean*. Cancer patients report themselves being 'dirty' or 'like an animal'; and, to throw in another dimension for a moment, male cancer patients describe themselves as 'feminized, weak, and fragile' (cf. Cassileth and Lief 1979:23). As we have seen, the associations between dirt, disease, and women go back centuries, if not millennia. In our present society both sexes are obsessed with baths, bidets, and showers, with antiperspirants and fragrances (all nonverbal euphemistic cover-ups for the simple fact that we stink). The obsession is partly motivated by the ritual aspect of cleanliness that we remarked upon in Ch. 3; and notice that women are expected to be more scrupulous than men in their personal freshness.

We have shown that highly r-rated bodily effluvia, and the body parts most closely associated with them, link up with decay and disease and dirt. Death is obviously drawn into this associative network through its connection with disease and decay. But as we pointed out way back in Ch. 1, fear has a very significant role to play in this story. For good reason, people fear disease and they fear death. They also, of course, fear other people, and we explored that aspect of euphemism and dysphemism in Chs. 2 and 5 when we discussed naming and addressing. Human beings fear losing control of their destinies, and this seems to be at the root of a lot of taboos: it is why madness featured in the discussion of euphemism. In normal nonclinical usage, madness is perceived as a lack of control, and fear of becoming insane has inspired some of the strongest linguistic taboos to be found in the general area of illness and disease. Similarly with bodily functions—there is no reasoning with the everyday demands of the body. The 'calls of nature' are not things over which we have a lot of control, and because being ruled by Nature is to be like an animal, and our view of animals is pretty dysphemistic, we despise those body parts, bodily functions, acts, and actions that answer to the call of nature. In no society can a human urinate or defecate with the freedom of an animal. Imagine, therefore, the distress of an ostomy patient. Ostomy is a form of cancer where normal excretory functions are distorted and feces and urine flow uncontrollably out of the body. Because such incontinence hits at one of our fundamental hangups, even nursing staff apparently have trouble concealing their distaste toward ostomy patients (cf. D'Orazio 1979). Disease is something that afflicts us; we are 'patients', we exercise no control. This adds to the horror of disease—which may have been even worse in times past when diseases were more mysterious and the fear of contagion much greater. The fastings, bleedings, diuretics, purgatives, and vomitories of earlier times were attempts to gain control over natural functions. Death is, of course, simply another unavoidable body process, as Mark Twain put it, 'no one gets out alive.' Although we can end our own life, we have no control over when Nature will rob us of it. Here is one more reason for death to be tabooed, and spoken of euphemistically.

We no longer impute supernatural powers to our effluvia, yet they remain a powerful source of verbal taboo. Whoever doubts it should utter a dirty word in an inappropriate context and observe the reaction. The doublethink that underpins euphemism and dysphemism perhaps explains paradoxical attraction–revulsion that dirty words inspire. As with most things forbidden, they have a special fascination for us; what is taboo is revolting, untouchable, filthy, unmentionable, dangerous, disturbing, thrilling, but above all powerful.

Nice Girls Say It, But Newspapers Won't Print It!

The Grace of Swearing has not obtain'd upon Good Manners to be a Mode yet
among the Women; *God damn ye*, does not sit well upon a Female Tongue; it
seems to be a Masculine Vice, which the Women are not arriv'd to yet; and I
wou'd only desire those Gentlemen who practice it themselves, to hear a Woman
swear: It has no Musik at all there, I am sure; and just as little does it become any
Gentleman, if he wou'd suffer himself to be judg'd by all the Laws of Sense or
Good Manners in the world.   (Daniel Defoe 1966/1697)

What nice girls say today, and what they used to say or not say, is very different.
Women have traditionally been presented as the bastions of etiquette and euphemism;
observations like Defoe's are common place. Early this century, Otto Jespersen wrote:
'There can be no doubt that women exercise a great and universal influence on
linguistic development through their instinctive shrinking from coarse and gross
expressions and their preference for refined and (in certain spheres) veiled and indirect
expressions [. . .] Among the things women object to in language must be specially
mentioned anything that smacks of swearing' (1967/1922:246). But even Jespersen
goes on to admit that female suffrage and economic changes created by World War I
led to many more women defying these prohibitions: 'many young ladies have begun
to imitate their brothers in that as well as in other respects.' (p. 248) In 1975 Robin
Lakoff wrote: 'Women are not supposed to talk rough [. . .] women don't use off-
color or indelicate expressions; women are the experts at euphemism.' (p. 55) It is not
clear how much of this is simply a popular characterization of women's speech which,
as with other stereotypes, may or may not match reality; as Jenny Coates (1987:108) is
forced to admit, most evidence remains impressionistic.

   People in so-called polite society avoid expressions associated with the speech of
the lower levels of society. In gender-mixed company, the most widespread social
convention is to avoid the scatological vocabulary stereotypically associated with
male groups. Conversely, working-class males may frankly avoid the euphemistic
expressions preferred by "polite society" so as not to seem effeminate and weak to
their peers. In this way, the use of conventionally tabooed language becomes a
desirable macho marker of sexual identity, just like fighting, or getting drunk, and
other antisocial behaviors. But these remarks oversimplify the differences in the
language used by men and women. For many young women today, 'freedom' is
freedom from euphemism, because—as we saw in Ch. 5—breaking any taboo gives
one a thrill, a sense of power. Many women now talk openly and frankly about
menstruation, orgasm, contraception, and so on. They are probably swearing more,
too. Certainly, none of the young and middle-aged women we know well is averse to
using expletives like *fuck* and *shit*, though most are averse to using *cunt* nonliterally.
The question is: Have standards changed? And the answer is probably that they
have, and the taboos of yesteryear are not the taboos of today. Or it may be that
researchers up to now have been just looking in the wrong place; at least this is what
is suggested by Barbara Risch's 1987 findings. Her study reveals that within a
subculture of women speakers, there exists a treasure trove of 'dirty' words, particularly ones used in reference to men. Some of these sexist gems are *jerk-off, hunk,
nice ass, candy ass, hard ass, ass wipe, jockhead, bulge head, cock sucker, hard shaft,*

bulge boy, fish-eater, piece of meat, dog meat, sweet meat, nice bulge, and surprisingly *whore, bitch,* and *slut,* which are traditionally applied only to females and gay males. What is more, most of the women Risch interviewed not only understood these forms, but also claimed to use them; moreover, they were from the middle class, supposedly the most "ladylike" women of all. It seems, then, that the stereotypical claim that foul language is the domain of men, whereas euphemism and politeness characterize women's language, needs revising. In reality the difference between the two sexes (genders) may not be so great after all. Risch offers evidence that in the private domain, at least, dysphemistic forms tabooed by social convention are as much a solidarity marker for women as for men. On the other hand, a number of researchers into the graffiti to be found in men's and women's toilets have found fundamental differences: men use more dirty words and dirty pictures, and are derogatory toward minority groups and women; the most frequent expressions are dysphemistic. Women, on the other hand, tend to be more romantic, supportive, and advisory; lesbians are not put down, although there are derogatory remarks about men (cf. Bruner and Kelso 1980).

Not all members of society observe the same taboos, just as not all people react in the same way to eating peas off a knife, or speaking with a mouth full of food. What is unspeakable for one social group is not necessarily so for another. Thus, some women find the expression *bird* meaning "girl, woman" offensive, whereas others don't. Some like to be addressed as *Ms.* and others don't. A couple of readers of earlier drafts of this text complained that serious discussion of terms like *prick, cunt, ass,* and *arsehole,* and a fortiori their occasional use within it, do not belong in a scholarly piece of work such as this purports to be; the overwhelming majority of readers have no such objection. Because the function of X-phemisms is to place whatever it is they denote in either a favorable or unfavorable light, euphemisms occur in all kinds of contexts in all walks of life. What determines whether a particular expression is euphemistic, dysphemistic, or neutral is a set of social attitudes or conventions that may vary significantly between dialect groups and ultimately between individual members of the same community. Individual Speakers use euphemistic and dysphemistic expressions in order to communicate an attitude both to Hearer and to what is spoken about. Given that there is such variety in people's opinions and attitudes, we are unlikely to ever find complete uniformity of judgment even between people of very similar social background; this was demonstrated in the responses to our questionnaire on tabooed body parts and effluvia, which gave rise to the revoltingness ratings reported in Ch. 3. There can probably be no such thing as a genuinely omnibus X-phemism (i.e., an expression that is everyone's euphemism or everyone's dysphemism in every context). In principle, judgments of the X-phemistic value of vocabulary items can be surveyed and tested in a given population, though we have not yet done that. The fact is that, although these judgments may be based on responses to particular expressions decontextualized and listed, they ultimately derive from the types of contexts in which they are typically uttered. More precisely, it is prejudiced social attitudes toward the contexts in which people BELIEVE a given expression would be uttered that determine its X-phemistic value. As Read (1934:277) put it 'A word is obscene not because the thing named is obscene, but because the speaker or hearer regards it, owing to the interference of a taboo with a sneaking, shamefaced, psychopathic attitude.' Although Read may be indulging in a little hyperbole here, it is for such reasons that we proposed the middle class politeness criterion, which characterizes an expression as euphemistic if, in order to be polite to a casual acquaintance of the opposite sex in a formal situation in a middle class environment, Speaker would normally be expected to use this expression rather than its dispreferred counterpart(s). The dispreferred counterpart(s) would be dysphemistic. We presume this to be the criterion used by Speaker in the public domain when the exact composition of the audience is not known; because it is good etiquette. We used the middle class politeness criterion, you will recall, to judge the intrinsic X-phemistic value of vocabulary items. Although some vocabulary items are intrinsically euphemistic or dysphemistic, euphemism and dysphemism are primarily determined from evaluating an expression within the particular context in which it is uttered. This is what gives rise to dysphemistic euphemisms on the one hand and euphemistic dysphemisms on the other.

Very few of our Western taboos now receive any sort of legal reinforcement. But as some taboos are relaxed, others come to replace them. There has been a gradual establishment of legal sanctions against -IST dysphemisms (see Ch. 5) concomitant with action to improve the lot of a heterogeneous group consisting of sections of the community who had been discriminated against until the 1960s or 1970s, a group somewhat euphemistically known as 'women and minorities.' In this era of self-congratulatory equality for all, the new taboos are ageism, racism, sexism, religiousism, and the like; so ageist, racist, sexist, and religiousist language is not only contextually dysphemistic, but it is also legally dysphemistic. These new legally recognized taboos serve as replacements for the relaxing of laws against blasphemy, profanity, and obscenity. In the final years of the 1980s, three separate obscene language charges were dismissed in different courts around Australia; all of them would have once led to conviction and possibly imprisonment of the accused. A *Sydney Morning Herald* article of May 17, 1989, reports one of them, saying that a Brisbane magistrate 'found that a commonly used four-letter word [*fuck*] had "well and truly" ceased to shock the public.' But it apparently depends who says it, and where; because in a remote western Queensland town, an Aboriginal woman was, in 1989, fined $70 for saying it. The magistrate chided her that 'obscene language is not nice, especially coming from a woman.' Some courts may have ruled that *fuck* is no longer obscene—which is sensible, since the word is neither infrequent in movies, nor on television in Britain and Australia, and it has even been heard on home grown American TV recently—but apparently the word maintains its evil power in print. Not one of the many newspaper articles reporting on each of these cases actually dared to quote any of the offensive expressions in full. And our editor at Oxford University Press firmly discouraged a suggestion that *f---* and even *sh-t* (never mind the full spellings) should occur within the title of this book. Such words have to be hidden away inside like either copulation or feces itself. Once again we see that context is a decisive factor in determining what is considered dysphemistic. Incidentally, the print media have come a long way: at least we can use these taboo words within the covers of the book. Read's 1934 article 'An obscenity symbol' is about 'the colloquial verb and noun, universally known by speakers of English, designating the sex act' (p. 267); not once does the word itself appear in any form anywhere in the article!

X-Phemisms Are Here to Stay

Immodest words admit of no defence,
For want of decency is want of sense.
(Wentworth Dillon 'Essay on
translated verse' 11.113f [C17])

Read (1934) described taboo words as symptoms of the 'diseased condition' of our language and 'a festering sore' in our minds. He, like D.H. Lawrence, advocated a simple cure for this condition: namely, abolish the taboo on dirty words. Like other utopian ideas, this one will not eventuate. History has shown time and time again that there will be many people who refuse to accept a directive to change their language usage: look at the ineffectiveness of attempts at language prescription and proscription by national language academies, by religious bodies, in schools, and so on. In any case, there will always be plenty more taboo terms to replace those that are upgraded to decency (cf. the -IST dysphemisms of our decade).

And what evidence is there that humans can survive without euphemism and dysphemism? None that we know of. Technology has made a lot of things in our lives easier, but it cannot help us greatly with the reality that, whether we like it or not, we are firmly rooted in Nature. Human beings are social animals, and face concerns are supremely important in human social interaction; that is why we need euphemisms and dysphemisms. Humankind would have to change beyond all recognition for X-phemisms to disappear.

Language as Shield and Weapon

He who stands on tiptoe is not steady.
He who strides cannot maintain the pace.
He who makes a show is not enlightened.
He who is self-righteous is not respected.
He who boasts achieves nothing.
He who brags will not endure.
According to the followers of the Tao,
    "These are extra food and unnecessary luggage."
They do not bring happiness.
Therefore followers of the Tao avoid them.
(Lao Tsu *Tao Te Ching* 24)

In the spirt of the Tao, we have tried to offer a balanced account of what we perceive euphemism and dysphemism to be. We shall not once more review our achievements in this book; the greatest achievement would, in any case, be your enjoyment of it as a reader. Enough to say again that we have seen language needs to be used as a shield against the feared, the offensive, the distasteful, even against out-groupers.

It is also necessary to use it as a weapon against those things and people that frustrate and annoy us, and whom we disapprove of, despise, dislike, or just plain hate. It is not for nothing that there are laws of libel and that repressive regimes resort to censorship: language is sometimes the only weapon against brute force. This is the heroic aspect of dysphemism. But there is another side to it, in the pleasurable effect of an expletive used to release pent up anger or frustration against a person, thing or malevolent fate, or to insult and wound another. This reveals the animal side of the human being, the side that euphemism so often strives to conceal.

Our investigation of X-phemism has demonstrated the poetic inventiveness of language users in the figures they create to construct euphemistic and dysphemistic expressions. We have seen the justification for Lakoff and Johnson's claim (1980:3) that the ordinary conceptual system of a human being is fundamentally metaphorical in nature. And we have seen that people's language is not divorced from the world of their perceptions and conceptions, but very firmly founded in it. Far from illustrating a diseased mind, euphemisms and dysphemisms are the product of the healthy, flourishing, superbly inventive, and subtle minds of ordinary people like you and us.

GLOSSARY

This book is a study of language and language use; and in the course of writing it, we, as professional linguists, made certain assumptions that may be neither familiar to nor shared by all our readers. In this Glossary we will try to identify and clarify these assumptions, most of which are straightforward; but a few require fairly detailed explanation. We hope that the explanation of our terminology, and the conventions we use, will facilitate understanding of the text that is to follow, so that most readers will not regret the time spent perusing this Glossary before hurrying on to more exciting parts of the book. The Glossary contains not only terms of art, but also some Australianisms and Britishisms that have puzzled some American readers. In the interests of wider understanding between peoples speaking different dialects of English, we will gloss them here (identified respectively as '[Oz]' and '[Br.]') rather than replace them in the text. Only relevant senses are included in the alphabetical listing below. First, though, some nonalphabetic conventions.

() "Parentheses": In addition to using parenthesis for parenthetical remarks, we use them to indicate optional items. For example, *have it (off)* should be interpreted as an abbreviation for both *have it* and *have it off.* More generally, *x(y)* should be interpreted "*x,* and also *x y*".

/ "Slash": With the exception of *s/he,* which is to be read "he or she", we use / "slash" to separate alternates. For example, from *up the creek in a barbed wire canoe / chickenwire boat,* we confidently rely on our readers to substitute 'chickenwire boat' for 'barbed wire canoe' so as to generate *up the creek in a chickenwire boat.* More generally, *x / y* should be interpreted "*x,* or alternatively, *y.*"

* "Asterisk": **x* should be interpreted "*x* is a hypothetical form". Either (1) *x* is a reconstructed form from some extinct or even hypothetical language such as, for example, (Proto-)Indo-European **bher-* "to bear"; or (2) *x* is an ungrammatical and therefore unacceptable form in some contemporary language, a form that the analyst has constructed for the sake of argument (e.g., **Ate Max up it*). In the latter case, *x* is said to have a 'stigma'; other stigmata are given next.

?x and ??y and ?*z "Stigmata": Respectively interpreted as follows: *?x* "the grammaticality/acceptability of *x* is somewhat dubious"; *??y* "the grammaticality/acceptability of *y* is certainly dubious"; *?*z* "the grammaticality/acceptability of *z* is extremely dubious, such that *z* is probably ungrammatical and therefore unacceptable."

acronyms versus **abbreviations**: The difference between acronyms and abbreviations is this: acronyms are proper words created from the initial letter or two of the words in a phrase, and they are pronounced like other words (cf. *snafu, radar, laser,* or *UNESCO*). By contrast, abbreviations do not form proper words, and so they are pronounced as strings of letters; for example, *S.O.B.*, *IOU*, *U.S.A.*, *MP*, *lp*, or *tv.*

bird *n* [*Br.*, *Oz*] "girl, woman."

bum *n* [*Br.*, *Oz*] "bottom, buttocks, fanny."

cactus *adj* [*Oz*] "ruined, useless; dead" (cp. *kaput*). The adjective probably derives from the figurative expression *in the cactus* "in difficulties, in trouble" (cf. the *Macquarie Dictionary*).

Ch. and **Chs.**: "chapter" and "chapters", respectively.

cognates: Words in different languages that are similar in form and meaning because the languages they belong to are (pre)historically related; for instance, Dutch *broeder*, Swedish *brodor*, German *Bruder*, English *brother* are cognates. All these languages derive (wholly or in part) from Proto-Germanic, and the reconstructed form from which they derive is **broÞer* (the symbol 'Þ' is called 'thorn', and in this case pronounced like the 'th' in English *brother*).

connotation: The connotations of a word or longer expression arise from our experiences, beliefs, and prejudices about the contexts in which the expression is used. Consequently, the connotations of an expression vary independently of its senses and denotations (qq.v.); they also vary between speech communities. Sets of words can have the same sense, but at the same time differ considerably in connotation. For example, each of the nouns *dog*, *mutt*, *hound*, *cur*, *dish-licker*, *bow-wow*, *whelp*, and *mongrel* typically denote a canine animal of either sex; but the lexemes (q.v.) have quite different connotations. *Hound* connotes a noble animal—"a dog for the chase." *Mutt*, *cur*, and *whelp* are pejorative—and therefore dysphemistic. *Dish-licker* smacks of dog racing slang, and *bow-wow* either racing slang or baby-talk (cp. *gee-gee*). *Dog*, however, connotes nothing in particular, being the unmarked lexeme among the others. Or take the different connotations of *Tom's dog killed Jane's rabbit* versus *Tom's doggie killed Jane's bunnie*: as Gazdar (1979:3) points out (without, incidentally, mentioning connotation), the speaker of the second of these 'is either a child, someone posing as a child, someone who thinks that they are addressing a child, or someone posing as someone who thinks that they are addressing a child.' And it is all due to the different connotations of the two sentences. Connotation makes it extremely difficult to find cases of absolute identity of meaning between vocabulary items: it is rare to find words that are identical in all the associations they have taken on from the various contexts in which they are used.

Connotation is very important to the subject matter of this book, because we find that, on the one hand, some vocabulary items will need to be classified within a dictionary as typically euphemistic, (e.g., *restroom*) in the sense "toilet"; and some items will be classified as typically dysphemistic because they are intrinsically offensive in almost all contexts (e.g., *cunt*). Such classifications are a function of both the senses and connotations of the lexemes.

context: We employ the term 'context' in pretty much its everyday meaning, but note that it can be interpreted in terms of three categories: **co-text, setting,** and **world spoken of,** qq.v. All three contribute significantly to the production and interpretation of utterances. Context is crucial in determining the meaning of an utterance; it is also critical in deciding what constitutes politeness. We will show that an expression will be understood as being either euphemistic (inoffensive) or dysphemistic (offensive) depending on the context (any one or more of the three categories) in which it is embedded.

cooperative principle: The cooperative principle comprehends three things. (1) A communicative presumption on Hearer's part that if Speaker appears to be using language, then s/he is doing so in order to communicate some message using the normal conventions of natural language. (2) A presumption that, in the prevailing context, Speaker has some reason for using the words s/he does in the manner s/he utters them, rather than remaining silent or uttering something different; that is what makes it worthwhile for Hearer to expend some effort figuring out what Speaker is trying to convey. (3) Speaker and Hearer observe the normal conventions pertaining to face effects—for details, see under **face,** and maxims of **quantity, quality, relation,** and **manner.** These maxims of cooperative behavior, based on work of H. Paul Grice (1975), are not laws to be obeyed, but reference points for language interchange—much as the points of the compass are conventional reference points for giving directions and locations over the surface of the earth. This allows Hearer to seek the communicative purpose behind apparent violations of the maxims; for instance, where a philosophy teacher uses *My brother is an only child;* or a man is warmly invited into his friend's house with *Come in you old bastard!;* or someone to whom you have given directions returns to say *That left turn at the traffic lights on High Street is a right turn;* or someone drops a plate and says *Clever!* (dropping a plate is not normally considered a clever thing to do). Were there no cooperative principle, systematic communication would be impossible. There would be no ground rules for deciding whether or not an utterance makes sense, nor what value should be put on it. Conversely, speakers would have no ground rules for getting their message across to the audience.

co-text: The term denotes the linguistic context of a given expression (i.e. the text that surrounds it). Utterances link up with their co-text by including devices to mark topic continuity; for example, pronouns and other anaphoric expressions (cf. Carter 1987 Ch. 2) versus full noun phrases, names, and the like; focusing devices like *As for NOUN PHRASE;* also sentence fragments whose gaps are filled using information provided by the co-text. Utterances not only take from their co-text, they also give to it: what we say or write at any point is probably going to have an important bearing how a text will continue.

denotation and reference: Denotation is the meaning relation that holds between language expressions and the world spoken of. In the act of denoting Speaker uses the senses of language-expressions to identify things in the world s/he is speaking of. It is important to identify WHAT THERE IS, and WHAT THERE ISN'T in the world spoken of; and so it divides up into **referring** expressions that do the former, and **nonreferring** expressions that do the latter. The world spoken of may be fictional (i.e., may not really exist) and yet there will be things in it that do, putatively, exist. For instance, take the imaginary world evoked in the television series/film 'Star Trek': speaking of that world, Speaker appropriately uses *Mr. Spock* to refer to a certain pointy-eared character in just the same way as s/he uses *a man Elizabeth Taylor twice married* to refer to the real-life (late) Richard

Burton, and to indirectly refer to actress Elizabeth Taylor, and to their two marriages. The use of referring expressions presents the denotatum as existing; it can therefore be denoted within what we linguists would call the complement noun phrase of an affirmative indicative existential clause of the type *There is* A CHARACTER CALLED 'MR. SPOCK' *in 'Star Trek'* —the referring expressions are in upper case. In this last example 'Star Trek' refers to a tv series (cf. *There is or was* A TV SERIES CALLED 'STAR TREK'). For the purposes of this book, we have no need to concern ourselves with nonreferring expressions (the interested reader could look for more details in Allan 1986a §1.6, §7.11.3, or Givón 1984 Ch. 11). To sum it up: saying *X denotes Y* makes no commitment as to the existence or nonexistence of **Y**; in other words, 'denoting Y' does not assure **Y**'s referentiality. Many authors (e.g., Bach 1987, Lyons 1977 Ch. 7) have a different notion of denotation and reference from the one described here; and yet others make no distinction between the two, conflating them under the one term, *reference*. We find these practices unsatisfactory when trying to account for nonreferring terms; but this is no place to pursue the matter.

dero *n* [*Oz*] "derelict person, typically homeless and suffering from alcohol abuse."

dunny *n* [*Oz*] "toilet."

discourse: See under **text**.

effluvia [*singular* **effluvium**]: This is the all-inclusive term we use for bodily emissions such as breath, sweat, snot, sperm, menstrual blood, urine, feces, fart, shed hair, nail-clippings, and the like. The Latin original of this noun denoted a variety of flows, oozings, leaks, emanations, and escapes of liquids, gases, and secretions; also hair falling out. If we have, out of necessity, extended the denotational scope of the term at all, we are delighted.

face: Face is a familiar enough concept; the figures of speech 'to save face' and 'to lose face' are current in everyday parlance. The 'face' that is referred to in both these expressions is essentially THE PUBLIC SELF-IMAGE that both Speaker and Hearer must have regard to in the speech situation. Strong feelings attach to face wants. If, for example, the outcome of an exchange means that a person's self-image is not sustained, if face has been lost or affronted, then s/he will generally feel bad, insecure, hurt, humiliated, and consequently become embarrassed, flustered, and even hostile. On the other hand, if the outcome is a (public) self-image that surpasses the norm (i.e., if face has been enhanced by the encounter), then the person will feel good, and perhaps confident and self-assured. But if the outcome is merely a self-image consistent with the usual, if face has been simply maintained, then the person is unlikely to feel any strong emotions one way or another.

There are two co-existing aspects to face. One, is the want of a person that his/her attributes, achievements, ideas, possessions, goals, and so on, should be desirable to at least certain others. This is called **positive face.** Attending to someone's positive face is to make that person feel good. It might involve showing a flattering interest in his/her activities, achievements, interests, or ideas; or it could mean simply "watching one's language" (by accommodating dialect, accent or style, for instance). The other aspect to face is the want of a person not to be imposed upon. This is usually referred to, following Brown and

Levinson (1987), as 'negative face,' a term that many people find misleading; we think **impositive face** is a more apt description, so that is the one we use. Paying attention to someone's impositive face includes taking account of nonverbal aspects of the encounter, such as eye contact (staring, or avoiding someone's gaze), physical distance (standing too close or too far away from someone), gesture (touching, nudging, gripping, even striking). It also includes straightforward impositions on a person's time, borrowing their possessions, or requesting help; or impositions caused when someone is required to expend unreasonable effort in order to understand the message because, for example, an utterance is either too loud or inaudible, incoherent or perhaps not even relevant—in other words, because it violates the politeness conditions enshrined in the cooperative principle. These two aspects of face (positive and impositive) can sometimes be in conflict, because satisfying one means infringing on the other; Oscar Wilde fingered this when he had Lord Henry Wotton remark with typical Wildean wit: 'There is only one thing in the world worse than being talked about, and that is not being talked about.' (*The Picture of Dorian Gray* 1926:2) There is additional discussion of face in the Introduction.

galah *n* 1 "Australian cockatoo (*Cacatua roseicapilla*), pale grey above, deep pink below; 2 [*nonliteral*] fool, berk, turkey."

Hearer: See under **Speaker and Hearer.**

homonyms: These are words that are either pronounced or spelled the same and yet have different meanings; occasionally, this may cause confusion:

> 'Mine is a long and sad tale!' said the Mouse, turning to Alice, and sighing.
> 'It *is* a long tail, certainly," said Alice, looking down with wonder at the Mouse's tail; 'but why do you call it sad?'
> (Lewis Carroll *Alice's Adventures in Wonderland* 1965:39)

illocution: See under **locution and illocution.**

impositive face: See under **face**.

innings *n* [*Br.*, *Oz*] despite appearances, a singular noun; in the game of cricket an innings is roughly counterpart to an *inning* in baseball (or so we believe, being regrettably ignorant of both games).

larrikin *n* [*Oz*] "man given to loutish behavior; a tearaway."

lexeme: A lexeme is almost a word, but not quite. For instance, *give, gives, gave, given* are separate words, but manifestations of the same lexeme GIVE: note that by convention, the lexeme is represented by the least marked form. Some lexemes consist of more than one word, for instance, a phrasal verb such as *put up with* constitutes one lexeme. It is lexemes, along with certain morphemes (q.v., that one finds listed in a dictionary).

loo *n* [*Br.*, *Oz*] "lavatory" [*euphemism*].

locution and illocution: In making an utterance, Speaker uses a form of words, the **locution,** to convey a certain message, the **illocution.** (The illocution is sometimes described as 'what Speaker is doing in making the utterance.') These terms were originally introduced by J.L. Austin in 1955 (see Austin 1975), and have undergone some slight variation in meaning since then (cf. Searle 1969; Bach and Harnish 1979; Allan 1986a). The same locution can be used to convey different messages, depending on the context of use. For instance *I will be late* can be used as a WARNING to one's spouse not to get anxious when one does not return at the usual hour; or it can be a PROMISE in reply to someone's request *Don't come early, 'cause I've got something I need to talk to Sammy about before you get there*; or it may be a resigned PREDICTION from the consequences of not setting the alarm clock the night before; and there are other messages it could convey, as well. Conversely, the same illocution can be achieved using different locutions: for instance there are all sorts of locutions to convey that Speaker wants to be informed of the time; for example, *Got the time?*; *Excuse me, but do you happen to know what time it is?*; *What's the time?*; *Gotta watch, mate?*; or *How late/early is it?*. In the course of this book we will frequently be referring to locutions and illocutions, and the difference between them should be borne in mind.

manner (maxim of): Where possible, Speaker's meaning should be presented in a clear, concise manner that avoids ambiguity, and avoids misleading or confusing Hearer through stylistic ineptitude. See the Introduction for an example.

metathesis: For this book, simply that the positions of phonemes (q.v.) within a word get swapped about at some point in history. For example, the /r/ in Old English *thridda* swapped places with the following vowel to give modern English *third*, which should be compared with the unmetathesized *three*.

morpheme: The smallest meaningful unit(s) within word structure (for reasons we don't have space to go into, phonemes (q.v.) aren't). A morpheme is typically a part of a word, nonetheless many words consist of a single morpheme. It is the function of a morpheme to bear meaning, and so it must not be confused with a syllable, which is a phonological entity: a syllable conveys no meaning unless it happens to realize a morpheme. For example, the monosyllabic word *cats* consists of two morphemes: one that doubles as a word, namely *cat* "feline quadruped" (!) and one that means "plural," namely *-s*. Another example is *unkindness*, which can be decomposed into morphemes *un-*, *kind*, and *-ness*.

noun phrase: Typically, a phrase headed by a noun, but also a phrase headed by (and often consisting solely of) a pronoun. Examples, with the heads bolded are: **John**; *another* **gardener**; *three very dilapidated late Victorian brick* **houses** *with the windows boarded up* (in this example there is a noun phrase 'the windows' within another noun phrase); *it;* **hers**; *the pink* **ones**.

phoneme: The phonemes of a language are the contrastive sounds in that language; English has between thirty-five and forty (dialects differ). Phonemic transcriptions are conventionally written between slashes. Double symbols are used only to indicate relative length. Since many phonemic symbols used in this book have easily recognizable values, we restrict ourselves to exemplifying only the others. A full inventory can be found in any good introductory book on linguistics intended for English readers. Some dictionaries also use them; for example, the *Macquarie Dictionary* has a full set, so does *O.E.D.* (1989) (2nd edn); *Webster's New Collegiate Dictionary* (1977) has a modified set. In this book we have simplifed the system to a minimum, which is why we yoke together British and Australian in the list below rather than dwell on irrelevant differences in pronunciation.

/ii/	/siin/	*seen, scene*
/ɪ/	/θɪn/	*thin*
/e/	/yel/	*yell*
/æ/	/šæm/	*sham*
/ʌ/	/čʌk/	*chuck*
/ə/	/ə'baut/	*about* (' is placed before a stressed syllable)
/əə/	/əən/	*urn, earn* (Br. & Oz, but N. Am. /ərn/)
/o/	/hot/	*hot* (In N. Am. this would be better represented /hat/)
/ʊ/	/wʊd/	*wood, would*
/uu/	/fuud/	*food*
/iə/	/iə/	*ear* (Br. & Oz, but N. Am. /ir/)
/eə/	/heə/	*hair, hare* (Br. & Oz., but N. Am. /her/)
/ei/	/hei/	*hay*
/ai/	/ai/	*I*
/au/	/šaut/	*shout*
/oi/	/boi/	*boy*
/ou/	/šou/	*show*
/θ/	/θɪn/	*thin*
/š/	/šæm/	*sham*
/č/	/čʌk/	*chuck*
/ʔ/		the glottal stop—infamous in Cockney speech, particularly in place of intervocalic and final ts; it is also common in this and other dialects before utterance initial vowels. In isolation, and therefore utterance initial and stressed, Cockney *hat* and *at* demonstrate both: /ʔæʔ/.
/ṭ/		this is a retroflex 't' (voiceless retroflex stop); it doesn't occur in English —except for the Asian Indian variety. The blade of the tongue is curled up so that the back of it closes the mouth at the teeth ridge.

phonestheme: A phonestheme is a cluster of sounds, located at either the onset of a word (e.g., *fl-* or *gl-*) or its rhyme (e.g., *-ash, -itter, -utter*), and which symbolize a certain meaning. In at least one sense of all of the following words with an *fl-* onset there is a suggestion of "sudden or violent movement": *flack, flag, flail, flame, flap, flare, flash, flay, flee, flick, flicker, flinch, fling, flip, flirt, flit, flood, flop, flounce, flounder, flourish, flow, fluent, flurry, flush, fluster, flutter, flux, fly*. Many verbs with the rhyme *-ash* denote "violent impact," for example, *bash, clash, crash, dash, flash, gash, gnash, lash, mash, slash, smash, thrash.* Phonesthemes are typically incomplete syllables, and their meaning is much more difficult to pin down than that of morphemes; furthermore, exactly the same cluster of sounds that constitute the phonestheme will appear in many vocabulary items which have no trace of the meaning associated with the phonestheme! Consequently, phonesthemes are often said to manifest a SUB-REGULARITY in the language; all languages seem to have them. (See Allan 1986 §4.9.3; Bolinger 1950.)

phonetics: (Study of) all the sounds humans use in languages.

phonology: (Study of) the sound system of one or more languages.

positive face: See under **face**.

prosody: The intonation patterns in a language that result from the pitch level used, the stresses employed, and the location of pauses, together with the tone of voice.

quality (maxim of): Speaker should be genuine and sincere. That is, Speaker should state as facts only what s/he believes to be facts; make offers and promises only if s/he intends carrying them out; pronounce judgments only if s/he is in a position to judge, and so forth. See the Introduction for an example.

quantity (maxim of): Speaker should make the strongest claim possible consistent with his/her perception of the facts, while giving no more and no less information than is required to make his/her message clear to Hearer. See the Introduction for an example.

reference/referentiality: See under **denotation and reference**.

reflexive-intention: Speaker will tailor the utterance to suit Hearer, taking into account what s/he knows or guesses about Hearer's ability to understand the message s/he wants to convey. Speaker is said to have a **reflexive-intention** toward Hearer: this is a speaker's intention to have a person in earshot recognize that the speaker wants him or her to accept the role of Hearer and therefore be the (or an) intended recipient of Speaker's message and consequently react to it. As we all know, it can happen that an utterance is overheard by someone when there was no original specific intention on Speaker's part that this should happen; to put it more precisely, Speaker has a reflexive-intention toward Hearer but not toward an overhearer. An overhearer may perchance understand the message in the same way as Hearer does but, because s/he is not necessarily party to the appropriate contextual information relevant to the proper interpretation of the utterance, it is possible that s/he may seriously misinterpret it.

relation (maxim of): In general, an utterance should not be irrelevant to the context in which it is uttered, because that makes it difficult for Hearer to comprehend; and we presume that Speaker has some reason for making this utterance in this context, in the particular form in which it occurs, rather than maintaining silence or uttering something different.

root *v,n* [*Oz*] "fuck"

sense: We ordinarily think of 'meaning' as something we look up in a dictionary: suppose we look up the meaning of *canine*; what we would find given in the dictionary are the **senses** of the word:

> ¹**canine** . . . *adj* [L *caninus*. fr. *canis* dog] **1:** of or relating to dogs or the family (Canidae) including the dogs, wolves, jackals, and foxes **2:** of, relating to, or resembling a dog
> ²**canine** *n* **1:** a conical pointed tooth; *esp:* one situated between the lateral incisor and the first premolar **2:** DOG
> (*Webster's New Collegiate Dictionary* 1977)

What this reveals is that there are two homonyms, ¹**canine** and ²**canine** (historically they are related, but we needn't go into that), and each of them has two senses. We can see that sense is a description of the informational content of a word (or combination of words) in terms of some other language expression; the latter may be in the same language as the word whose sense is being given, or it may be in some other language, even a formal language of semantic primitives, perhaps. For example, the senses of the translation equivalents *ájá* (Yoruba), *cane* (Italian), *dog*, *Hund* (German), *kare* (Hausa), *mbwa* (Kiswahili), *pies* (Polish) are identical with one another and also with *canine quadruped*—it is exactly that fact that makes them translation equivalents. It should be noted that they are translation equivalents independently of any particular context of use: the significant characteristic of sense is that it is decontextualized meaning, abstracted from innumerable occurrences of the word in texts. Looking back at the previously quoted dictionary entries for *canine*, note that you have to decide which of the four senses given is relevant to a particular context in which the word is used. Each sense of an expression, whether word or sentence, can be thought of as a property of the expression itself, rather than one that is induced by context on a particular occasion of that expression's use. For example, the sense of, *I totaled my car yesterday* is (virtually) unchanging: "Speaker did irreparable damage to his or her car the day before this sentence was uttered." However, the **denotation** (q.v.) will depend on WHO makes the utterance— which determines between 'his' or 'her' car, and WHEN it was uttered—which dates 'yesterday'. What Speaker does in the act of denoting is to use the senses of language-expressions to identify things in the world s/he is speaking of.

setting: Setting, also known as 'the situation of utterance,' is one of the three categories of context; the other two are co-text and world spoken of. It defines the spatial and temporal location of utterances; that is, the place and the time in which an utterance is made and the place and time in which it is heard or read. So, it can be thought of as 'the world spoken **in**.' Setting determines the choice of what are collectively known as deictic expressions: tense, personal pronouns (identifying Speaker and Hearer, among others); demonstratives (*this, that, yonder*, etc.); choices of adverbs and directional verbs relative to the location of Speaker and Hearer (e.g., whether Speaker chooses *come, go, bring, come up, come down, come over*, etc.). The setting will also have a role to play in determining the topic and the linguistic register or jargon—that is, the variety of language associated with a particular occupational, institutional, or recreational group: for instance, legalese, medicalese, cricketese, baseballese, criminalese, linguisticalese, and so forth. The setting is also where and when paralanguage such as gesture, facial expression, the positions and postures of interlocutors, take place (see Argyle 1988).

shiela *n* [*Oz*] "girl, woman."

Speaker and Hearer: We use 'Speaker' as shorthand for "speaker or writer", and 'Hearer' for "audience" (i.e., for "hearer or reader"). We do this for the sake of convenience to ourselves and our readers, because it avoids, to the greatest extent possible, the ghastly stylistic effects of being required to write *speaker or writer*, and so forth, in order to be accurate. We are confident that you will take this idiosyncrasy in your stride and still be able to enjoy the book. Unless otherwise stated, we shall assume that Speaker and Hearer are NOT particular living individuals, each of whom we know to have their own peculiar ways of behaving; instead we regard Speaker and Hearer as model persons, with the sort of judgment and capabilities assigned to the 'reasonable man' in law; they adopt whatever practice is customary under the prevailing circumstances; they have no abnormal idiosyncrasies; they are not omniscient, nor clairvoyant, nor especially foolish. We shall also assume that there is only the one Speaker at a time—which is in any case the norm. Hearer is normally individually identifiable to Speaker in face to face conversation, when speaking on the phone, and when writing a personal letter, but rarely in other varieties of written discourse. In all cases Speaker will normally tailor the utterance to suit Hearer (see further under **reflexive-intention**), taking into account what s/he knows or guesses about Hearer's ability to understand the message s/he wants to convey; this is governed by the cooperative principle, q.v. Even though there is commonly more than the one Hearer at a time, we assume in our discussion that there is only one.

The notion of face (q.v.) is useful in distinguishing between types of Hearers and overhearers. We define Hearer as anyone whom, at the time of utterance, Speaker reflexively intends should recognize the illocutionary point of the utterance (i.e., Speaker's message). Clark and Carlson (1982) distinguish between Hearer as a 'direct addressee,' and Hearer as a 'ratified participant,' the latter being a member of the audience participating in the speech act (cf. Goffman 1981:131). An **addressee** is someone who cannot reject the role of Hearer without serious affront to Speaker's face. A **ratified participant** can reject the Hearer role more freely than an addressee and with less of an affront to Speaker's face. Any other person hearing the utterance is an **overhearer** (i.e., either a bystander or an eavesdropper). A **bystander** within earshot was not originally intended as a Hearer and may, depending on the circumstances, accept or reject the role of Hearer without loss of face. An **eavesdropper** can only admit to listening in at the expense of their own positive face, because it makes them look bad, and sometimes at the expense of Speaker's impositive face, because Speaker feels affronted by the intrusion (see Allan 1986b for further discussion).

speech act: An act of speaking or writing.

style: Successful communication requires Speaker to pay careful attention to the style of language s/he uses. In making an utterance, Speaker will select particular forms in response to the degree of formality, informality, and familiarity appropriate to the context of the utterance. Joos (1961) identified five levels of formality: **frozen, formal, consultative, casual,** and **intimate.** Needless to say, 'intimate' style is less formal than 'casual,' 'casual' less formal than 'consultative,' and so forth; for that reason we often list the five levels in the following way: frozen > formal > consultative > casual > intimate. There are no fixed boundaries between each of the different styles and any one person's language will reflect a wide range between the extremes of frozen and intimate style. Style varies according to who we are and whom we are communicating with; whether we are speaking or writing; where we are and when the utterance takes place; what we are talking about; and how we feel about the whole situation. If any one of these factors is changed, the style may well change accordingly. For any given utterance, there exists a wide variety of stylistic choices possible: not only lexical choices, although these are the most obvious, but also grammar, pronunciation, and paralinguistic features like gesture and facial expression. Usually, Speaker's stylistic choices are tuned to create just the impression s/he wants to create: where Speaker wants to avoid offense s/he will be euphemistic and choose a style appropriate to that end; where s/he wants to be offensive s/he will choose a style of language that is dysphemistic. Thus, euphemism and dysphemism (which are themselves defined in Ch. 1) are intimately bound up with style, in ways we will examine in Chs. 2 and 8.

text: Usually, a speaker or writer makes a number of utterances to an audience that are linked by some sort of arrangement of related topics. These constitute what is known as the 'text' or 'discourse.' There are many varieties of texts. Some involve the utterances of a single individual: for example, lectures, speeches, sermons, recitations, narratives, and jokes. Others, like arguments, negotiations, interviews, conversations, and debates are typically made up from the utterances of more than one individual; indeed, they can involve any number of people, who all share certain expectations about the structure and flow of the talk-exchange. Even in casual conversations not everyone can talk at once; and there are certain sequencing conventions and cues which govern when and how the interlocutors take turns at speaking. Hearers become speakers, while speakers in turn become hearers, and so forth. There are also ritual utterances, like those used in greetings and partings, or to show that one is paying attention to the speaker holding the turn, and so on. Anyone who does not observe these sorts of conventions is considered uncooperative and even rude (see under **cooperative principle**).

utterance: In ordinary usage, utterances occur only in spoken language, but here we understand them to include segments of either speech or writing. They can be complete sentences or, in the case of speech, they are frequently just fragments of sentences. An utterance is brought into being by an act of speaking or writing, by what is more succinctly called a **speech act**—a term applied even to the act of producing a written utterance! We presume that the utterance is the proper source for linguistic data.

vagina *n*: This word turns up several times in the book. We, like most people, use *vagina* to denote "those parts of the female genitalia used in sexual intercourse" or, to put it into the vernacular, "cunt." This is its normal meaning in all dialects of English. Thus, for instance, Meigs (1978) writes that a Papuan people, the Hua, 'traditionally wore, and to a large extent still do wear, only loin cloths, which leave their genitals open and exposed to objects beneath them' (p. 305); and they believe that 'the sight of a woman's vagina can cause sores to erupt on a boy's face' (p. 311). Here, Meigs is clearly referring to what a boy might catch sight of, which is, if one needed to be absolutely specific, the **vulva**. Today, the word *vagina* is polysemous: its original meaning, "the passage between the vulva and the cervix," has been semantically extended in lay usage to include the vulva, giving rise to what is now the primary everyday meaning. That is why Meigs, most ordinary folk, and we in this book, all use *vagina* to denote "those parts of the female genitalia used in sexual intercourse." It is precious to claim, as do Ash (1980) and Hankey (1980), that people like Meigs and ourselves

MISUSE *vagina* to denote (among other bits of the body) the vulva; *vulva* is a specialist term, a piece of jargon, to be used when it is communicatively required to distinguish this denotatum from the rest of the female genitalia. We doubt that Ash or Hankey would claim of a man who says *I've just been bitten by a dog* that he is MISUSING the word *dog* when his attacker is in fact known to be a *chihuahua bitch*. Specialist terms, or jargon, are not appropriate to every context—as we shall clearly demonstrate in Ch. 8.

whinge *v* [*Oz*] a pejorative term meaning "to complain, whine about something."

wombat *n* Native Australian burrowing marsupial, not unlike a small bear—heavily built with short legs and a rudimentary tail.

world spoken of: The most important category of context is **the world spoken of** (and that includes the world written of). The world spoken of is the mental model of the world that we all construct in order to be able to produce or understand a text. Evidence for the constructive nature of text understanding is: **(A)** the proven use of inferences and speculations in the course of language understanding, which enables Hearer to predict what is likely to happen next in a story; **(B)** the effect that titles and headings have on the way a text is interpreted; **(C)** experimental evidence for the realignment of scrambled stories in both summaries and recall; and the replacement of abnormal by normal events but not vice versa in recall situations (cf. Van Dijk and Kintsch 1983). The world we construct can be our own version of the real world or perhaps a deliberately evoked fictional world—as in a futuristic novel, for example. A text is judged coherent where this world is internally consistent and generally accords with accepted human knowledge; even an imaginary world is necessarily interpreted in terms of the world of our experience. This is the most crucial category of context, because the two other categories are mostly only relevant by reference to it. The co-text identifies entities, events, and such within the world spoken of; and the setting is most significant to language understanding when it is spoken of: the exception is in those matters of politeness (and hence, as we shall see, of euphemism and dysphemism) that are determined by relationships between participants irrespective of the world being spoken of.

wowser *n* [*Oz*] "prude, killjoy, teetotaller," said to be an acronym from 'We Only Want Social Evils Remedied.'

zero-derivation: A zero-derivation is where, for instance, a noun like *waitress* is used as a verb, with the regular verb suffixes (cf. *she waitresses / waitressed/ has been waitressing*, etc.). It is called ZERO-derivation because there is no additional derivational morpheme such as the *-ize* on *computerize* or *atomize*. In English we readily zero-derive nouns from verbs and verbs from nouns (e.g., *to doctor* from *a doctor*). There are also other zero-derivations (e.g., *But me no buts*), but they are rarer.

REFERENCES

Abbey, Edward. 1968. *Desert Solitaire*. New York: Ballantine.

Abbey, Edward. 1975. *The Monkey Wrench Gang*. New York: Avon Books.

Adams, Robert M. 1985. Soft Soap and Nitty Gritty. In Enright D.J. (ed.), *Fair of Speech: The Uses of Euphemism*. Oxford: Oxford University Press. pp. 44–55.

Adler, Max K. 1978. *Naming and Addressing: A Sociolinguistic Study*. Hamburg: Helmut Buske Verlag.

Aesop. 1913. *Fables of Aesop and Others*. London: Dent, New York: Dutton.

Aitchison, Jean. 1983. *The Articulate Mammal: An Introduction to Psycholinguistics*. New York: Universe Books.

Allan. Keith. 1977. Classifiers. *Language* 53:281–311.

Allan, Keith. 1981. Interpreting from context. *Lingua* 53:151–73.

Allan, Keith. 1984. The component functions of the high rise terminal contour in Australian declarative sentences. *Australian Journal of Linguistics* 4:19–32.

Allan, Keith. 1986a. *Linguistic Meaning*, 2 *Vols*. London: Routledge & Kegan Paul.

Allan, Keith. 1986b. Hearers, overhearers, and Clark & Carlson's informative analysis. *Language* 62:509–17.

Allan, Keith. 1990. Some English terms of insult invoking sex organs: evidence of a pragmatic driver for semantics. *Meanings and Prototypes: Studies in Linguistic Categorization*, Savas L. Tsohatzidis (ed.). London: Routledge & Kegan Paul.

Allen, Edward. 1989. *Straight Through the Night*. New York: Soho Press.

Allen, Wendy. 1987. *Teenage Speech: The Social Dialects of Melbourne Teenagers*. Unpublished B.A. Honors Thesis, Linguistics Department, La Trobe University.

Aman, Reinhold. 1984/5. Offensive language via computer. *Maledicta* 8:105f.

Aman, Reinhold, and Grace Sardo. 1982. Canadian sexual terms. *Maledicta* 6:21–28.

Anderson, Elaine S. 1978. Lexical universals of body-part terminology. *Universals of Human Language, Volume 3: Word Structure*, Joseph H. Greenberg (ed.), pp. 335–68. Stanford: Stanford University Press.

Anttila, Raimo. 1972. *An Introduction to Historical and Comparative Linguistics*. New York: Macmillan.

Arbeitman, Yoël. 1980. Look ma, what's become of the sacred tongues. *Maledicta* 4:71–88.

Argyle, Michael. 1988. *Bodily Communication*. London & New York: Methuen (2nd edition).

Aries, Phillippe. 1974. *Western Attitudes towards Death*. Baltimore: John Hopkins University Press.

Ash, Mildred. 1980. The vulva: a psycholinguistic problem. *Maledicta* 4:213–19.

Atkinson, Martin, David Kilby, and Iggy Rocca. 1988. *Foundations of General Linguistics*. London: Unwin Hyman (2nd edition).

Aurilianus, Caelius. 1950. *On Acute Diseases and on Chronic Diseases*. I.E. Drabkin (trans.), Chicago: University of Chicago Press.

Bach, Kent. 1987. *Thought and Reference*. Oxford: Clarendon Press.

Bach, Kent, and Robert M. Harnish. 1979. *Linguistic Communication and Speech Acts*. Cambridge, MA: MIT Press.

Bailey, Lee A., and Lenora A. Timm. 1976. More on women's and men's expletives. *Anthropological Linguistics* 18:438–49.

Baird, Jonathan. 1976. The funeral industry in Boston. In Shneidman, Edwin S. (ed.), *Death: Current Perspectives*. Palo Alto: Mayfield. pp. 82–91.

Baird, Lorrayne Y. 1981. *O.E.D.* cock 20: the limits of lexicography of slang. *Maledicta* 5:213–26.

Barker, Ronnie. 1979. *Fletcher's Book of Rhyming Slang*. London: Pan.

Barolsky, Paul. 1978. *Infinite Jest: Wit and Humor in Italian Renaissance Art*. Columbia: University of Missouri Press.

Baruch, Joel. 1976. Combat death. In Shneidman, Edwin S. (ed.), *Death: Current Perspectives*. Palo Alto: Mayfield. pp. 92–98.

Baugh, Albert C., and Thomas Cable. 1978. *A History of the English Language*. London: Routledge & Kegan Paul (Third edition).

Becker, Ernest. 1973. *The Denial of Death*. New York: The Free Press.

Behagel, Otto. 1923. *Deutsche Syntax, Band IV*. Heidelberg: Carl Winter.

Benson, Robert W. 1985. The end of legalese: the game is over. *Review of Law and Social Change* 13:519–73.

Birke, Lynda. 1980. From zero to infinity: scientific views of lesbians. *Alice Through the Microscope*, The Brighton Women & Science Group (eds.), pp. 108–23. London: Virago.

Black, David. 1986. *The Plague Years: A Chronicle of AIDS*. London: Picador.

Black, Maggie. 1977. *A Heritage of British Cooking*. London: Charles Letts.

Bloomfield, Leonard. 1927. On recent work in general linguistics. *Modern Philology* 25: 211–30.

Boase, T.S.R. 1966. King Death. *The Flowering of the Middle Ages*, Joan Evans (ed.), pp. 203–44. London: Thames & Hudson.

Boch, Oscar, and Walther von Wartburg. 1975. *Dictionnaire Etymologique de la Language Française*. Paris: Presses Universitaires (6th edition).

Boec van medicinen in Dietsche. Ca 1300. *Een Middelnederlandse compilatie van medisch-farmaceutische literatuur*, W. F. Daems (ed.). Thesis, Leiden, 1967.

Bolinger, Dwight. 1950. Rime, assonance, and morpheme analysis. *Word* 6:117–36.

Bolinger, Dwight. 1975. *Aspects of Language*. New York: Harcourt, Brace, Jovanovich (2nd edition).

The Book of Common Prayer, and Administration of the Sacraments, and Other Rites and Ceremonies of the Church, according to the use of the United Church of England and Ireland. 1852. London: Eyre & Spottiswoode.

T Bouck van Wondre. 1513. H.G. Frencken (ed.). Thesis, Rijksuniversiteit Leiden, 1943.

Boyle, T. Coraghessan. 1984. *Budding Prospects*. New York: Penguin.

Braekman, Willy. L. (ed.). 1970. *Middelnederlandse geneeskunkige recepten: Een bijdrage tot de geschiedenis van de vakliteratuur in de Nederlanden*. Gent: Koninklijke Vlaamse Academie voor Taal- en Letterkunde.

Braekman, Willy. L. (ed.). 1975. *Medische en technische middelnederlandse recepten: Een tweede bijdrage tot de geschiedenis van de vakliteratuur in de Nederlanden*. Gent: Koninklijke Vlaamse Academie voor Taal- en Letterkunde.

Braekman, Willy. L. 1987. Een merkwaardige collectie Sectreten uit de vijftiende eeuw. *Verslagen en Mededelingen van de Koninklijke Academie voor Nederlandse Taal- en Letterkunde* 2:270–87.

Brain, James L. 1979. *The Last Taboo: Sex and the Fear of Death*. New York: Anchor/Doubleday.

Brophy, John, and Eric Partridge. 1931. *Songs and Slang of the British Soldier: 1914–1918.* London: Routledge & Kegan Paul. (3rd edition)
Brown, Penelope, and Stephen C. Levinson. 1987. *Politeness: Some Universals in Language Usage.* Cambridge: Cambridge University Press.
Brown, Roger, and M. Ford. 1961. Address in American English. *Journal of Abnormal and Social Psychology* 62:375–85.
Brown, Roger, and A. Gilman 1960. The pronouns of power and solidarity. *Style in Language,* Thomas A. Sebeok (ed.), 253–76. Cambridge MA: MIT Press.
Brown, Roger, and A. Gilman 1989. Politeness theory and Shakespeare's four major tragedies. *Language in Society* 18:159–212.
Bruner, Edward M., and Jane P. Kelso. 1980. Gender differences in graffiti: a semiotic perspective. *Women's Studies International Quarterly* 3:239–52.
Burchfield, Robert. 1985. An Outline History of Euphemism in English. In Enright, D.J. (ed.), *Fair of Speech: The Uses of Euphemism.* Oxford: Oxford University Press, pp. 13–31.
Burridge, Kate. [forthcoming.] *Syntactic Change in Germanic with Particular Reference to Dutch.* Amsterdam: John Benjamins.
Camporesi, Piero. 1988. *The Incorruptible Flesh: Bodily Mutation and Mortification in Religion and Folklore.* Cambridge: Cambridge University Press.
Carter, David. 1987. *Interpreting Anaphors in Natural Language Texts.* Chichester: Ellis Horwood.
Cassileth, Barrie R. (ed.). 1979. *The Cancer Patient: Social and Medical Aspects of Care.* Philadelphia: Lea & Febiger.
Cassileth, Peter A., and Barrie R. Cassileth. 1979. Learning to care for cancer patients: the student dilemma. In Cassileth, Barrie R. (ed.), *The Cancer Patient: Social and Medical Aspects of Care.* Philadelphia: Lea & Febiger. pp. 301–18.
Cassileth, Barrie R., and Jane Hamilton. 1979. The family with cancer. In Cassileth, Barrie R. (ed.), *The Cancer Patient: Social and Medical Aspects of Care.* Philadelphia: Lea & Febiger. pp. 233–49.
Cassileth, Barrie R., and Harold I. Lief. 1979. Cancer: a biopsychosocial model. In Cassileth, Barrie R., (ed.), *The Cancer Patient: Social and Medical Aspects of Care.* Philadelphia: Lea & Febiger. pp. 17–32.
Catullus. 1969. *The Poems of Catullus.* A bilingual edition, James Michie (trans.), New York: Random House.
Chafin, Mary. 1979. *Mary Chafin's Original Country Recipes from a Dorset family cookery book of the 17th century.* London: Macmillan.
Chappell, Hilary, and William Mcgregor (eds.). [forthcoming.] *The Grammar of Body Parts.* Berlin: Mouton de Gruyter.
Charrow, Veda R. 1982. Linguistic theory and the study of legal and bureaucratic language. *Exceptional Language and Linguistics,* Loraine K. Obler and Lise Menn (eds.), New York: Academic Press. pp. 81–101.
Charrow, Robert P., and Veda R. Charrow. 1979. Making legal language understandable: A psycholinguistic study of jury instructions. *Columbia Law Review* 79:1306–74.
Chomsky, Noam. 1966. *Cartesian Linguistics.* New York: Harper & Row.
Cicero. 1959. *Letters to His Friends (Epistulae ad Familiares,* W. Glynn Williams, trans.). London: Heineman.
Clark, Herbert H., and Thomas B. Carlson. 1982. Hearers and speech acts. *Language* 58:332–73.
Cleland, John. 1985 [1748–9]. *Fanny Hill or Memoirs of a Woman of Pleasure,* (Peter Wagner, ed.). Harmondsworth: Penguin.

Clemens, Samuel. 1968. *Mark Twain's 1601, A Conversation as it was by the Social Fireside in the Time of the Tudors*. Hackensack: Privately Printed.

Clyne, Michael. 1986. The role of linguistics in peace and conflict studies. *Australian Review of Applied Linguistics* 10:76–97.

Clyne, Michael. 1987. Language and Racism. *Prejudice in the Public Arena: Racism*, Andrew Markus and Radha Rasmussen (eds.), pp. 35–44. Melbourne: Centre for Migrant and Intercultural Studies, Monash University.

Coates, Jennifer. 1986. *Women, Men and Language: A Sociolinguistic Account of Sex Differences in Language*. London: Longman.

Cove, John J. 1978. Ecology, structuralism, and fishing taboos. *Adaptation and Symbolism: Essays on Social Organization*, Karen A. Watson-Gegeo and S. Lee Seaton (eds.), pp. 143–54. Hawaii: East-West Center, University of Hawaii.

Crawley, Ernest. 1927/1960. *The Mystic Rose: A Study of Primitive Marriage and of Primitive Thought in its Bearing on Marriage*. New York: Meridian Books (2nd edition).

Crowley, Terry. 1989. Expansion in Bislama. *English World-Wide* 10:85–118.

Crowley, Terry. [forthcoming.] Body parts and part-whole constructions in Paamese grammar. In Chappell, Hilary, and William McGregor (eds). *The Grammar of Body Parts*. Berlin: Mouton deGruyter.

Crownover, Richard. 1976. *Lover's Language Guide to Japan*. Phoenix: Phoenix Books.

Crystal, David, and Derek Davy. 1969. *Investigating English Style*. London: Longman.

Cutts, Martin, and Chrissie Maher. 1984. *Gobbledygook*. London: George Allen & Unwin.

Danat, Brenda. 1980. Language in the legal process. *Law and Society Review* 14:445–564.

Dauzat, Albert. 1938. *Dictionnaire Étymologique de la Langue Française*. Paris: Librairie Larousse.

Dawidowicz, Lucy. 1975. *The War Against the Jews, 1939–1945*. New York: Bantam.

Deetz, James. 1977. *In Small Things Forgotten: The Archeology of Early American Life*. New York: Anchor Press.

Defoe, Daniel. 1966. Of Academics. *The English Language: Essays by English and American Men of Letters 1490–1839*, W.F. Bolton (ed.), pp. 91–101. Cambridge: Cambridge University Press.

Dixon, Robert M. W. 1972. *The Dyirbal Language of North Queensland*. Cambridge: Cambridge University Press.

Dixon, Robert M. W. 1980. *Languages of Australia*. Cambridge: Cambridge University Press.

Dixon, Robert M. W. 1982. *Where Have All the Adjectives Gone? And Other Essays in Semantics and Syntax*. Berlin: Mouton.

D'Orazio, Michael L. 1979. Rehabilitation for ostomy patients. In Cassileth, Barrie R. (ed.), *The Cancer Patient: Social and Medical Aspects of Care*, Philadelphia: Lea & Febiger, pp. 185–99.

Douglas, Mary. 1966. *Purity and Danger: An Analysis of Concepts of Pollution and Taboo*. London: Routledge & Kegan Paul.

D'Urfey, Thomas. 1719. *Wit and Mirth: or Pills to Purge Melancholy*. (6 vols). London.

Durkheim, Emile. 1963 [1897]. *Incest: The Nature and Origin of the Taboo*. New York: Lyle Stuart.

Eagleson, Robert D. 1982. Aboriginal English in an urban setting. *English And the Aboriginal Child*. Robert D. Eagleson, Susan Kaldor, Ian G. Malcolm (eds.), pp. 113–162. Canberra: Curriculum Development Centre.

Eckler, A. Ross. 1986/7. A taxonomy for taboo-work studies. *Maledicta* 9:201–3.

Een middelnederlandse versie van de "Circa Instans" van Platearius (1387), L.J. Vanderwiele (ed.). Oudenaarde: Sanderus, 1970.

Elder, Bruce. 1988. *Blood on the Wattle: Massacres and Maltreatment of Australian Aborigines since 1799*. French's Forest, NSW: Child & Associates.

Eliot, Gil. 1976. Agents of death. In Shneidman, Edwin S. (ed.) *Death: Current Perspectives*. Palo Alto: Mayfield. pp. 110–33.

Eliot, Thomas S. 1958. *Collected Poems 1910–1935*. London: Faber & Faber.

Ellis, Albert. 1963. *The Origins and the Development of the Incest Taboo*. New York: Lyle Stuart.

English 18th Century Cooking. (n.d.) Romania: Roy Bloom

Enninger, Werner, John Hostetler, Joachim Raith, Karl-Heinz Wandt. 1989. Rules of speaking and their mediation: the case of the Old Order Amish. *Studies on the Languages and the Verbal Behavior of the Pennsylvania Germans II*, Enninger, W., Raith, J., and Wandt K-H. (eds.), pp. 137–168. Stuttgart: Steiner Verlag.

Enright, D. J. (ed.) 1985. *Fair of Speech: The Uses of Euphemism*. Oxford: Oxford University Press.

Epstein, Joseph. 1985. Sex and euphemism. In Enright, D.J. (ed.), *Fair of Speech: The Uses of Euphemism*. Oxford: Oxford University Press. pp. 56–71.

Ervin-Tripp, Susan. 1969. Sociolinguistics. *Advances in Experimental Social Psychology*, Vol. 4, ed. by Leon Berkowitz, pp. 93–107.

Ervin-Tripp, Susan, et al. 1984. Language and power in the family. In Kramarae, C., M. Schulz, and W. M. O'Barr (eds.), *Language and Power*, Beverly Hills: Sage Publications. pp. 116–35.

Eschholz, Paul, Alfred Rosa, and Virginia Clark (eds). 1982. *Language Awareness*. New York: St. Martin's Press.

Farb, Peter. 1974. *Word Play: What Happens When People Talk*. New York: Knopf.

Farmer, John S., and W.E. Henley. 1890–1904. *Slang and Its Analogues (7 Vols)*. London. Reprint by Arno Press, New York, 1970.

Field, David. 1976. The social definition of illness. In Tuckett, David (ed.) *An Introduction to Medical Sociology*. London: Tavistock Publications. pp. 334–66.

Fleming, John, and Hugh Honour. 1977. *The Penguin Dictionary of Decorative Arts*. London: Penguin.

Florio, John. 1611 [1598]. *Queen Anna's New World of Words or Dictionarie of the Italian and English Tongues*. London. (Scolar Press Facsimile. Menstone, 1968.)

Folb, Edith. 1980. *Runnin' Down Some Lines: The Language and Culture of Black Teenagers*. Cambridge, MA: Harvard University Press.

Fox, Barbara A. 1987. *Discourse Structure and Anaphora*. Cambridge: Cambridge University Press.

Frank, Francine W., and Paula A. Treichler (eds). 1989. *Theoretical Approaches and Guidelines for Nonsexist Usage*. New York: Modern Language Association of America.

Frantz, David O. 1989. *Festum Voluptatis: A Study of Renaissance Erotica*. Columbus: Ohio State University Press.

Frazer, Sir James G. 1911. *The Golden Bough Part II: Taboo and The Perils of the Soul*. London: Macmillan (3rd edition).

Freud, Sigmund. 1953 [1913]. *Totem and Taboo*. London: Hogarth Press

Fromkin, Victoria et al. 1984. *An Introduction to Language*. Australian Edition. Sydney: Holt, Rinehart & Winston.

Fryer, Peter. 1963. *Mrs. Grundy: Studies in English Prudery*. London: Dennis Dobson.

Gandour, Jackson T. 1978. Talking backwards about sex (etc.) in Thai. *Maledicta* 2:111–14.

Gillis, Lynn. 1972. *Human Behaviour in Illness: Psychology and Interpersonal Relationships*. London: Faber & Faber.

Givón, Talmy. 1984. *Syntax: A Functional Typlogical Introduction. Volume 1.* Amsterdam: John Benjamins.

Goffman, Erving. 1955. On face-work: an analysis of ritual elements in social interaction. *Psychiatry* 18:213–31. Reprinted in *Communication in Face to Face Interaction*, John Laver and Sandy Hutcheson (eds.), 1972:319–63. Harmondsworth: Penguin.

Goffman, Erving. 1981. *Forms of Talk.* Philadelphia: University of Pennsylvania Press.

Gordon, Benjamin L. 1959. *Medieval and Renaissance Medicine.* New York: Philosophical Library.

Gordon, David P. 1983. Hospital slang for patients: crocks, gomers, gorks, and others. *Language in Society* 12:173–85.

Gorer, Geoffrey. 1965. *Death, Grief and Mourning in Contemporary Britain.* London: Cresset Press.

Gottfried, Robert S. 1983. *The Black Death: Natural and Human Disaster in Medieval Europe.* New York: The Free Press

Gowers, Sir Ernest. 1970. *The Complete Plain Words.* Harmondsworth: Penguin.

Graves, Robert. 1972 [1927]. *Lars Porsena, or The Future of Swearing and Improper Language.* London: Martin Brian & O'Keeffe.

Grey, Sir George. 1841. *Journals of Two Expeditions of Discovery in North West and Western Australia.* London: T. and W. Boone.

Grice, H. Paul. 1975. Logic and conversation. *Syntax and Semantics 3: Speech Acts*, Peter Cole and Jerry L. Morgan (eds.), pp. 41–58. New York: Academic Press.

Griffin, Jasper. 1985. Euphemisms in Greece and Rome. In Enright, D. J. (ed.), *Fair of Speech: The Uses of Euphemism*, Oxford: Oxford University Press. pp. 32–43.

Grose, (Captain) Francis. 1811 [1783]. *Dictionary of the Vulgar Tongue.* London.

Gross, John. 1985. Intimations of Mortality. In Enright, D. J. (ed.) *Fair of Speech: The Uses of Euphemism.* Oxford: Oxford University Press. pp. 203–19.

Guthrie, Douglas. 1945. *A History of Medicine.* London: Thomas Nelson.

Haiman, John. 1980. The iconicity of grammar: isomorphism and motivation. *Language* 56:515–40.

Hankey, Clyde. 1980. Naming the vulvar part. *Maledicta* 4:220–22.

Healey, Tim. 1980. A new erotic vocabulary. *Maledicta* 4:181–201.

Heine, Bernd, Ulrike Claudi, and Friederike Hünnemeyer. [forthcoming.] From cognition to grammar—evidence from African languages. To appear in Elizabeth C. Traugott and Bernd Heine (eds.) *Grammaticalization*, Amsterdam: John Benjamins.

Henderson, Jeffrey. 1975. *The Maculate Muse: Obscene Language in Attic Comedy.* New Haven: Yale University Press.

Hinton, John. 1976. Speaking of death with the dying. In Schneidman, Edwin S. (ed.), *Death: Current Perspectives.* Palo Alto: Mayfield. pp. 303–14.

Hock, Hans H. 1986. *Principles of Historical Linguistics.* Berlin: Mouton.

Hodgett, Gerald, A.J. Hodgett, and Delia Smith. (n.d) *Stere Htt Well: Medieval Recipes and Remedies from Samuel Pepys's Library.* [Sic] Adelaide: Mary Martin Books.

Holbein, Hans. 1538. *Dance of Death: A complete facsimile of the original 1538 edition of Les simulachres & historiees faces de la mort (with a new introduction by Werner L. Gundersheimer).* New York: Dover Publications, 1971.

Hollis, Anthony C. 1905. *The Masai: Their Language and Folklore.* Oxford: Clarendon Press.

Holy Bible, New International Version. London: Hodder & Stoughton. (1980)

Holzknecht, Susanne. 1988. Word taboo and its implications for language change in the Markham family of languages, PNG. *Language and Linguistics in Melanesia* 18:43–69.

Horvath, Barbara. 1985. *Variation in Australian English*. Cambridge: Cambridge University Press.

Hostetler, John A. 1980. *The Amish Society*. Baltimore: Johns Hopkins University Press (3rd edition).

Hudson, Joyce, and Eirlys Richards. 1978. *The Walmatjari: An Introduction to the Language and Culture*. Darwin: Summer Institute of Linguistics.

Hudson, Keith. 1978. *The Jargon of the Professions*. London: Macmillan.

Huizinga, Johan. 1924. *The Waning of the Middle Ages*. New York: Doubleday Anchor.

Hüsken, Wim N. 1987. *Noyt Meerder Vreucht*. Deventer: Sub Rosa.

Ide, Sachiko. 1982. Japanese sociolinguistics: politeness and women's language. *Lingua* 57:357–85.

Irving, John. 1979. *The World According to Garp*. London: Corgi.

Jaworksi, Adam. 1984/5. A note on sexism and insulting, with examples from Polish. *Maledicta* 8:91–94.

Jespersen, Otto. 1969 [1922]. *Language: Its Nature Development and Origin*. London: George Allen & Unwin.

Joffe, Natalie F. 1948. The vernacular of menstruation. *Word* 4:181–86.

Johnson, Diane, and John F. Murray. 1985. Do Doctors Mean What They Say? In Enright, D. J. (ed.), *Fair of Speech: The Uses of Euphemism*. Oxford: Oxford University Press. pp. 151–58.

Jongh, E. de. 1968/9. Erotica in Vogelperspekticf. *Simiolus* 3:22–74.

Jonson, Ben. *Ben Jonson. Vol. III*. C. H. Herford, Percey Simpson, and Evelyn Simpson. Oxford: Clarendon Press.

Joos, Martin. 1961. *The Five Clocks*. New York: Harcourt, Brace & World.

Kachru, Braj. 1984. The alchemy of English: social and functional power of non-native varieties. In Kramarae C., M. Schulz, and W. M. O'Barr (eds.), *Language and Power*. Beverly Hills: Sage Publications. pp. 176–93.

Keesing, Nancy. 1982. *Lily on the Dustbin: Slang of Australian Women and Families*. Ringwood: Penguin.

Keesing, Roger M., and Jonathan Fifi?i. 1969. Kwaio word tabooing in its cultural context. *Journal of the Polynesian Society* 78:154–77.

Kerr, Walter C. (translator). 1961 [1919]. *Martial: Epigrams*. (Loeb Classical Library). London: Heinemann.

Kirshenblatt-Gimblett, Barbara (ed.). 1976. *Speech Play*. Philadelphia: University of Pennsylvania Press.

Knipe, Edward E. 1984. *Gamrie: An Exploration in Cultural Ecology*. Lanham, New York, London: University Press of America.

Knipe, Edward E., and David G. Bromley. 1984. Speak no evil: word taboos among Scottish fishermen. *Forbidden Fruits: Taboos and Tabooism in Culture*, Ray B. Browne (ed.), pp. 183–92. Bowling Green: University of Ohio Popular Press.

Kobjitti, Chart. 1983. *The Judgement* (Tr. by Laurie Maund). Bangkok: Laurie Maund

Kramarae, Cheris, Muriel Schulz, and William M. O'Barr. 1984. *Language and Power*. Beverly Hills: Sage Publications.

Kubler-Ross, Elizabeth. 1969. *On Death and Dying*. New York: Macmillan

La Fontaine, Jean de. 1875. *Fables de La Fontaine*. Paris: Garnier Frères.

Lakoff, George. 1987. *Women, Fire, and Dangerous Things*. Chicago: University of Chicago Press.

Lakoff, George, and Mark Johnson. 1980. *Metaphors We Live By*. Chicago: Chicago University Press.

Lakoff, Robin. 1975. *Language and Woman's Place*. New York: Harper & Row.

Lao Tsu. 1972. *Tao Te Ching*. Gia-Fu Feng and Jane English (trans.). New York: Vintage.

Lawrence, D. H. 1961. *Lady Chatterley's Lover*. Harmondsworth: Penguin.

Laycock, Donald C. 1972. Towards a typology of ludlings, or play-languages. *Linguistic Communications* (Working Papers of the Linguistic Society of Australia) 6:61–113.

Leach, Edmund. 1964. Anthropological aspects of language: animal categories and verbal abuse. *New Directions in the Study of Language*, ed. by Eric H. Lenneberg, pp. 23–63. Cambridge, MA: MIT Press.

Leech, Geoffrey N. 1983. *Principles of Pragmatics*. Longman: London & New York.

Lehrer, Adrienne. 1986. English classifier constructions. *Lingua* 68:109–48.

Lehrer, Adrienne. [forthcoming.] As American as apple pie—and sushi and bagels: the semiotics of food and drink. *The Semiotic Web*, Thomas A. Sebeok and Jean Umiker-Sebeok (ed.). New York: Mouton de Gruyter.

Lewis, Charlton T., and Charles Short. 1975 [1879]. *A Latin Dictionary*. Oxford: Clarendon Press.

Lifton, Robert J. & Eric Olson. 1976. The nuclear age. In Shneidman, Edwin S. (ed.), *Death: Current Perspectives*. Palo Alto: Mayfield. pp. 99–109.

Lloyd, Geoffrey E.R. 1983. *Science, Folklore and Ideology—Studies in the Life Sciences in Ancient Greece*. Cambridge: Cambridge University Press.

Lodge, David. 1989. *Nice Work*. Harmondsworth: Penguin.

Lynn, Jonathon, and Anthony Jay. 1986. *Yes Prime Minister: The Diaries of the Right Honourable James Hacker. Volume I*. London: BBC Publications.

Lynn, Jonathan, and Antony Jay. 1987. *Diary 1988 of Sir Humphry Appleby*. Sydney: Doubleday.

Lyons, John. 1977. *Semantics (Vols. 1 & 2)*. Cambridge: Cambridge University Press.

MacLaury, Robert E. 1989. Zapotec body-part locatives: prototypes and metaphoric extensions. *International Journal of American Linguistics* 55:119–54.

Macquarie Dictionary. St Leonards, NSW: Macquarie Library.

MacWhinney, Brian, Janice M. Keenan, and Peter Reinke. 1982. The role of arousal in memory for conversation. *Memory & Cognition* 10:308–17.

Massinger, Philip. 1870. *The Plays of Philip Massinger*, Lieut. Colonel F. Cunningham (ed.). London: Albert Crocker.

Massinger, Philip. 1631. The Emperor of the East. *The Plays and Poems of Philip Massinger Vol. 3*, Philip Edwards and Colin Gibson (eds.). Oxford: Clarendon Press. 1976.

May, Derwent. 1985. Euphemisms and the media. In Enright, D. J. (ed.), *Fair of Speech: The Uses of Euphemism*. Oxford: Oxford University Press. pp. 122–34.

McDonald, James. 1988. *A Dictionary of Obscenity, Taboo and Euphemism*. London: Sphere Books.

McFate, Patricia A. 1979. Ethical issues in the treatment of cancer patients. In Cassileth, Barrie R. (ed.), *The Cancer Patient: Social and Medical Aspects of Care*. Philadelphia: Lea & Febiger. pp. 59–74.

Meigs, Anna S. 1978. A Papuan perspective on pollution. *Man* 13:304–18.

Middelnederlandsch Handwoordenboeck. 1911.

Middelnederlandsch Woordenboeck. 1885–1952. E. Verwijs and J. Verdam (eds.). s-Gravenhage: Martinus Nijhoff.

Milne, A. A. 1948. *The House at Pooh Corner*. London: Methuen.

Mitford, Jessica. 1963. *The American Way of Death*. London: Hutchinson.

Mitford, Nancy. 1979. *The Pursuit of Love and Love in a Cold Climate*. New York: Modern Library.

Montagu, Ashley. 1967. *The Anatomy of Swearing*. New York: Macmillan.

Neaman, Judith S., and Carole G. Silver. 1983. *Kind Words: A Thesaurus of Euphemisms*. New York: Facts on File.

Nohain, Jean & F. Caradec. 1985. *Le Petomane, 1857–1945*. New York: Bell.

O'Barr, Jean F. 1984. Studying power in literary texts. In Kramarae, C., M. Schulz, and W. M. O'Barr (eds.), *Language and Power*. Beverly Hills: Sage Publications. pp. 218–26.

O.E.D. = *The Oxford English Dictionary*, q.v.

Ortner, Sherry B. 1974. Is female to male as nature is to culture? *Woman, Culture, and Society*, Michelle Z. Rosaldo and Louise Lamphere (eds.), pp. 67–87.

Orwell, George. 1946. *Animal Farm*. New York: Harcourt, Brace & World.

Orwell, George. 1946. Politics and the English Language. *Shooting an Elephant and Other Essays*. New York: Harcourt, Brace & World.

Orwell, George. 1987 [1949]. *Nineteen Eighty-Four*. London: Secker & Warburg.

Osgood, Charles E., George J. Suci & Percy H. Tannenbaum. 1957. *The Measurement of Meaning*. Urbana: University of Illinois Press.

Oxford English Dictionary. 1971. Oxford, New York, Melbourne: Oxford University Press. Oxford: Clarendon Press. 1989 (2d edition).

Palmer, Roy. 1973. *The Water Closet: A New History*. Wellington, NZ: A.H. & A.W. Reed.

Pannick, David. 1985. The law. In Enright, D. J. (ed.), *Fair of Speech: The Uses of Euphemism*. Oxford: Oxford University Press. pp. 135–50.

Paros, Lawrence. 1984. *The Erotic Tongue: A Sexual Lexicon*. Seattle: Madrona.

Partridge, Eric. 1955. *Shakespeare's Bawdy*. London: Routledge & Kegan Paul.

Partridge, Eric. 1961. *A Dictionary of Slang and Unconventional English, Volume 1*. London: Routledge & Kegan Paul (5th edition).

Partridge, Eric. 1970. *A Dictionary of Slang and Unconventional English, Volume 2, The Supplement*. London: Routledge & Kegan Paul (7th edition).

Partridge, Eric. 1984. *A Dictionary of Slang and Unconventional English*. Paul Beale (ed.). London: Routledge & Kegan Paul (8th edition).

Patterson, James T. 1987. *The Dread Disease: Cancer and Modern American Culture*. Cambridge, MA: Harvard University Press.

Peterson Alan. 1986. *Word Words Words: The Use and Misuse of English in Australia Today*. Sydney: The Fairfax Library.

Picoche, Jacqueline. 1979. *Dictionnaire Étymologique du Français*. Paris: Le Robert.

Pinero, Arthur W. 1926. *The Gay Lord Quex*. London: Heineman.

Pound, Louise. 1936. American euphemisms for dying, death, and burial. *American Speech* 11:195–202.

Price, Reynolds. 1990. *The Tongues of Angels*. New York: Atheneum.

Pulan, Stephen G. 1983. *Word Meaning and Belief*. London: Croom Helm.

Pyles, Thomas. 1971. *The Origins and Development of the English Language*. New York: Harcourt, Brace, Jovanovich.

Quang Phuc Dong. 1971. English sentences without overt grammatical subject. *Studies Out in Left Field: Defamatory Essays Presented to James D. McCawley*, Arnold Zwicky, Peter Salus, Robert Binnick, and Anthony Vanek (eds.), pp. 3–18. Edmonton: Linguistic Research Inc.

Rawson, Hugh. 1981. *A Dictionary of Euphemisms and Other Doubletalk*. New York: Crown.

Rawson, Hugh. 1989. *Wicked Words*. New York: Crown.

Read, Allen W. 1934. An obscenity symbol. *American Speech* 9:264–78.

Read, Allen W. 1977. *Classic American Graffitti: Lexical Evidence from Folk Epigraphy in*

Western North America. Waukesha, WI: Maledicta Press. (First published in Paris, 1935)

Richards, Peter. 1977. *The Medieval Leper*. Cambridge: D.S. Brewer.

Risch, Barbara. 1987. Women's derogatory terms for men: That's right, "dirty" words. *Language in Society* 16:353–58.

Rosten, Leo. 1968. *The Joys of Yiddish*. New York: McGraw-Hill

Rose, Debbie. 1988. Obituary: Eric Michaels. *Australian Aboriginal Studies* 2:117f.

Rusbridger, Alan. 1986. *A Concise History of the Sex Manual*. London: Faber & Faber.

Scellinck, Thomaes. 1343. Het "Boeck van Surgien" van Meester Thomaes Scellinck van Thienen. *Opuscula Selecta Neerlandicorum de Arte Medica*, ed. by E.D. Leersum. Amsterdam, 1928.

Schmidt, Casper G. 1984/5. AIDS jokes, or, *Schadenfreude* around an epidemic. *Maledicta* 8:69–75.

Schneidman, Edwin S. 1976. *Death: Current Perspectives*. Palo Alto: Mayfield.

Searle, John. 1969. *Speech Acts*. London: Cambridge University Press.

Searle, John. 1979. *Expression and Meaning: Studies in the Theory of Speech Acts*. Cambridge: Cambridge University Press.

Shakespeare, William. 1963. *Hamlet*. Horace H. Furness (ed.). New York: Dover.

Shakespeare, William. 1987. *Henry IV, Part 1*. David Bevington (ed.). Oxford: Clarendon Press.

Shakespeare, William. 1966. *The Second Part of King Henry IV*. A.R. Humphreys (ed.). (Arden edition). London: Methuen & Cambridge, MA: Harvard University Press.

Shakespeare, William. 1947. *King Henry V*. John Dover Wilson (ed.). Cambridge: Cambridge University Press.

Shakespeare, William. 1986. *The Tragedy of Julius Caesar*. William Rosen and Barbara Rosen (eds.). New York: Signet.

Shakespeare, William. 1963. *The Tragedy of King Lear*. Russell Fraser (ed.). New York: Signet.

Shakespeare, William. 1987. *The Tragedy of Macbeth*. Sylvan Barnet (ed.). New York: Signet.

Shakespeare, William. 1964. *Measure for Measure*. S. Nagarajan (ed.). New York: Signet.

Shakespeare, William. 1971. *The Merry Wives of Windsor*. H.J. Oliver (ed.). (Arden edition). London: Methuen.

Shakespeare, William. 1981. *Much Ado About Nothing*. A.R. Humphreys (ed.). (Arden edition). London & New York: Methuen.

Shakespeare, William. 1986. *The Tragedy of Othello, The Moor of Venice*. Alvin Kernan (ed.). New York: Signet. 1986.

Shakespeare, William. 1980. *Romeo and Juliet*. Brian Gibbons (ed.). (Arden edition). London & New York: Methuen.

Shakespeare, William. 1988. *The Sonnets*. New York: Harper & Row.

Shakespeare, William. 1981. *The Taming of the Shrew*. Brian Morris (ed.). (Arden edition). London & New York: Methuen.

Shakespeare, William. 1970. *The Life of Timon of Athens*. G. R. Hibbard (ed.). Harmondsworth: Penguin.

Shakespeare, William. 1986. *The History of Troilus and Cressida*. Daniel Seltzer (ed.). New York: Signet.

Shaw, George B. 1946. *Pygmalion*. Harmondsworth: Penguin Books.

Shem, Samuel. 1978. *The House of God*. New York: Dell.

Shipley, Joseph T. 1977. The origin of our strongest taboo-word. *Maledicta* 1:23–29.

Silverstein, Theodore. 1971. *Medieval English Lyrics*. London: Edward Arnold.

Simons, Gary F. 1982. Word Taboo and comparative austronesian linguistics. *Papers from the Third International Conference on Austronesian Linguistics, Vol. 3. Accent on Variety*, Amran Halim, Lois Carrington, and Stephen A. Wurm (eds.), pp. 157–226. Canberra: Pacific Linguistics.

Smith, Michael A. 1968. Process technology and powerlessness. *British Journal of Sociology* 19:76–88.

Smith, Philip M. 1985. *Language, The Sexes, and Society*. Oxford: Basil Blackwell.

Solt, John. 1982. Japanese sexual maledicta. *Maledicta* 6:75–81.

Sontag, Susan. 1979. *Illness as Metaphor*. New York: Vintage books.

Sontag, Susan. 1989. *AIDS and Its Metaphors*. New York: Farrar, Strauss, Giroux.

Spears, Richard A. 1982. *Slang and Euphemism*. New York: Signet.

Spender, Dale. 1984. Defining reality: a powerful tool. In Kramarae, C., M. Schulz, and W. M. O'Barr (eds.), *Language and Power*. Beverly Hills: Sage Publications. pp. 194–205.

Sperber, Dan & Deirdre Wilson. 1986. *Relevance: Communication and Cognition*. Oxford: Basil Blackwell.

Stephenson, John. S. 1985. *Death, Grief, and Mourning: Individual and Social Relations*. New York: The Free Press.

Stern, Gustaf. 1965 [1931]. *Meaning and Change of Meaning (with Special Reference to the English Language)*. Bloomington: Indiana University Press.

Stiles, Henry R. 1934 [1871]. *Bundling: its origin, progress and decline in America*. New York: Book Collectors Association.

Strom, K. A. (ed.) 1984. *The Best of Attack! and National Vanguard Tabloid*. Arlington: National Alliance.

Style Manual for Authors, Editors and Printers. 1988. Fourth edition. Canberra: Australian Government Publishing Services.

Swift, Jonathan. 1958 [1735]. *Gulliver's Travels and Other Writings by Jonathan Swift*, Ricardo Quintana (ed.). New York: Random House.

Thackeray, William M. 1962 [1848]. *Vanity Fair*. New York: Signet.

The CCH Macquarie Concise Dictionary of Modern Law. 1988. Sydney: CCH Australia Limited.

Tuckett, David (ed.). 1976. *An Introduction to Medical Sociology*. London: Tavistock Publications.

Tuckett, David, and Joseph M. Kaufert (eds.). 1978. *Basic Readings in Medical Sociology*. London: Tavistock Publications.

Valenstein, Edward, and Kenneth M. Heilman. 1979. Emotional disorders resulting from lesions of the central nervous system. *Clinical Neuropsychology*, Heilman and Valenstein (eds.) pp. 413–38. New York: Oxford University Press.

Van Dijk, Teun A., and Walter Kintsch. 1983. *Strategies of Discourse Comprehension*. New York: Academic Press.

Vigilans. 1952. *Chamber of Horrors: A glossary of Official Jargon both English and American*. (Introduction by Eric Partridge.) London: Andre Deutsch.

Wardhaugh, Ronald. 1986. *Introduction to Sociolinguistics*. Oxford: Basil Blackwell.

Webster's New Collegiate Dictionary. 1977. Springfield, MA: G. & C. Merriam.

Weisman, Avery D. 1976. Denial and middle knowledge. In Shneidman, Edwin S. (ed.) *Death: Current Perspectives*. Palo Alto: Mayfield. pp. 452–68.

Wentworth, Harold, and Stuart B. Flexner. 1975. *Dictionary of American Slang, Second Supplemented Edition*. New York: Thomas Crowell.

Wilde, Oscar. 1926. *The Picture of Dorian Gray & De Profundis*. New York: The Modern Library.

Wilde, Oscar. 1988. *The Complete Plays of Oscar Wilde*. London: Methuen.

Williams, Joseph M. 1981a. The English language as use-governed behaviour. *Style and Variables in English*, Timothy Shopen and Joseph M. Williams (eds.), pp. 27–60. Cambridge MA: Winthrop.

Williams, Joseph M. 1981b. Literary style: the personal voice. *Style and Variables in English*, Timothy Shopen & Joseph M. Williams (eds.), pp. 117–216. Cambridge, MA: Winthrop.

Woordenboek der Nederlandsche Taal. 1882–. M. de Vries and L. A. Te Winkel (eds.). s-Gravenhage: Martinus Hijhoff.

Wright, Lawrence. 1960. *Clean and Decent*. New York: Viking.

Yeats, William B. 1965. *The Collected Poems of W. B. Yeats*. London: Macmillan.

INDEX